Anaerobic Microbiology

The Practical Approach Series

SERIES EDITORS

D. RICKWOOD
Department of Biology, University of Essex
Wivenhoe Park, Colchester, Essex CO4 3SQ, UK

B. D. HAMES
Department of Biochemistry and Molecular Biology, University of Leeds
Leeds LS2 9JT, UK

Affinity Chromatography
Anaerobic Microbiology
Animal Cell Culture
Animal Virus Pathogenesis
Antibodies I and II
Biochemical Toxicology
Biological Membranes
Biosensors
Carbohydrate Analysis
Cell Growth and Division
Cellular Neurobiology
Cellular Calcium
Centrifugation (2nd edition)
Clinical Immunology
Computers in Microbiology
Crystallization of Nucleic Acids and Proteins
Cytokines
Directed Mutagenesis
DNA Cloning I, II, and III
Drosophila
Electron Microscopy in Molecular Biology
Electron Microscopy in Biology
Essential Molecular Biology I and II
Fermentation
Flow Cytometry
Gel Electrophoresis of Nucleic Acids (2nd edition)
Gel Electrophoresis of Proteins (2nd edition)
Genome Analysis
HPLC of Small Molecules
HPLC of Macromolecules
Human Cytogenetics
Human Genetic Diseases
Immobilised Cells and Enzymes
Iodinated Density Gradient Media
Light Microscopy in Biology
Liposomes
Lymphocytes
Lymphokines and Interferons
Mammalian Development

Mammalian Cell Biotechnology
Medical Bacteriology
Medical Mycology
Microcomputers in Biology
Microcomputers in Physiology
Mitochondria
Molecular Neurobiology
Mutagenicity Testing
Neurochemistry
Nucleic Acid and Protein Sequence Analysis
Nucleic Acid Hybridisation
Nucleic Acids Sequencing
Oligonucleotides and Analogues
Oligonucleotide Synthesis
PCR
Photosynthesis: Energy Transduction
Plant Cell Culture
Plant Molecular Biology
Plasmids
Post-Implantation Mammalian Embryos
Prostaglandins and Related Substances
Protein Architecture
Protein Function
Protein Purification Applications
Protein Purification Methods
Protein Sequencing
Protein Structure
Proteolytic Enzymes
Radioisotopes in Biology
Receptor Biochemistry
Receptor–Effector Coupling
Ribosomes and Protein Synthesis
Solid Phase Peptide Synthesis
Spectrophotometry and Spectrofluorimetry
Steroid Hormones
Teratocarcinomas and Embryonic Stem Cells
Transcription and Translation
Virology
Yeast

Anaerobic Microbiology

A Practical Approach

Edited by

P. N. LEVETT

University of the West Indies
Faculty of Medical Sciences
Queen Elizabeth Hospital
Barbados

OXFORD UNIVERSITY PRESS
Oxford New York Tokyo

Oxford University Press
Walton Street, Oxford OX2 6DP

Oxford is a trade mark of Oxford University Press

Published in the United States
by Oxford University Press, New York

A catalogue record for this book is
available from the British Library

Library of Congress Cataloging in Publication Data
Anaerobic microbiology: a practical approach/edited by P. N. Levett.
p. cm. —(Practical approach series)
1. Diagnostic microbiology. 2. Anaerobic bacteria—
Identification. I. Levett, Paul N. (Paul Nigel), 1957–
II. Series.
[DNLM: 1. Bacteria, Anaerobic—isolation & purification.
2. Bacteriological Techniques. QW 25 A5314]
QR.67.A53 1991
589.9'5—dc20
DNLM/DLC

ISBN 0–19–963204–9 (hbk.)
ISBN 0–19–963262–6 (pbk.)

Typeset by Cotswold Typesetting Ltd, Gloucester
Printed in Great Britain by
Information Press Ltd, Eynsham, Oxford

For Dr Ken Phillips
Our trusted colleague and close friend

Preface

For many years anaerobic microbiology was regarded by microbiologists as being a difficult area in which to work. Many potential anaerobic bacteriologists were deterred from the study of these diverse and interesting organisms because of the perceived difficulties in cultural methods. The pioneering work by Robert Hungate on the bacterial flora of the rumen using roll-tube methods facilitated the isolation of some of the most oxygen-sensitive of anaerobes. This advance was followed by the application of these methods to the human gut flora and studies of non-sporing anaerobes in human disease. Taxonomic advances were facilitated by the use of gas–liquid chromatography for the analysis of metabolic end-products. The wide availability of anaerobic cabinets led to a much broader study of anaerobes and their activities in a considerable variety of environments. More recently, advances in molecular biological techniques have been applied to anaerobes and are yielding results of fundamental interest.

Despite the production of several excellent monographs and laboratory manuals on various aspects of anaerobic bacteriology (particularly in relation to the clinical laboratory) there has been no attempt to provide a ready source of reference to methods which are common to all anaerobic microbiologists in addition to those which may be of value in the study of specific habitats or groups of organisms (including the anaerobic eukaryotes). It was with this need in mind that this volume was prepared.

This is a practical book, the contributors to which are all active researchers in their fields, but within a volume of this size it is impossible to cover all the methods used for all groups of anaerobes. However, an attempt has been made to provide a broad coverage of the most useful methods. The coverage of some chapters naturally overlaps, but duplication has been kept to a minimum. Where space has precluded the inclusion of particular methods, source references are included. Although several chapters include tables of characters intended to facilitate identification of particular anaerobic taxa, this is not intended solely as an identification manual. Other publications, such as *Bergey's Manual of Systematic Bacteriology* (1), *The Prokaryotes: a Handbook on Habitats, Isolation, and Identification of Bacteria* (2), and the *VPI Anaerobe Laboratory Manual* (3), should also be consulted for that purpose.

There remain an enormous number of challenges for those interested in the study of anaerobic micro-organisms, whether in the fields of taxonomy, ecology, physiology, or molecular biology. It is the hope of all concerned with this volume that it will stimulate the interdisciplinary approach to these fascinating organisms.

I sincerely wish to thank all the authors who contributed chapters to this volume, for without their efforts it would not have been possible. Finally I must express my gratitude to the publishers and staff at Oxford University Press for their assistance and their enthusiasm for the subject.

References

1. Krieg, N. R. and Holt, J. G. (ed.) (1984–90). *Bergey's Manual of Systematic Bacteriology*, 4 vols. Williams and Wilkins, Baltimore, MD.
2. Starr, M. P., Stolp, H., Truper, H. G., Balows, A., and Schlegel, H. G. (ed.) (1981). *The Prokaryotes, a Handbook on Habitats, Isolation, and Identification of Bacteria*, 2 vols. Springer-Verlag, Berlin.
3. Holdeman, L. V., Cato, E. P., and Moore, W. E. C. (1977). *Anaerobe Laboratory Manual* (4th edn). Virginia Polytechnic Institute and State University, Blacksburg, VA.

Barbados
January 1991

P. N. Levett

Contents

List of contributors xix

Abbreviations xxi

1. Anaerobic culture methods 1

A. T. Willis

1. Introduction 1

2. Shake cultures and fluid cultures 1

Reducing agents 2

3. The anaerobic jar 3

Operation of the jar by evacuation-replacement 3

Operation of the jar by internal gas generation 5

Microenvironment (commonly disposable) systems 6

4. Anaerobic cabinets 6

Comment 8

5. Pre-reduced anaerobically sterilized (PRAS) roll-tube techniques 8

Preparation of oxygen-free gas 9

Anaerobic media 9

Comment 11

References 11

2. Isolation of anaerobes from clinical material 13

P. N. Levett

1. Introduction 13

2. The specimen 13

3. Processing samples for anaerobes in the laboratory 14

Smell 14

Fluorescence 15

Gram stain 15

Culture 15

Identification and sensitivity testing 18

4. Blood cultures 18

Blood culture media 18

Collection of blood for culture 19
Incubation 20
Detection of growth 20
Subculture 20

5. Detection and isolation of anaerobes from faeces 20
Clostridium perfringens food poisoning 21
Antibiotic-associated diarrhoea and pseudomembranous colitis 22
Botulism 24

References 26

3. Identification of clinically important anaerobes 29

P. N. Levett

1. Introduction 29

2. Initial examination of anaerobic isolates 29
Atmospheric requirements for growth 29
Colonial morphology 30
Microscopic morphology 31

3. Rapid presumptive identification using disc resistance tests 31

4. Identification using commercial identification systems 33

5. Conventional identification methods for clinically important anaerobes 34
Identification using PRAS media 34
Agar plate identification method 35
Egg yolk agar 38
Miniaturized methods in conventional identification schemes 41

6. Gas-chromatographic analysis of end-products of anaerobic metabolism 43
Media for gas-chromatographic analysis 43
GLC conditions 46
Other end-products detectable by GLC 46
GLC of blood cultures 48
GLC of clinical material 48

References 49

4. Susceptibility testing of anaerobic bacteria 51

Jon E. Rosenblatt

1. Introduction 51

2. Antimicrobial resistance of anaerobes 52

3. Mechanisms of antimicrobial resistance in anaerobes 54

4. Development of new antimicrobials 54

5. Methods for antimicrobial susceptibility testing of anaerobes 57
General 57
Antimicrobial breakpoints 58
Quality control procedures 58
Specific testing methods 60
Additional testing methods 61
Detection of β-lactamase activity using nitrocefin (chromogenic cephalosporin) 62

References 62

5. Chemotaxonomic methods 65

H. N. Shah, S. E. Gharbia, and P. A. Lawson

1. Introduction 65

2. Acid end-products 65
High-performance liquid chromatography 66
Enzymic determination of L(+)- and D(−)-lactic acid 67

3. Electrophoretic methods 68
SDS-PAGE patterns 69
Isoelectric focusing 71
Enzyme electrophoresis 73

4. Peptidoglycan composition 74
Preparation of cell walls 75
Chromatographic methods 78

5. Lipids as major chemotaxonomic markers 80
Polar lipids 80
Fatty acids 84
Menaquinones 86

6. DNA base composition, DNA–DNA hybridization, and ribosomal RNA gene restriction patterns 89
Isolation and purification of DNA 89
DNA–DNA hybridization 91
16S ribosomal RNA gene restriction patterns 94

References 98

6. Immunochemical methods 101

I. R. Poxton

1. Introduction 101

2. Historical applications of immunological methods 101

3. Antigens of anaerobic bacteria 102

4. Preparation of some relevant antigens 102
Cell breakage 104
Preparation of the outer membrane of Gram-negative bacteria 105
EDTA extraction of Gram-positive surface antigen 106
Secondary cell-wall carbohydrate antigens 106
Lipocarbohydrate antigens 107
Lipopolysaccharide 107
Exopolysaccharides 108
Preparation of appendages 109

5. Production of antisera 109
Whole-bacteria vaccine 110
Subcellular or soluble vaccine 110

6. Immunological methods 111
Precipitation in gels 112
Enzyme-linked immunosorbent assay (ELISA) 115
Immunoblotting 116

7. Conclusions 119

References 119

7. Methods for biochemical studies 121

S. J. Forsythe

1. Introduction 121

2. Cell culture 121
Preparation of media for anaerobes 121
Formulation of media 121

3. Harvesting anaerobic cultures 122
Manipulation of cell suspensions under anaerobic conditions on open bench tops 122
Preparation of anaerobic buffers 122

4. Cell fractionation 124
Methods of cell fractionation 124
Separation of oxygen-sensitive cell components 125

5. Biochemical studies of whole and perforated cells 126
Dehydrated enzyme kits 126
Fluorogenic substrates 126
Enzyme specific activity 128
Thunberg anaerobic tubes and cuvettes 128
Artificial electron carriers 129
Electrophoresis of enzymes 130

6. Oxidoreductase enzymes 131
Fumarate reductase 131
Pyruvate dehydrogenase 132

Nitrogen oxide reductases 133
Nitroreductase activity of strict anaerobes 135
Methanogens 136

7. Membrane potential generation 137
Production of proton motive force by methanogens 137
Sodium-transport decarboxylases 138

8. Proteolytic enzymes 139

9. Cellulose degradation 139
Cellulose activity detected using carboxymethylcellulase 140

10. Measurement of microbial activity *in situ* 141
Methods of sampling intestinal contents 141

References 143

8. Genetics and molecular biology 145

B. W. Wren, P. Mullany, and F. I. Lamb

1. Introduction 145

2. Extraction and purification of genomic bacterial DNA 146
Extraction of bacterial genomic DNA by SDS lysis 146
Extraction of bacterial DNA using guanidine hydrochloride 147
Extraction of plasmid DNA from anaerobic bacteria 148
Purification of bacterial DNA 150

3. Development of vectors for anaerobic bacteria 150
Shuttle plasmids 151
Potential transposon-based vector systems 151

4. Methods for the introduction of genetic material into anaerobic bacteria 153
Conjugation 153
Transformation of anaerobic bacteria 154

5. Application of the polymerase chain reaction (PCR) to anaerobes 155
Diagnostic applications of PCR 157
PCR with mixed oligonucleotide primers (PCRMOP) 159

Acknowledgements 161

References 161

9. Toxins of anaerobes 163

David M. Lyerly and Tracy D. Wilkins

1. Introduction 163

2. Clostridial toxins 163
C. botulinum toxins 163

C. tetani toxins 168
C. perfringens toxins 169
C. difficile toxins 172
Toxins produced by 'gas-gangrene' clostridia 178
3. *Bacteroides fragilis* enterotoxin 179
4. Cloning of toxin genes from clostridia 179
5. Conclusions 180
Acknowledgements 180
References 180

10. Methods for the study of anaerobic microflora 183
B. S. Drasar and April K. Roberts
1. Introduction 183
2. Sampling 184
Preservation of samples 185
3. Microscopy 186
4. Bacterial cultivation 187
Culture media 188
The culture of a sample using an anaerobic cabinet 189
5. Bacterial metabolism 195
The metabolism of xenobiotics 197
Metabolic indicators of the bacterial flora 198
References 199

11. Sulphate-reducing bacteria 201
George T. MacFarlane and Glenn R. Gibson
1. Introduction 201
2. Growth media 201
General description 201
Media for the growth of desulfovibrios and desulfotomacula 202
Media for the growth of other SRB 203
Enumeration of SRB 209
3. Identification of SRB 212
4. Methods for studying SRB activity 213
Radiotracer analysis of sulphate reduction rates 214
Colorimetric determination of H_2S 216
Chemical measurement of sulphate 217
Ecological studies with sodium molybdate 218
References 220

12. Methanogens 223

Mahendra K. Jain, J. Gregory Zeikus, and Lakshmi Bhatnagar

1. Introduction 223

2. Habitat 223

3. Isolation 224
- Culture requirements 225
- Enrichment 228
- Gas chromatographic analysis 230
- Microscopic analysis 231
- Colony isolation 232
- Enumeration 234

4. Taxonomic characterization 235

5. Large-scale cultivation 236

6. Preparation of cell extracts 239

7. Maintenance and preservation 241

8. Culture sources 243

References 243

13. Isolation and cultivation of anoxygenic phototrophic bacteria 247

B. J. Tindall

1. Introduction 247

2. Enrichment methods 251
- General considerations 251

3. Enrichment in natural samples 259
- Preparation of Winogradsky columns 259

4. Isolation of pure cultures 260
- Purification in agar deeps 261
- Use of agar plates or roll tubes 267
- Purification in liquid culture 269
- Checking for purity 269

5. Media suitable for enrichment and cultivation 270
- Special equipment 270
- Preparation of media 271

6. Storage of pure cultures 274

7. Summary 275

References 275

14. Anaerobic protozoa and fungi 279

G. S. Coleman

1. Introduction 279

2. Cultivation of anaerobic protozoa 280

Cultivation of rumen entodiniomorphid protozoa 280
Cultivation of rumen flagellate protozoa 285
Cultivation of flagellates from the termite hind gut 285
Cultivation of *Metopus striatus* 288
Cultivation of *Metopus contortus* 289
Cultivation of *Pelomyxa palustris* 290

3. Cultivation of anaerobic fungi 290

Cultivation of rumen fungi 290
Cultivation of other anaerobic fungi 296

References 296

Appendix

A1 Manufacturers and suppliers of specialist items 297

Index 299

Contributors

LAKSHMI BHATNAGAR
Michigan Biotechnology Institute, 3900 Collins Road, PO Box 27609, Lansing, MI 48909, USA.

G. S. COLEMAN
AFRC Institute of Animal Physiology, Babraham, Cambridge CB2 4AT, UK.

B. S. DRASAR
Department of Clinical Sciences, London School of Hygiene and Tropical Medicine, Keppel Street, London WC1E 7HT, UK.

S. J. FORSYTHE
Department of Life Sciences, Nottingham Polytechnic, Clifton Lane, Nottingham NG11 8NS, UK.

S. E. GHARBIA
Department of Oral Microbiology, Dental School, London Hospital Medical College, Turner Street, London E1 2AD, UK.

GLENN R. GIBSON
MRC Dunn Clinical Nutrition Centre, 100 Tennis Court Road, Cambridge CB2 1QL, UK.

MAHENDRA K. JAIN
Michigan Biotechnology Institute, 3900 Collins Road, PO Box 27609, Lansing, MI 48909, USA.

F. I. LAMB
Department of Medical Microbiology, St. Bartholomew's Hospital Medical College, West Smithfield, London EC1A 7BE, UK.

P. A. LAWSON
Department of Oral Microbiology, Dental School, London Hospital Medical College, Turner Street, London E1 2AD, UK.

P. N. LEVETT
Department of Surgery and Pathology, Faculty of Medical Sciences, University of the West Indies, Queen Elizabeth Hospital, Barbados.

DAVID M. LYERLY
Department of Anaerobic Microbiology, Virginia Polytechnic Institute and State University, Blacksburg, VA 24061, USA.

Contributors

GEORGE T. MACFARLANE
MRC Dunn Clinical Nutrition Centre, 100 Tennis Court Road, Cambridge CB2 1QL, UK.

P. MULLANY
Department of Medical Microbiology, St. Bartholomew's Hospital Medical College, West Smithfield, London EC1A 7BE, UK.

I. R. POXTON
Department of Medical Microbiology, University Medical School, Edinburgh EH8 9AG, UK.

APRIL K. ROBERTS
Public Health Laboratory Service Centre for Applied Microbiology and Research, Porton Down, Salisbury SP4 0JG, UK.

JON E. ROSENBLATT
Department of Laboratory Medicine, Mayo Clinic, Rochester, MN 55905, USA.

H. N. SHAH
Department of Oral Microbiology, Dental School, London Hospital Medical College, Turner Street, London E1 2AD, UK.

B. J. TINDALL
DSM-Deutsche Sammlung von Mikroorganismen und Zellkulturen GmbH., Mascheroder Weg 1b, D-3300 Braunschweig, Germany.

TRACY D. WILKINS
Department of Anaerobic Microbiology, Virginia Polytechnic Institute and State University, Blacksburg, VA 24061, USA.

A. T. WILLIS
Public Health Laboratory, Luton and Dunstable Hospital, Luton LU4 0DZ, UK.

B. W. WREN
Department of Medical Microbiology, St. Bartholomew's Hospital Medical College, West Smithfeld, London EC1A 7BE, UK.

J. GREGORY ZEIKUS
Michigan Biotechnology Institute, 3900 Collins Road, PO Box 27609, Lansing, MI 48909, USA.

Abbreviations

ATCC	American Type Culture Collection
BAC	*Bacteroides* agar
BHF	brain–heart infusion agar + volatile fatty acids
BHI	brain–heart infusion
BIF	*Bifidobacterium* agar
BSA	bovine serum albumin
CA	clavulanic acid
CD	*Clostridium difficile* agar
c.f.u.	colony-forming units
CIE	crossed immunoelectrophoresis
CMB	cooked meat broth
CMC	carboxymethylcellulose
CPE	cytopathic effect
CSC	concentrated saline–citrate
CTAB	cetyl trimethylammonium bromide
CVA	crystal violet erythromycin agar
DMACA	*p*-dimethylaminocinnamaldehyde
DMSO	dimethyl sulphoxide
DNase	deoxyribonuclease
DSC	dilute saline-citrate
DSM	Deutsche Sammlung von Mikroorganismen und Zellkulturen
EDTA	ethylenediamine-tetraacetic acid
Eh	redox potential
ELISA	enzyme-linked immunosorbent assay
FAA	fastidious anaerobe agar
FAB	fastidious anaerobe broth
FAME	fatty acid methyl esters
GCl	guanidine hydrochloride
GFF	glass fibre filters
GPT	glutamate-pyruvate transaminase
Hepes	*N*-(2-hydroxyethyl)piperazine-*N'*-(2-ethanesulphonic acid)
IU	international units
KV	kanamycin–vancomycin agar
LDH	lactate dehydrogenase
LP	Braun's lipoprotein
LPS	lipopolysaccharide
LR	London resin
MDH	malate dehydrogenase

MESNA	2-mercaptoethanesulphonic acid
MIC	minimum inhibitory concentration
MLS	macrolide–lincosamide–streptogramin
MPN	most probable number
MTT	thiazolyl blue tetrazolium
MUC	methyl umbelliferyl-β-D-cellobioside
NCCLS	National Committee for Clinical Laboratory Standards
NVFA	non-volatile fatty acids
OCM	Oregon Collection of Methanogens
OFN	oxygen-free nitrogen
OP	over proof
P	protein
PABA	*p*-aminobenzoic acid
PAGE	polyacrylamide gel electrophoresis
PBP	penicillin-binding proteins
PBS	phosphate-buffered saline
PBS–T	phosphate-buffered saline–Tween 20
PBS–T–RS	phosphate-buffered saline–Tween 20–neutral rabbit serum
PCR	polymerase chain reaction
PCRMOP	polymerase chain reaction with mixed oligonucleotide primers
PL	phospholipid
PMS	phenazine methosulphate
PRAS	pre-reduced anaerobically sterilized
PTFE	polytetrafluoroethylene
PY	peptone–yeast extract
R_f	movement relative to solvent front
RFLP	restriction fragment length polymorphism
R_g	movement relative to glucose standard
RI	refractive index
RIF	rifampicin agar
RNase	ribonuclease
RPLA	reversed passive latex agglutination
SDS	sodium dodecyl sulphate
SPS	sodium polyanethol sulphonate
SRB	sulphate-reducing bacteria
SSC	saline–citrate
TBE	Tris–boric acid–EDTA
TBS	Tris-buffered saline
TE	Tris–EDTA
TENS–PK	Tris–EDTA–saline–SDS–proteinase K
TES	Tris–saline–EDTA
TCA	trichloracetic acid
TEMED	*N,N,N′,N′*-tetramethylethylenediamine
TLC	thin layer chromatography

TPP^+	tetraphenylphosphonium chloride
Tris	tris(hydroxymethyl)-aminomethane
SUL	sulbactam
V	*Veillonella* agar
VFA	volatile fatty acids
WCA	Wilkins–Chalgren agar
WH	Willis and Hobbs agar

1

Anaerobic culture methods

A. T. WILLIS

1. Introduction

A variety of methods is available for the culture of anaerobic organisms.

(a) Exclusion of oxygen from part of the medium is the simplest method, and is effected by growing the organisms *within* the medium as a shake or fluid culture.

(b) When an anaerobic atmosphere is required for obtaining surface growths, anaerobic jars and cabinets provide the routine methods of choice.

(c) A more sophisticated method for the surface culture of anaerobes is the pre-reduced anaerobically sterilized (PRAS) roll-tube technique (1) and various modifications of it. This technique utilizes specialized equipment, is time-consuming, and requires specially trained staff and specialized medium-preparation facilities. It provides the most meticulous anaerobic conditions, and is appropriately used for the isolation and study of exacting anaerobic species that are highly sensitive to oxygen (2). The method, however, is too demanding in time and space for routine use in most diagnostic laboratories where anaerobic work is but one facet of a service commitment.

Comparative studies have shown that PRAS methods are not superior to the anaerobic jar and cabinet for the recovery of relevant anaerobes in diagnostic settings.

2. Shake cultures and fluid cultures

Exclusion of oxygen from part of the medium is the simplest way of growing anaerobes and is effected by growing the organisms *within* the culture medium, which may be either solid or fluid. In solid media growth is obtained as a shake culture in a tube or bottle, or as a pour-plate culture. Fluid media in deep tubes are extensively used in anaerobic bacteriology since most anaerobes grow readily in this type of culture and manipulations with them are easy. Before use, media for shake and deep fluid culture are steamed or heated in boiling water to drive off dissolved oxygen, and then cooled quickly before inoculation. When incubated in air, oxygen tends to diffuse back into shake and fluid cultures, so that anaerobic growth appears some distance below the surface; strict anaerobes grow only in the depths of the medium, and highly exacting species may not grow at all; less oxygen-intolerant organisms show growth almost to the surface. The deeper parts of shake cultures remain anaerobic much longer than those of fluid cultures, since convection currents in fluid media help to distribute oxygen rapidly throughout the medium. For this reason it is helpful to thicken fluid media by the addition of a small amount of agar (0.05–0.1%). These 'sloppy' agar media are still fluid, but convection currents in them are minimized. Growth of anaerobes in such media is greatly enhanced by the addition of reducing substances. Although it is a common practice to incubate shake and fluid cultures under ordinary atmospheric conditions, some strict anaerobes may fail to grow; it is better to incubate all cultures in the anaerobic jar or cabinet.

2.1 Reducing agents

Although the addition of reducing agents to fluid and shake cultures of some anaerobes is not necessary, it is essential for the growth of the more exacting species. Many workers add reducing substances routinely to all media (including surface plating media) used for anaerobic work. The following are some of the reducing agents commonly used.

2.1.1 Thioglycollic acid

Thioglycollic acid or its sodium salt may be added to media in the proportion of 0.01–0.2%. Sodium thioglycollate is inhibitory to the growth of some clostridia, and gradually becomes slightly toxic towards a variety of anaerobes during storage.

2.1.2 Glucose

Glucose is usually incorporated in media in the proportion of 0.5–1.0%. It is a good reducing agent, is non-toxic, and serves as a nutrient for bacterial growth.

2.1.3 Ascorbic acid

Ascorbic acid may be used in the proportion of 0.1%; it is inhibitory towards some strains of non-sporing anaerobic bacilli.

2.1.4 Cysteine

Cysteine is a valuable reductant if used in concentrations not exceeding 0.05%; higher concentrations may be inhibitory. In neutral and slightly alkaline solution oxidation of cysteine in the presence of atmospheric oxygen proceeds rapidly, so that during pouring, setting, and drying of plates, loss of cysteine is likely to be appreciable. Cysteine in culture media is protected from oxidation by ascorbic acid, thioglycollate, and dithiothreitol (3). In addition to its non-specific reducing effect cysteine may be a growth requirement of some anaerobes, e.g. *Clostridium novyi* types C and D (4).

2.1.5 Metallic iron

Metallic iron in the form of wire, filings, nails, and steel wool may be used as a reducing agent in fluid media. Ordinary 'tin tacks' or screws are convenient as they are quickly sterilized by flaming before being added to sterile stock media. Metallic iron greatly favours the growth of *C. novyi* types B, C, and D on solid media (5), an effect that appears to be specific for these organisms.

2.1.6 Cooked meat medium

Cooked meat is widely used for the culture of anaerobes. The cooked sterile muscle tissue contains reducing substances, particularly glutathione, which enable the growth of many strict anaerobes without the use of other anaerobic methods (6).

3. The anaerobic jar

Modern anaerobic jars use a room-temperature catalyst (D catalyst, Engelhard) which consists of pellets of alumina coated with finely divided palladium. This catalyst removes oxygen in the presence of excess hydrogen. The hydrogen may be provided either from an external source (evacuation-replacement method), or may be generated within the jar (Oxoid gas generating kit; Becton Dickinson GasPak generator).

3.1 Operation of the jar by evacuation-replacement

3.1.1 Hydrogen source

The most convenient source of hydrogen is a cylinder of the compressed gas mixed with 10% carbon dioxide. Because the mixed gases must be supplied to the jar at a low pressure, the cylinder should be fitted with a reducing valve and the gas delivered initially to a low-pressure reservoir, such as a rubber football bladder, interposed between the reducing valve and the jar. Because carbon dioxide is a growth requirement of some anaerobes, and because none is adversely affected by it, carbon dioxide should be added to the anaerobic jar routinely; a concentration of 7–10% is suitable. Cylinders of mixed gases (90% hydrogen + 10% carbon dioxide; or 10% hydrogen + 10% carbon dioxide + 80% nitrogen) are available commercially. Although I prefer the 90% hydrogen gas mixture, low hydrogen content mixtures are popular because they reduce the risk of minor explosions inside the jar. Low hydrogen content mixtures require more complete preliminary evacuation of the jar (to at least −600 mm Hg) for satisfactory anaerobiosis to develop, a procedure which itself causes some problems including difficulties in manometric monitoring of catalysis.

Protocol 1. Setting up the anaerobic jar

1. Using a 90% hydrogen + 10% carbon dioxide gas mixture, place cultures in the jar and secure the lid.
2. Attach the jar to a vacuum pump fitted with a vacuum gauge, and evacuate until there is a negative pressure in the jar of about −300 mm Hg. Much more evacuation (to −600 mm Hg) is required if a gas mixture of low hydrogen content is used.
3. Close the valve on the jar, disconnect the vacuum pump, and connect the low-pressure gas source to the jar.
4. Open the valve to admit the gas mixture, which is drawn in rapidly from the football bladder as the jar loses its partial vacuum.
5. Close the inlet valve. Catalysis now proceeds for 10–15 min and induces a substantial secondary vacuum (−200 to −300 mm Hg) if the jar is functioning efficiently.[a]
6. Equilibrate the pressure in the jar to atmospheric pressure by adding further gas mixture. The jar is now ready for incubation.

[a] Monitoring the pressure changes within the jar is important since jar integrity and adequate catalyst activity are necessary for the development and maintenance of anaerobiosis.

3.1.2 Catalyst

The optimum amount of catalyst pellets is 1.0 g/l of jar volume (7). The catalyst may be reversibly inactivated by moisture (reactivate by heating at 160°C for 90 min), or it may be irreversibly poisoned by products of bacterial metabolism, notably hydrogen sulphide. Failure of the catalyst, whether from moisture inactivation or poisoning may occur between consecutive uses, thus emphasizing the importance of observing catalyst activity each time the jar is set up. If the anticipated degree of secondary vacuum is not achieved, the catalyst must be changed, and the jar checked for leaks. Note that damaged catalyst sachets may allow escape of crushed catalyst powder into the body of the jar, leading to ignition of hydrogen and explosion.

A catalyst sachet marketed by Oxoid Ltd contains standard palladium catalyst pellets wrapped in perforated aluminium foil and encased in wire gauze. The aluminium foil protects the pellets from physical damage and acts as a heat sink during exothermic catalysis, thus minimizing the risk of ignition of gases.

3.1.3 Carbon dioxide

This is essential for, or stimulatory to, the growth of many anaerobes and is a well established supplement to the anaerobic atmosphere at a concentration of 5–10% carbon dioxide (8–10). Although a precise concentration of carbon dioxide is not critical for the growth enhancement of anaerobes, it assumes greater importance when controlled atmospheres are required, e.g. in some antimicrobial sensitivity testing (11).

3.1.4 Jar design

Various designs of anaerobic jar are commercially available, most of which are suitable for use by the evacuation-replacement method. All jars should be vented, for which Schraeder valves are convenient and reliable. Vents are not only necessary for operation by evacuation-replacement, but are essential for monitoring the internal pressure.

3.1.5 Care of the jar

After removal of cultures the jar should be cleaned, dried, and stored (lid unattached) in a warm dry place.

3.1.6 Indicators of anaerobiosis

With all types of anaerobic jar, it is important to include some system to verify the development of anaerobic conditions.

- Methylene blue indicator solution is suitable and is easily prepared in the laboratory (12).
- Disposable methylene blue indicator strips are available commercially from Becton Dickinson.
- Resazurin strips, which perform the same function, are available from Oxoid Ltd.
- Biological indicators can also be used; failure of an exacting anaerobe such as *Clostridium tetani* to grow, or growth of a strict aerobe such as *Pseudomonas aeruginosa*, indicate a failure to achieve satisfactory anaerobiosis.

Note that chemical and biological methods are retrospective indicators and should be considered as adjuncts only to the essential prospective control afforded by pre-incubation manometry.

For further guidance on the use of evacuation-replacement methods for anaerobic jars see Willis (12). A recently introduced device (the Anoxomat) for the automatic operation of jar evacuation-replacement was described and evaluated by Brazier and Smith (13).

3.2 Operation of jar by internal gas generation

Brewer *et al.* (14) described a technique for producing hydrogen inside the closed jar, using moist sodium borohydride which slowly liberates hydrogen in the presence of a cobalt catalyst. Commercial anaerobic jars operating on this principle are produced by Becton Dickinson (GasPak System) and by Oxoid Ltd. These systems incorporate a cold catalyst, the hydrogen being generated from a disposable hydrogen–carbon dioxide envelope which is activated by the addition of water. The apparatus is convenient and efficient, although the unit cost per jar is high compared with the use of cylinders of compressed gas mixture.

3.3 Microenvironment (commonly disposable) systems

Miniaturization has for long been a preoccupation of microbiologists; single tube and plate culture devices for the isolation of anaerobes, many of which were reviewed by Hall (15), date from the earliest days of anaerobic bacteriology. Modern microenvironment anaerobic systems (16) include:

- the Bio-Bag, Type A (Marion Scientific)
- the GasPak Pouch (Becton Dickinson)
- Anaerocult P (Merck)

Among these recent innovations my experience is confined to the GasPak pouch.

3.3.1 GasPak pouch

This disposable system consists of a clear plastic laminate bag containing reagents for the removal of atmospheric oxygen, and with accommodation for three culture plates. After activation of the reagents the pouch is sealed with a snap-on clip and anaerobiosis is created and maintained for up to 72 h; a redox potential (Eh) indicator is included in the pouch. This device is most effective and should be useful for small laboratories, in field work, and in emergency back-up situations.

4. Anaerobic cabinets

An anaerobic cabinet, equipped with glove ports and a rigid airlock for transfer of materials, provides an oxygen-free environment in which conventional bacteriological techniques can be applied to the isolation and manipulation of obligate anaerobes in conditions of strict and continuous anaerobiosis.

The development of efficient, commercially available anaerobic cabinets has provided modern diagnostic and research laboratories with the cultural method of choice, although properly used modern anaerobic jars yield isolation rates of anaerobes comparable with those obtained with anaerobic cabinets (17). Anaerobic cabinets are not a substitute for poorly performed conventional anaerobic techniques; those who fail with anaerobic jars will also fail with cabinets. Moreover, for laboratories whose anaerobic workload is small, an anaerobic cabinet may not be cost-effective compared with an efficient anaerobic jar procedure. Anaerobic incubators, as distinct from anaerobic cabinets, do not seem to offer any special advantage (18).

There is a number of specifications of anaerobic cabinet available in the United Kingdom (19); the Whitley models (Don Whitley Scientific) and the Microflow cabinet (MDH) are recommended (20, 21).

Anaerobiosis in these cabinets is achieved with palladium catalyst pellets and a non-explosive gas mixture (10% hydrogen + 10% carbon dioxide + 80% nitrogen) maintained at a slight positive pressure. The interchange is programmed automatically to evacuate and flush with gas mixture prior to access being gained to the main working chamber; traces of oxygen entering the cabinet from the interchange are rapidly removed by catalysis. The whole working area is maintained at incubation temperature (37°C).

Anaerobic cabinets should always be used with a redox indicator such as resazurin in order to monitor the current status of the chamber atmosphere.

Protocol 2. Preparation of resazurin solution for use in anaerobic chambers

1. Dissolve 0.1 g resazurin in 100 ml distilled water. Store at 4°C.

2. Dissolve the following in 19 ml distilled water

• Tris(hydroxymethyl)aminomethane	4 g
• Glucose	0.2 g
• 0.1% (w/v) aqueous resazurin	1 ml

3. Boil in a Universal bottle until the solution is decolourized. Then remove from the waterbath, tighten the cap, and transfer quickly to the anaerobic chamber.

4. The clear solution will turn pink when the oxygen concentration is about 300 ppm. The intensity of coloration is roughly proportional to the oxygen concentration.

5. Resazurin is also a pH indicator; when the resazurin solution begins to turn yellow (because of the CO_2 in the chamber atmosphere), it should be replaced by freshly boiled solution.

A notable feature of modern anaerobic cabinets is their facility for 'bare hands' operation which avoids the impaired manual dexterity and discomfort of gloved manipulations in the chamber. This is achieved by the use of latex rubber sleeves which are evacuated and flushed with anaerobic gas mixture prior to entry to the chamber via sealable ports.

Important among the range of commercial anaerobic cabinets is the WISE anaerobic work station (Don Whitley Scientific), which is simpler in concept and design than those referred to earlier (22). It consists of three interconnected chambers: (a) a fan-purged interchange chamber leading to (b) a 'low-oxygen' chamber purged with nitrogen which is used for manipulating cultures, and (c) a heated anaerobic chamber for incubation in a standard anaerobic atmosphere. This simple low-cost cabinet is valuable in the diagnostic laboratory.

It is important to economize on gas usage in anaerobic cabinets, because the special gas mixtures used by most cabinets are expensive. Inadequate integrity of the door and port gaskets is a common problem; unless these seals are tight, excessive gas is used, leading to rapid exhaustion of the cylinders. A heavy demand for gas occurs if the entry interchange is operated frequently; one of the great advantages of the evacuable latex sleeves is that small items, e.g. four to six plate cultures, can be moved in and out of the cabinet via the sleeves without using the airlock.

Gas leaks may develop at many points in this type of complex equipment and may be very difficult to trace; a highly sensitive electronic hydrogen leak detector is provided by Don Whitley Scientific, and is regarded as essential ancillary hardware.

Excessive humidity may cause problems in any anaerobic environment produced by the removal of oxygen by catalysis with hydrogen. This is especially true of anaerobic cabinets, as water condensate is derived not only from continuous catalysis but also from the large bulk of constantly changing media. The problem is overcome in Whitley cabinets by the incorporation of a humidistat which can be set between 20 and 80%. The humidistat controls the operation of an external circulating system through which the atmosphere of the cabinet is pumped. Incorporation of an atmospheric scrubber into this unit prolongs the life of the catalyst by removing toxic metabolites, particularly hydrogen sulphide. The MDH cabinet incorporates a refrigeration–dehumidification unit for control of condensate.

4.1 Comment

Previously, anaerobic cabinets were used chiefly for research purposes, and specialist workers in the field usually designed and constructed their own systems. The availability of commercial units enables diagnostic laboratories to take advantage of the benefits offered by a continuity of anaerobiosis which is much more efficient than the cyclic operation of multiple anaerobic jars. Cultures can be inspected at any time without the disadvantages of exposure to air and the temperature drop associated with examination of anaerobic jar cultures. Prolonged incubation of cultures is easier in the cabinet and may reveal an unexpected anaerobic bacterial presence in specimens at first thought not to contain significant organisms. The use of commercially available anaerobic cabinets is becoming more common in diagnostic laboratories as their cost-effectiveness and other advantages are appreciated.

5. Pre-reduced anaerobically sterilized (PRAS) roll-tube techniques

Hungate (1) first developed this technique for the isolation of oxygen-intolerant anaerobes from rumen fluid and sewage. Since then the method has undergone many modifications and improvements (23) and has been adapted in various ways to suit the particular requirements of other anaerobic disciplines (24, 25).

The method has been used mainly as a research tool for the isolation and study of sewage, gastrointestinal, and oropharyngeal bacteria; it has not been widely used in diagnostic work (26, 27). The brief notes which follow are offered only as an introduction to the technique; those who wish to learn more should consult specialized publications (28–30).

In essence, the PRAS method involves the use of pre-reduced anaerobically sterilized media which are prepared in stoppered roll tubes. Exposure of bacteria and culture media to oxygen is avoided by displacing the air in culture tubes with an oxygen-free gas such as carbon dioxide or nitrogen. When the stoppers are removed from the tubes for manipulation of the culture, anaerobiosis in the tubes is maintained by continuous flushing with an oxygen-free gas.

5.1 Preparation of oxygen-free gas

The inert gas (carbon dioxide or nitrogen) from a cylinder is rendered oxygen-free by passing it through a vertical column of heated copper turnings; the heated copper is periodically reactivated (reduced) by passing hydrogen through the column. Alternatively, the inert gas may be rendered oxygen-free with a Model D laboratory Deoxo Purifier. The purifier, which contains Deoxo cold catalyst (palladinized alumina) is inserted in the gas line from a cylinder containing 95% carbon dioxide and 3% hydrogen. The purifier lasts indefinitely and requires no reactivation.

5.2 Anaerobic media

Protocol 3. Preparation of PRAS media

1. Weigh out the dry ingredients and place in a conical flask.
2. Add distilled water, haemin, mineral salts solution (30), rumen fluid or volatile fatty acid mixture, and resazurin solution (to a final concentration of 1 mg/l).
3. Place a polypropylene magnet in the flask, fit a chimney to the flask, and heat on a magnetic stirrer.
4. Boil the medium to drive off the dissolved oxygen, until the resazurin turns from pink to colourless.
5. Remove the flask from the heat and remove the chimney. Immediately pass a stream of oxygen-free CO_2 through the medium; the flow rate should be sufficient to cause gentle bubbling.
6. After cooling the medium add the reducing agents L-cysteine hydrochloride (0.25 g/l) and sodium formaldehyde–sulphoxylate (0.3 g/l).
7. Adjust the pH of the medium.

Protocol 4. Anaerobic dispensing of medium

1. Tube the cooled medium under oxygen-free gas without exposure to air by flushing both the flask of bulk medium and the culture tube continually with oxygen-free gas.
2. Transfer the medium with a medium pump.
3. Close the culture tube with a butyl rubber stopper, withdrawing the gassing cannula as the stopper is inserted.

Protocol 5. Sterilization of medium

1. Wire or clamp the rubber closures firmly in place; alternatively use a combined screw-cap and rubber closure.
2. Sterilize tubes of medium in the autoclave at 121°C for 15 min.
3. After sterilization the autoclave pressure may be reduced rapidly, but do not remove the clamps holding the rubber closures until the tubes have cooled to a safe temperature, e.g. below 80°C.
4. Store tubes at 4°C until required for use as either roll-tube dilution or roll-tube streak cultures.

Protocol 6. Preparation of roll-tube streak cultures

1. Melt the medium.
2. Spin the stoppered tubes of molten medium about their long axes in a roll-tube apparatus; centrifugal force spreads a thin even layer of the agar medium over the wall of the tube, where it hardens.
3. Hold the tube of medium for inoculation vertically on a rotating platform (about 70 r.p.m.).
4. Inoculate the medium with a charged stainless steel or platinum loop by moving the loop upwards over the revolving agar surface; a spiral inoculum streak results. Carry out this procedure under conditions of continuous flushing of the vessels with oxygen-free gas to maintain anaerobiosis.
5. Flame the gas cannulae and necks of the tubes to prevent contamination before and after inoculation.
6. Replace the stopper and place the tube in the incubator.

Protocol 7. Preparation of roll-tube dilution cultures

1. Prepare dilutions of material for culture in an anaerobic diluent by the anaerobic dispensing method.
2. Inoculate appropriate volumes of these dilutions into tubes of molten medium (using the same anaerobic technique) to produce shake cultures.
3. Spin the tubes on the roll-tube apparatus until the agar has set.
4. Incubate the tubes.

Inoculation of PRAS tubes may also be carried out by a closed method in which a syringe and needle are used to inoculate the tube through the rubber seal (23).

5.3 Comment

The re-stoppered roll tube is thus its own anaerobic culture chamber. Culture tubes are incubated without further treatment and they may be examined individually at any time without exposing the cultures to air. That anaerobiosis is maintained in any individual tube is indicated by the resazurin remaining colourless; any trace of oxygen entering the tube rapidly turns the Eh indicator pink. Uninoculated tubes have a relatively long shelf life; they are satisfactory for use until they become oxidized (the resazurin turns pink). Tube breakage is a particular problem during medium preparation (especially at stopper insertion). Chain mail gloves must be worn by workers to avoid the possibility of serious injury when tubing media.

References

1. Hungate, R. E. (1950). *Bact. Rev.* **14**, 1.
2. Barnes, E. M. and Impey, C. S. (1970). *Br. Poultry Sci.* **11**, 467.
3. Moore, W. B. (1968). *J. Gen. Microbiol.* **53**, 415.
4. Walker, P. D., Harris, E., and Moore, W. B. (1971). In *Isolation of Anaerobes* (ed. D. A. Shapton and R. G. Board), pp. 25–38. Academic Press, London.
5. Collee, J. G., Rutter, J. M., and Watt, B. (1971). *J. Med. Microbiol.* **4**, 271.
6. Lepper, E. and Martin, C. J. (1929). *Br. J. Exp. Pathol.* **10**, 327.
7. Watt, B., Hoare, M. V., and Collee, J. G. (1973). *J Gen. Microbiol.* **77**, 447.
8. Willis, A. T. (1964). *Anaerobic Bacteriology in Clinical Medicine* (2nd edn). Butterworths, London.
9. Watt, B. (1973). *J. Med. Microbiol.* **6**, 307.
10. Reilly, S. (1980). *J. Med. Microbiol.* **13**, 573.
11. Jansen, J. E. and Bremmelgaard, A. (1986). *Acta Pathol. Microbiol. Scand.* **B94**, 319.
12. Willis, A. T. (1977). *Anaerobic Bacteriology: Clinical and Laboratory Practice* (3rd edn). Butterworths, London.
13. Brazier, J. S. and Smith, S. A. (1989). *J. Clin. Pathol.* **42**, 640.
14. Brewer, J. H., Heer, A. A., and McLaughlin, C. B. (1955). *Appl. Microbiol.* **3**, 136.

15. Hall, I. C. (1929). *J. Bacteriol.* **17**, 255.
16. Slifkin, M. and Engwall, C. (1988). *Clin. Lab. Sci.* **1**, 114.
17. Collee, J. G. (1980). *Infection* **8** (Suppl. 2), S150.
18. Berry, P. L., Taylor, E., and Phillips, I. (1982). *J. Clin. Pathol.* **35**, 1158.
19. Brazier, J. (1987). *Med. Tech.* **17**, 14.
20. Phillips, K. D. and Willis, A. T. (1981). *J. Clin. Pathol.* **34**, 1110.
21. Brazier, J. S. (1985). *Med. Tech.* **15**, 8.
22. Sisson, P. R., Ingham, H. R., and Byrne, P. O. (1987). *J. Clin. Pathol.* **40**, 286.
23. Hungate, R. E. (1969). In *Methods in Microbiology*, Vol. 3B (ed. R. Norris and D. W. Ribbons), pp. 117–132. Academic Press, London.
24. Smith, P. H. (1966). *Develop. Indust. Microbiol.* **7**, 156.
25. Berg, J-O. and Nord, C-E. (1973). *Scand. J. Dent. Res.* **81**, 163.
26. McMinn, M. T. and Crawford, J. J. (1970). *Appl. Microbiol.* **19**, 207.
27. Killgore, G. E., Starr, S. E., Del Bene, V. E., Whaley, D. N., and Dowell, V. R. (1973). *Am. J. Clin. Pathol.* **59**, 552.
28. Moore, W. E. C. (1966). *Int. J. Syst. Bacteriol.* **16**, 173.
29. Holdeman, L. V. and Moore, W. E. C. (1972). *Am. J. Clin. Nutrit.* **25**, 1314.
30. Holdeman, L. V., Cato, E. P., and Moore, W. E. C. (1977). *Anaerobe Laboratory Manual* (4th edn). Virginia Polytechnic Institute and State University, Blacksburg, VA.

2

Isolation of anaerobes from clinical material

P. N. LEVETT

1. Introduction

With the notable exceptions of the clostridial diseases such as tetanus and botulism, infections caused by obligate anaerobes are endogenous in origin. Such infections are invariably caused by non-sporing anaerobes derived from the normal flora of the large intestine, the female genital tract, or the oropharynx (1). Because of their origin these are usually mixed infections involving several species of anaerobes in addition to facultative anaerobes such as coliforms. Experience throughout the world during the last 20 years has shown that obligate anaerobes are usually the principal components of such infections and that effective therapy depends upon eradication of the anaerobes (2).

The essential question to which the clinician needs an answer is 'are anaerobes present or not?'. It is often unnecessary and almost always wasteful of both time and resources to attempt isolation and identification of all anaerobes present in a specimen. The difficulties of anaerobic work and methods of functioning adequately within these constraints were discussed succinctly by Finegold and Edelstein (3).

2. The specimen

The approach to successful clinical anaerobic microbiology begins with clinical awareness, which in turn leads to the collection of appropriate specimens. Whenever possible it is desirable to examine samples of pus, other exudates, or tissue, rather than swabs. Many studies have shown that anaerobes are better protected from the harmful effects of oxygen in pus than they are on the surface of a cotton wool swab in transport medium (4, 5). Moreover, the recovery of any organism from a swab is likely to be reduced by a factor of about 90%, simply because bacteria adhere to the swab material (4, 6).

Expensive and complex transport tubes and media are neither necessary nor desirable for routine diagnostic work. Indeed, they may encourage complacency among those handling the specimen *en route* to the laboratory bench. If it is impossible or impractical to examine pus, a plain swab in Stuart's transport medium will suffice if transported to the laboratory and processed promptly.

Specimens which routinely should be cultured for anaerobes include the following:

- infected lesions associated with the gastrointestinal tract or the female genital tract;
- deep-seated sepsis (peritonitis, subphrenic abscess);
- blood from septicaemic patients;
- chronic upper respiratory tract infections (sinusitis, otitis media, and mastoiditis);
- suppurative lesions in the lower respiratory tract (particularly aspiration pneumonia);
- cerebrospinal fluid in cases of meningitis associated with chronic upper respiratory tract infection;
- brain abscess;
- dental sepsis;
- superficial abscesses (breast, perianal, Bartholin's);
- other superficial lesions (such as balanitis, paronychia, bite wounds);
- infected wounds (whether surgical or traumatic).

This is not an exhaustive list; further indications are discussed elsewhere (1, 7, 8). Other specimens may be cultured for anaerobes after consultation between clinician and microbiologist. In general, it seves little purpose to culture specimens from sites with an anaerobic flora, such as faeces (but see Section 5), high vaginal swabs from patients without obvious sepsis, chronic leg ulcers, and pressure sores.

3. Processing samples for anaerobes in the laboratory

A number of rapid direct tests are available which indicate the presence of anaerobes in a high proportion of specimens from suppurative infections. In general, these tests are both simple and cheap, yet yield results at least as useful as more expensive methods using some of the newer molecular biological techniques.

3.1 Smell

Smell the sample if it is pus or tissue. Malodorous lesions are ALWAYS anaerobic; occasional samples will contain anaerobes but will not smell.

Gas–liquid chromatography (GLC) can be used to detect the short-chain volatile fatty acids responsible for the smell, but GLC is no faster or more sensitive than the human nose and costs a great deal more to use. Once again it is only necessary to note the presence or absence of a fetid odour.

3.2 Fluorescence

Examine the specimen under long-wave (365 nm) ultraviolet light (Wood's lamp). This applies to all specimens of pus and tissue. Red fluorescence denotes the presence of the black-pigmented *Bacteroides* and *Porphyromonas* species. However, these organisms are almost invariably part of a mixed anaerobic flora, so they are seldom recovered in pure culture; thus red fluorescence can be relied upon as a further marker of anaerobic infection.

3.3 Gram stain

Examine a Gram-stained smear from the specimen. Note the presence of pus cells and organisms, particularly any pleiomorphic Gram-negative rods, small Gram-positive cocci, or Gram-positive bacilli.

3.4 Culture

3.4.1 Media

It is of paramount importance that media for anaerobic culture should be:

(a) sufficiently nutritious to support the growth of fastidious organisms;

(b) adequately reduced, so that oxygen-sensitive anaerobes are not inhibited by oxygen dissolved in the medium.

There is a range of basal media which are useful for the preparation of media for anaerobes. This includes brain–heart infusion, Columbia agar base, Schaedler medium, Wilkins–Chalgren medium, fastidious anaerobe agar (FAA), fastidious anaerobe broth (FAB), and GLC broth. Of these, brain–heart infusion, FAB, and GLC broth are excellent broth media (see Section 4.1). FAA, FAB, and GLC broth are available only from LabM; the other media are available from all major manufacturers of culture media. It is not necessary to use all the basal media listed; in practice one or two will suffice.

Most anaerobic media contain general growth-enhancing and detoxifying compounds, such as starch, sodium bicarbonate, sodium succinate, cysteine hydrochloride, and sodium pyruvate, in addition to compounds which enhance the growth of specific anaerobes (see *Table 1*). Reducing agents may be added to media for clinical microbiology (see Chapter 1). However, note that some compounds (thioglycollate, cysteine, and ascorbate) may become inhibitory to anaerobes on storage. Therefore, media containing reducing agents should be used as fresh as possible.

Isolation of obligate anaerobes from clinical material is simplified by the use of selective media (see *Table 2*), of which there are many variations (9–13). Many of the antibiotic combinations included in *Table 2* are available commercially. The preparation of one commonly used selective medium is described in *Protocol 1*.

Table 1. Medium additives

Compound	Concentration	Stimulatory for
Starch	1 g/l	Anaerobes in general
Sodium bicarbonate	0.4 g/l	Anaerobes in general
Sodium succinate	0.5–2.5 g/l	Anaerobes in general
Sodium pyrovate	1 g/l	Anaerobes in general
Cysteine hydrochloride	0.5 g/l	*Bacteroides fragilis* and *Fusobacterium necrophorum*
Tween 80	1 ml/l	Gram-positive non-sporing anaerobes
Haemin[a]	5–10 mg/l	Black-pigmented *Porphyromonas* and *Bacteroides* spp.
Menadione (vitamin K)[a]	0.5–1 mg/l	*Bacteroides* spp.
Arginine	1 g/l	*Eubacterium* spp.
Pyruvate	1 g/l	*Veillonella* spp.
Pyrophosphate	0.25 g/l	*Porphyromonas asaccharolyticus, Porphyromonas gingivalis,* and *Bacteroides intermedius*

All additives listed (with the exception of Tween 80) are included in FAA (LabM Lab 90).
[a] Haemin/menadione solution is available from Gibco, and is added to media after autoclaving, at a concentration of 10 ml/l.

Table 2. Selective agents for obligate anaerobes

Organism(s)	Selective agents
Obligate anaerobes from clinical material	1. Neomycin (70 mg/l) 2. Nalidixic acid (10 mg/l)
Actinomyces spp.	Metronidazole (5 mg/l)
Bacteroides spp. and *Fusobacterium* spp.	Nalidixic acid (10 mg/l) +vancomycin (2.5 mg/l)
Bacteroides ureolyticus	Nalidixic acid (10 mg/l) +teicoplanin (20 mg/l)
Clostridium difficile	Cycloserine (250 mg/l) +cefoxitin (8 mg/l)
Fusobacterium spp.	1. Rifampicin (50 mg/l) 2. Neomycin (100 mg/l) +josamycin (3 mg/l) +vancomycin (5 mg/l)

Protocol 1. Preparation of nalidixic acid–vancomycin medium for Gram-negative anaerobes

1. Dissolve in 465 ml distilled water
   - Columbia agar base 19.5 g
   - Yeast extract 2.5 g
2. Autoclave at 121°C for 15 min. Cool to 55°C in a water bath.
3. Rehydrate 1 vial of Oxoid GN Anaerobe Supplement (Oxoid SR108) with 10 ml sterile distilled water. This contains

Protocol 1. *Continued*

- Haemin 2.5 mg
- Menadione 0.25 mg
- Sodium succinate 1.25 g
- Nalidixic acid 5 mg
- Vancomycin 1.25 mg

4. Aseptically add to the molten medium:
   - Supplement SR108 10 ml
   - Sterile defibrinated blood 25 ml
5. Swirl to mix and dispense 20 ml aliquots into sterile Petri dishes.
6. Dry the plates in an incubator before inoculation.
7. Plates may be pre-reduced if desired, by incubating in an anaerobic cabinet or an anaerobic jar overnight.

3.4.2 Inoculation

Plate the specimen on to blood agar and MacConkey agar for aerobic incubation, and on to blood agar and selective medium for anaerobic incubation. After streaking out the inoculum, place a 5-μg metronidazole disc near the junction of the inoculum and the first set of streaks on each plate for anaerobic incubation (*Figure 1*).

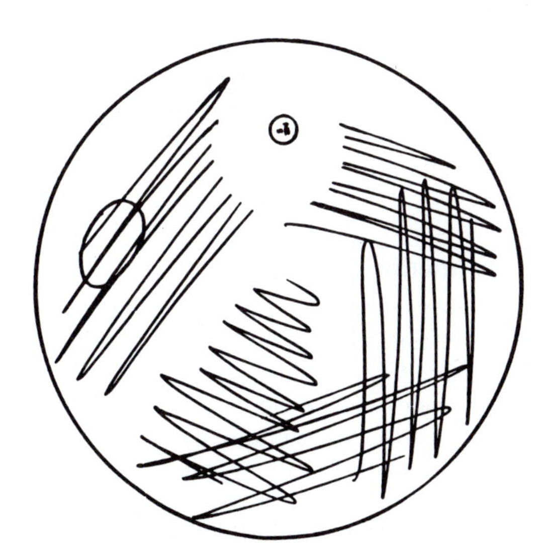

Figure 1. Placement of a 5-μg metronidazole disc on each anaerobic plate after inoculation facilitates early recognition of obligate anaerobes.

3.4.3 Incubation

Incubate in an anaerobic atmosphere containing 10% CO_2 for 48–72 h. Such an atmosphere is conveniently obtained in anaerobe jars or cabinets (see Chapter 1). Reincubate for a further 48–72 h if no growth is evident.

3.4.4 Inspection of plates

After removing the plates from the jar or chamber inspect them under long-wave UV light for red fluorescence as described in Section 3.2. The presence of black-pigmented *Bacteroides* again can be taken as a marker of infection with other anaerobes. Inspect the plates in natural light and compare the growth with that obtained on the plates incubated aerobically. Growth of obligate anaerobes is indicated by a zone of inhibition around the 5-μg metronidazole disc. CO_2-dependent organisms will grow anaerobically only but are resistant to metronidazole.

3.5 Identification and sensitivity testing

Identification of clinical isolates of anaerobes is described in Chapter 3. Susceptibility to antimicrobial agents is determined using standard methods (Chapter 4).

4. Blood cultures

Blood cultures are among the most important investigations performed in diagnostic laboratories. The need for speed in isolation and reporting of significant isolates is apparent, yet so also is the ability to recover the most fastidious organisms which may be present in blood in very low numbers and in the presence of antibiotics.

4.1 Blood culture media

Any blood culture medium must be able to support the rapid growth of a wide range of fastidious organisms from very small inocula and must contain inhibitors of antibiotics.

The traditional approach to blood cultures has been to include at least two media, one specifically for anaerobic culture. In most cases this has been cooked meat broth (CMB) prepared in the laboratory (see *Protocol 2*). However, the lack of clarity associated with cooked meat particles may make visible detection of growth difficult. A clear broth may therefore be preferable (see *Table 3*), prepared in the same volume as the CMB. For paediatric use, blood culture sets should be prepared in Universal bottles, containing 20 ml medium.

Table 3. Recipe for clear broth blood culture medium

Fastidious anaerobe broth (LabM Lab 79)[a]	29.7 g
Sodium polyanethol sulphonate (Liquoid)[b]	0.3 g
Gelatin[c]	1.0 g
p-aminobenzoic acid[d]	0.05 g
Distilled water	1 l

[a] Contains reducing agents sodium thioglycollate (0.5 g/l) and cysteine hydrochloride (0.5 g/l), Eh indicator resazurin (1 mg/l), and haemin/menadione (see *Table 1*).

[b] SPS is an anti-coagulant, anti-phagocytic agent, which neutralizes human serum and also partially inhibits polymyxin B and aminoglycoside antibiotics.

[c] SPS inhibits some strains of anaerobic cocci. Gelatin neutralizes the inhibitory activity.

[d] Competitively inhibits sulphonamides.

Protocol 2. Preparation of cooked meat broth blood culture bottles

You will need

- 100 ml bottles with metal caps that have rubber septa
- Foil covers
- Balance

1. Weigh out and add to each bottle 0.5 g cooked meat granules (LabM Lab 24).
2. Prepare brain–heart infusion containing yeast extract (5 g/l). Heat to dissolve, stirring well.
3. Add 70 ml broth to each bottle.
4. Replace and tighten caps; cover each cap with a foil cover. These can be colour-coded for each medium used in the blood culture set.
5. Autoclave at 121°C for 15 min.

Alternatively, blood culture media may be purchased ready to use. Among the best of such products are Thiol broth (Difco) and supplemented FAB (LabM). Commercially prepared media are supplied with a CO_2 headspace. Growth detection systems such as Bactec (Becton-Dickinson) necessitate the use of commercially prepared media.

4.2 Collection of blood for culture

The correct procedure for collection of blood cultures is described in detail elsewhere (14, 15). The volume of blood cultured is the subject of compromise between an adequate inoculum and necessary dilution of antibacterial activity. The optimum dilution lies between 1/10 and 1/20. 5–10 ml is an appropriate inoculum for a 75-ml culture volume. In cases of endocarditis a larger volume (up to 15 ml) is desirable because of the extremely low number of bacteria usually present. The number of cultures to perform is also open to question. It is not often necessary to perform more than three sets of culture.

4.3 Incubation

Blood cultures are incubated at 37°C as soon as possible after collection. It is not necessary to vent anaerobic bottles.

4.4 Detection of growth

Growth in broth media may be detected visually or by automated growth detection systems. Turbidity produced by anaerobic growth may be difficult to detect if CMB media are used. The Oxoid Signal system consists of a sterile chamber which is attached to the blood culture bottle after receipt in the laboratory. Organisms growing in the medium produce enough gas to force the medium up into the additional chamber, producing a positive 'signal'. Automated systems detect CO_2 produced in the headspace either by radiolabelling or by infra-red spectroscopy.

4.5 Subculture

All blood cultures should be subcultured within 24 h of collection and again after 48 h. A terminal subculture is performed before discarding cultures, after 5–7 days. Routine subculture of all positive cultures after 7 days may yield a significant number of anaerobic isolates (16) since these often occur in cases of polymicrobial septicaemia.

Subcultures are made, preferably by syringe and needle, after disinfecting the rubber septum in the cap. All subcultures from anaerobic bottles are inoculated on to blood agar for aerobic incubation, chocolate agar for incubation aerobically in 10% CO_2, and an enriched blood agar (such as FAA) for anaerobic incubation as described in Section 3.4.3.

5. Detection and isolation of anaerobes from faeces

There are few indications for anaerobic culture of faeces. They include:

- suspected *Clostridium perfringens* food-poisoning
- antibiotic-associated diarrhoea or pseudomembranous colitis
- botulism

Faecal specimens and food can be plated directly on to selective media or may be cultured after heat or alcohol shock. A combination of methods usually yields the highest isolation rates.

5.1 *Clostridium perfringens* food poisoning

This condition is characterized by the onset of nausea, abdominal pain, and diarrhoea, usually 8–24 h after the ingestion of food containing large numbers of vegetative cells of *C. perfringens* type A. The symptoms are relatively mild and are of short duration (<24 h). Because of the food-borne nature of the disease, outbreaks often occur in large institutions such as hospitals, schools, canteens, restaurants, and hotels. Sporadic cases occur in elderly hospital patients. Symptoms result from the action of an enterotoxin released at the time of sporulation.

In order to confirm the diagnosis it is necessary to demonstrate the presence of enterotoxin in faeces. Isolation of *C. perfringens* type A is of less significance, since many individuals carry the organism in their normal gut flora.

Enterotoxin may be detected by a variety of methods, including gel diffusion, tissue culture, ELISA, and reversed passive latex agglutination (see Chapter 9). Of these methods latex agglutination is the most convenient for use in diagnostic laboratories and is available commercially (Oxoid TD930).

5.1.1 Isolation of *Clostridium perfringens* from food and faeces

In order for culture to be of value it is necessary to isolate *C. perfringens* type A in large numbers ($> 10^4$ colony-forming units (c.f.u.)/g) from both the suspect food and from as many patients as possible. Serotyping of isolates is best performed by reference laboratories. Many isolates are non-typable; bacteriocin typing may then be of assistance in demonstrating the identity of outbreak isolates.

Since both heat-sensitive and heat-resistant strains of *C. perfringens* type A may cause food poisoning it is necessary to employ isolation methods for both types (see *Protocols 3* and *4*). Identification of *C. perfringens* is described in Chapter 3.

Protocol 3. Isolation of *C. perfringens* from faeces

1. Prepare a 10% (w/v) suspension of faeces in 1/4-strength Ringer's solution.
2. Prepare serial 10-fold dilutions in Ringer's solution.
3. Plate 0.2 ml aliquots on to neomycin blood agar plates (see *Table 2*) for a total count.
4. Heat the original 10% faecal suspension in a water bath at 80°C for 10 min to activate heat-resistant spores.
5. Prepare dilutions and plate out as described in steps 2 and 3 for a total spore count.
6. Inoculate 1–2 g of faeces into cooked meat broth in a Universal bottle. Heat at 100°C for 60 min. Incubate overnight at 37°C, then subculture 20 μl on to a neomycin blood agar plate for isolation of heat-resistant *C. perfringens*.

Protocol 3. *Continued*

7. Incubate all plates anaerobically at 37°C overnight. Inspect for typical colonies of *C. perfringens* and calculate total counts and spore counts.

Protocol 4. Isolation of *C. perfringens* from food

1. Homogenize, using a stomacher, 10 g of food in 90 ml 0.1% peptone water.
2. Prepare serial 10-fold dilutions in peptone water.
3. Plate 0.2 ml aliquots on to neomycin blood agar.
4. Inoculate 1–2 g of food into cooked meat broth. Incubate overnight at 43°C. Subculture 20 μl on to a neomycin blood agar plate.
5. Incubate all plates anaerobically and enumerate *C. perfringens* as described in *Protocol 3*.

5.2 Antibiotic-associated diarrhoea and pseudomembranous colitis

These conditions represent a spectrum of disease ranging from mild diarrhoea, which resolves spontaneously upon the cessation of antibiotic therapy, to fulminant colitis accompanied by pseudomembrane formation.

Symptoms follow infection of the large bowel by *Clostridium difficile* and subsequent toxin production. A variety of methods has been applied to the detection of *C. difficile* in faeces. These include detection of toxins A and B (see Chapter 9), culture, counter-immunoelectrophoresis, gas–liquid chromatography (17), and latex agglutination (18). However, few methods are more sensitive than culture, and detection of the toxin(s) of *C. difficile* remains the most important diagnostic test.

5.2.1 Isolation of *Clostridium difficile*

Faeces may be inoculated directly on to a selective medium (see *Protocol 5*) or subjected to alcohol shock prior to plating out (see *Protocol 6*). After incubation plates are inspected for the typical colonial morphology of *C. difficile*. Colonies of *C. difficile* exhibit yellow-green fluorescence under long-wave ultraviolet light (365 nm). However, this fluorescence is medium-dependent (19); basal media which support fluorescence include Columbia agar, brain–heart infusion agar, Wilkins–Chalgren agar, fastidious anaerobe agar (LabM Lab90), and Iso-sensitest agar (Oxoid CM471). Commercial *C. difficile* selective agar bases do not support fluorescence.

Cultures of *C. difficile* on media containing *p*-hydroxyphenylacetic acid produce a characteristic odour of *p*-cresol (20). This odour is a further aid to the recognition and identification of *C. difficile*. Again this is a medium-dependent characteristic (21). GLC can be used to identify *p*-cresol production by *C. difficile* (see Chapter 3).

Protocol 5. Preparation of selective medium for *Clostridium difficile*

1. Dissolve the following in 930 ml distilled water.
   - Columbia agar base 39 g
   - Yeast extract 5 g
   - *p*-hydroxyphenylacetic acid 1 g
   - Sodium cholate 500 mg
2. Autoclave at 121°C for 15 min. Cool to 50°C in a water bath and then add aseptically
   - Cycloserine 250 mg[a]
   - Cefoxotin 8 mg[a]
   - Sterile defibrinated blood 70 ml
3. Pour 20-ml aliquots into sterile Petri dishes.
4. Dry the surface of the plates in an incubator before inoculating.

[a] Cycloserine/cefoxitin mixtures are available lyophilized in vials containing 250/8 mg from Oxoid, Difco, Gibco, and LabM. Some manufacturers recommend concentrations of 500/16 mg/l; however, lower concentrations increase isolation rate significantly (22).

Protocol 6. Alcohol shock method for isolation of *C. difficile* from faeces

1. Prepare a 50% (w/v) suspension of faeces in absolute alcohol[a] in a bijou bottle.
2. Mix well on a vortex mixer.
3. Incubate at room temperature for 1 h. Invert the bottle occasionally to remix the specimen.
4. Inoculate 4 drops (100 μl) of the suspension on to the following media
   - Blood agar
   - Selective medium (see *Protocol* 5)
   - Cooked meat broth
5. Incubate plates anaerobically at 37°C for 48 h; then inspect for growth and fluorescence.
6. Incubate cooked meat broth at 37°C for 3–5 days; then subculture on to blood agar and selective medium (as in steps 4 and 5).

[a] Lab alcohol (74 OP) may be substituted.

Several methods for enrichment of *C. difficile* have been described. The choice of enrichment culture lies between non-selective enrichment in cooked meat broth, following alcohol shock, or direct inoculation of selective enrichment medium (23–25). For non-selective enrichment, five drops of the alcohol/stool mixture (as described in *Protocol 6*) are inoculated into a universal container containing cooked meat broth. Broths are incubated at 37°C for 3–5 days before subculture on to blood agar and selective medium. For selective enrichment see *Protocol 7*.

Protocol 7. Selective enrichment for *C. difficile*

1. Prepare brain–heart infusion containing
   - Cycloserine 250 mg/l
   - Cefoxitin 8 mg/l
2. Dispense in 5-ml aliquots and pre-reduce in an anaerobic cabinet or jar prior to use.
3. Inoculate with one drop (0.02 ml) of fluid stool, or two drops of a 50% (w/v) suspension of stool in saline.
4. Incubate anaerobically at 37°C for 48 h.
5. Subculture on to selective medium (see *Protocol 5* and steps 4 and 5 of *Protocol 6*).

Growth of *C. difficile* in selective enrichment medium may be detected by use of a latex agglutination reagent (Serobact *C. difficile* latex agglutination kit; Disposable Products Pty Ltd). One drop of the culture is mixed on a black tile with one drop of *C. difficile* latex reagent. Agglutination within 2 min of mixing indicates the presence of *C. difficile*. Other clostridia, such as *C. bifermentans* and *C. sordellii*, give rise to false positive results; thus it is necessary to confirm a positive latex agglutination result by subculture.

Both non-selective and selective enrichment may increase the isolation rate of *C. difficile* from stools by 50–100% (23, 24). Selective enrichment combined with the use of latex agglutination would appear slightly more sensitive (25). The increased sensitivity offered by enrichment methods is particularly valuable in studies of carriage of *C. difficile* (26), particularly in hospital patients, among whom cross-infection may occur.

5.3 Botulism

The diagnosis of botulism is a clinical one. However, laboratory investigations are indicated for confirmation and for epidemiological purposes. Faeces and food are examined for *Clostridium botulinum* and its toxins; toxin may also be detected in serum from affected patients. Detection of botulinum toxins is accomplished by mouse inoculation and neutralization by type-specific antitoxins (see Chapter 9). Other methods, such as ELISA, lack sensitivity in comparison with the mouse bioassay.

In cases of infant botulism examination of food is not usually helpful and, in the rare case of wound botulism, faecal cultures are also unlikely to be successful. Recently, two cases of infant botulism have been described in which the toxigenic isolates were not *C. botulinum*, but *C. butyricum* and *C. baratii*, producing botulinum toxin of types E and F respectively. Therefore, it is imperative that laboratory investigations include toxin assays. Moreover, any clostridia isolated from stools of infants with botulism in the absence of *C. botulinum* should be screened for toxin production.

5.3.1 Isolation of *C. botulinum* from food and faeces

It is necessary to examine as many foods remaining from a suspect meal as possible, in particular canned or preserved foods. Empty cans or lids should be retained for culture and toxin assays, and relevant information about the products recorded. For isolation of *C. botulinum* from food see *Protocol 8*, and from faeces see *Protocol 9*.

Protocol 8. Isolation of *C. botulinum* from food

1. Record carefully condition of food and any packaging. Record pH and examine a Gram-stained film.
2. Inoculate 2 g portions of suspect foods into each of three pre-reduced CMB in Universal bottles (prepared as described in *Protocol 2*).
3. Heat one CMB at 80°C for 10 min and a second at 60°C for 10 min.
4. Incubate all three CMB at 30°C for 5 days.
5. Subculture on to pre-reduced agar, egg-yolk agar, and selective medium (see *Table 4*).
6. Incubate plates anaerobically at 37°C for 24–48 h; then inspect for growth of lipase-producing clostridia.
7. Pick several colonies to blood agar and egg yolk agar for identification (see Chapter 3) and inoculate CMB for toxin produciton (see Chapter 9).

Table 4. Composition of selective medium for *Clostridium botulinum*

Columbia agar base	39 g
Yeast extract	5 g
50% fresh egg yolk emulsion	60 ml
Cycloserine[a]	250 mg
Sulphamethoxazole[a]	76 mg
Trimethoprim[a]	4 mg
Thymidine phosphorylase[a]	100 IU

[a] Add these components aseptically to molten medium at 55°C.

Protocol 9. Isolation of *C. botulinum* from faeces

1. Inoculate faeces lightly on to selective medium (see *Table 4*) and into CMB (steps 2–4 of *Protocol 8*).
2. Perform alcohol shock (see *Protocol 6*); then inoculate selective medium and CMB as described in step 1.
3. Incubate all plates anaerobically at 37°C for 24–48 h; then follow steps 6 and 7 of *Protocol 8*.

5.3.2 Identification of *Clostridium botulinum*

Any isolate of lipase-producing clostridia should be identified. Isolates from food, in particular, should be assayed for toxigenicity by the methods described in Chapter 9.

References

1. Finegold, S. M. and George, W. L. (ed.) (1989). *Anaerobic Infections in Humans*. Academic Press, San Diego.
2. Willis, A. T. (1977). *Anaerobic Bacteriology, Clinical and Laboratory Practice* (3rd edn). Butterworths, London.
3. Finegold, S. M. and Edelstein, M. (1988). In *Anaerobes Today* (ed. J. M. Hardie and S. P. Borriello), pp. 1–10. John Wiley, Chichester.
4. Collee, J. G., Watt, B., Brown R., and Johnson, S. (1974). *J. Hyg.* **72**, 339.
5. Gargan, R. A. and Phillips, I. (1979). *Med. Lab. Sci.* **36**, 159.
6. Ross, P. W., Cumming, C. G., and Lough, H. (1982). *J. Clin. Pathol.* **35**, 223.
7. Willis, A. T. and Phillips, K. D. (1988). *Anaerobic Infections*. Public Health Laboratory Service, London.
8. Bushell, A. C. (1989). In *Medical Bacteriology: A Practical Approach* (ed. P. M. Hawkey and D. A. Lewis), pp. 91–137. IRL Press, Oxford.
9. Eley, A., Clarry, T., and Bennett, K. W. (1989). *Eur. J. Clin. Microbiol. Infect. Dis.* **8**, 83.
10. Livingston, S. J., Kominos, S. D., and Yee, R. B. (1978). *J. Clin. Microbiol.* **7**, 448.
11. Lyznicki, J. M., Busch, E. L., and Blazevic, D. J. (1982). *J. Clin. Microbiol.* **15**, 123.
12. Morgenstein, A. A., Citron, D. M., and Finegold, S. M. (1981). *J. Clin. Microbiol.* **13**, 666.
13. Wren, M. W. (1980). *J. Clin. Pathol.* **33**, 61.
14. Shanson, D. C. (1989). *Microbiology in Clinical Practice* (2nd edn). Wright, London.
15. Freeman, R. (1989). In *Medical Bacteriology: A Practical Approach* (ed. P. M. Hawkey and D. A. Lewis), pp. 21–42. IRL Press, Oxford.
16. Spencer, R. C. and Nicol, C. D. (1985). Polymicrobial septicaemia. *Lancet* **i**, 1210.
17. Levett, P. N. (1984). *J. Clin. Pathol.* **37**, 117.
18. Borriello, S. P., Barclay, F. E., Reed, P. J., Welch, A. R., Brown, J. D., and Burdon, D. W. (1987). *J. Clin. Pathol.* **40**, 573.
19. Levett, P. N. (1985). *Lett. Appl. Microbiol.* **1**, 75.

20. Phillips, K. D. and Rogers, P. A. (1981). *J. Clin. Pathol.* **34**, 642.
21. Levett, P. N. (1987). *Lett. Appl. Microbiol.* **5**, 71.
22. Levett, P. N. (1985). *J. Clin. Pathol.* **38**, 233.
23. Carroll, S. M., Bowman, R. A., and Riley, T. V. (1983). *Pathology* **15**, 165.
24. Levett, P. N. (1984). *Microbios Lett.* **25**, 67.
25. Riley, T. V., Brazier, J. S., Hassan, H., Williams, K., and Phillips, K. D. (1987). *Epidem. Inf.* **99**, 355.
26. George, R. H. (1986). *J. Antimicrob. Chemother.* **18** (Suppl. A) 47.

3

Identification of clinically important anaerobes

P. N. LEVETT

1. Introduction

Definitive identification of anaerobic isolates in clinical laboratories is not usually justified. Important exceptions are isolates recovered from blood cultures, cerebral abscesses, and some isolates recovered in pure culture from other specimens. Unfortunately, the most reliable methods of identifying anaerobic bacteria are time-consuming and do not fit readily into the routine of a busy diagnostic laboratory, in which anaerobic organisms may not always seem to be of the greatest importance. A number of approaches may be adopted in order to provide clinically useful information without the costs and potential delays associated with conventional identification methods.

2. Initial examination of anaerobic isolates

A great deal of information can be gained from a few simple tests which should be made on all isolates. The results of these investigations usually give some indication of the genus of the isolate, making possible a more rational choice of future tests. The initial work-up of anaerobic isolates is outlined in *Table 1*. These tests represent the absolute minimum required. They can be performed in any laboratory, including those lacking the facilities for further identification. It can hardly be overemphasized that, before any identification can be attempted, a pure culture must be obtained. The confusion created by the use of contaminated cultures was discussed eloquently by Willis (1). It may be necessary to subculture repeatedly in order to obtain a pure culture, particularly when handling some clostridia.

2.1 Atmospheric requirements for growth

Many Gram-positive cocci initially isolated only from anaerobic cultures will be found to grow equally well on subculture in an aerobic atmosphere enriched with CO_2. Such CO_2-dependent streptococci are often misreported as metronidazole-resistant anaerobes. Many non-sporing Gram-positive bacilli are similarly found

Table 1. Initial investigations on all anaerobic isolates

Growth on blood agar
In air
In air+10% CO_2
In an anaerobic atmosphere
Colonial morphology
Haemolysis
Swarming
Pitting
Pigmentation
Odour
Microscopic morphology
Gram stain
Malachite green spore stain

to grow aerobically on subculture. These organisms are usually found to be lactobacilli or propionibacteria, only some species of which are obligate anaerobes. However, some clostridia also grow aerobically, albeit without much vigour. These 'facultative aerobes' include *Clostridium tertium*, *C. carnis*, and *C. histolyticum*. They can be distinguished from the genus *Bacillus* by their colonial morphology and by their failure to form spores under aerobic conditions.

2.2 Colonial morphology

Colonial morphology can be most helpful, particularly for some of the clostridia. Many clostridia produce swarming growth, whereas few, if any, of the non-sporing anaerobes do so. It is relatively simple to distinguish the fine swarming of *C. tetani* or *C. septicum* from the coarse swarming of *C. sporogenes*. To the experienced worker the colonies of many clostridial species are quite distinctive. Many illustrations are found in the reviews by Mitsuoka (2) and by Walker *et al.* (3).

Haemolysis is also a valuable characteristic, but erythrocytes from different species vary in their susceptibility to haemolysins. Many clostridia, including several important pathogens, produce β-haemolysis on horse- or sheep-blood agar. Among the non-sporing anaerobes β-haemolysis is less common. However, some of the black-pigmented *Bacteroides* spp. and, in particular, *Fusobacterium necrophorum* produce intense β-haemolysis.

Few bacterial species produce pitting of agar surrounding the colonies, but many strains of *Bacteroides ureolyticus* do so. This species may be distinguished from the CO_2-dependent *Eikenella corrodens* because the latter produces oxidase but does not produce urease.

A number of *Bacteroides* spp. produce black-pigmented colonies, and this property can be used as an aid to their presumptive identification. Colonies of these organisms exhibit red fluorescence under long-wave ultraviolet light before they become pigmented. Thus red fluorescence can be used to presumptively identify these organisms after only 24–48 h growth. Fluorescence may also be detected in clinical material (see Chapter 2). Several of the black-pigmented *Bacteroides* (*B. asaccharolyticus*, *B. endodontalis*, and *B. gingivalis*) were recently transferred to a new genus, *Porphyromonas* (4). While taxonomic rearrangement of the genus *Bacteroides* was overdue, clinical laboratories should continue to report black-pigmented Gram-negative rods as the '*B. melaninogenicus* group' in order to avoid the inevitable confusion of clinical staff that would result from following taxonomic changes to the letter.

2.3 Microscopic morphology

Every isolate should be Gram-stained. The microscopic morphology of anaerobes is often characteristic, but can be misleading. Many clostridia, for example, lose the Gram-reaction in relatively young cultures; some species fail to produce spores in regular cultures and may then appear similar to *Bacteroides* or *Fusobacterium*. Isolates of Gram-positive cocci also commonly appear Gram-negative in all but the youngest cultures. Over-decolorization can be avoided by using alcohol rather than acetone in the Gram-stain method. Some strains of *Fusobacterium* and *Bacteroides* exhibit extreme pleiomorphism, producing both long filamentous cells and coccoid forms. Further confusion is possible among the non-sporing Gram-positive bacilli, where branching may be present or absent. Experience and familiarity with a wide range of anaerobes make interpretation of Gram-stained films somewhat easier.

If spores are not seen in a Gram stain made from a culture of Gram-positive bacilli at least 48 h old, a malachite green spore stain should be performed. Using this method (5) spores stain green while vegetative cells stain red. Most strains of *C. perfringens* sporulate poorly, if at all, in the usual culture media. However, spore formation can be demonstrated when cultures are grown on a specially developed medium (6).

3. Rapid presumptive identification using disc resistance tests

A presumptive identification can be made by determining the resistance of an isolate to a number of antimicrobial agents. The agents and disc strengths employed are shown in *Table 2*. Discs are available commercially (Mastring-S, Mast Laboratories; An-ident discs, Oxoid).

Table 2. Identification of anaerobes using antibiotic[a] disc resistance[b] tests

	Ery	Rif	Col	Pen	Kan	Vanc	Vic	Gent
Bacteroides fragilis	S	S	R	R	R	R	R	S
Bacteroides oralis/ Bacteroides melaninogenicus	S	S	v	S	R	R	v	S
Bacteroides ureolyticus	S	S	S	S	S	R	R	S
Fusobacterium spp.	R	R	S	S	S	R	R	R
Veillonella	S	S	S	S	S	R	—	—

[a] Ery, erythromycin, 60 μg; Rif, rifampicin, 15 μg; Col, colistin, 10 μg; Pen, penicillin, 2 units; Kan, kanamycin, 1000 μg; Vanc, vancomycin, 5 μg; Vic, Victoria blue; Gent, gentian violet.
[b] S, sensitive (zone of inhibition); R, resistant (no zone); v, variable (some strains S, some R).

Protocol 1. Presumptive identification using disc resistance tests

1. Prepare a suspension of the isolate in sterile brain–heart infusion.
2. Inoculate the whole surface of a blood agar plate to obtain confluent growth.
3. Place a Mastring-S or Oxoid An-ident discs upon the surface of the plate.
4. Incubate anaerobically at 37°C for 48 h, or until confluent growth is visible.
5. Record the sensitivity of the isolate to the antimicrobials. Identify the isolate using the resistance patterns shown in *Table 2*. (*This is not a therapeutic sensitivity test—do not report these sensitivities.*)

Other useful agents for disc resistance tests include Liquoid/SPS (sodium polyanethol sulphonate). A disc containing 100 μg SPS will inhibit *Peptostreptococcus anaerobius*, while other anaerobic cocci are resistant. Resistance to a 5-μg novobiocin disc was similarly used to presumptively identify *Peptococcus* spp. However, this approach was made redundant by changes in the taxonomy of the Gram-positive anaerobic cocci, all but one species of which are now regarded as *Peptostreptococcus*.

Tolerance to certain dyes may also be useful for the presumptive identification of Gram-negative anaerobic bacilli, and can be used to reinforce the results obtained from anti-microbial resistance tests (7).

Protocol 2. Preparation of discs for dye susceptibility tests

You will need

- 1/18 000 aqueous solution of Victoria blue 4R (pH 7.4)
- 1/10 000 aqueous solution gentian violet (pH 7.4)
- Blank filter-paper discs (Whatman A/A discs)

1. Add 2 ml Victoria blue solution to 100 discs in a bijou bottle. Sterilize by autoclaving at 121°C for 20 min.

Protocol 2. *Continued*

2. Similarly add 2 ml gentian violet to 100 discs and sterilize by autoclaving.

3. Determine tolerance or inhibition using the method described in *Protocol 1*.

4. Expected patterns of tolerance are shown in *Table 2*.

---

4. Identification using commercial identification systems

A considerable variety of miniaturized identification kits is commercially available. They are of two types: those which rely on conventional biochemical reactions, principally fermentation of carbohydrate substrates; and those which detect the presence of pre-formed enzymes in cultures of anaerobes. The former are exemplified by the API 20A and Minitek systems, while the latter include the RapID-ANA, AN-Ident, API ATB32A, and MicroScan systems. The API ZYM system can be used in the same way, but no data base is available for this otherwise useful product.

All commercial kits which rely on fermentation tests have shortcomings, because many clinically important obligate anaerobes are asaccharolytic. Thus such kits often require the performance of several supplementary tests (usually including GLC) in order to provide accurate identifications. Moreover, incubation for at least 48 h is invariably necessary. Some of their potential benefits for diagnostic laboratories without anaerobic expertise are therefore diminished. These kits have been largely superseded by systems employing enzyme detection tests.

Kits which rely on the detection of pre-formed enzymes have several potential advantages over other identification methods. They offer rapid identification of anaerobes without the need for preparation of complex media and in the absence of expensive anaerobic apparatus. An aerobic, 4-h incubation period is preferable to the 24–48 h required for the previous generation of commercial identification systems.

Comparative studies of the newer systems and conventional PRAS methods have demonstrated correct identification to species level, without further tests being necessary, of 70–90% of isolates. In some studies there was considerable variation between laboratories. The main disadvantages of these systems are their cost, their lack of flexibility when compared to the conventional approach, and, in some cases, the inadequacy of the data base supporting the system. It is important when using such kits that methods are standardized as far as possible, since enzyme profiles vary with age of the culture tested, the composition of the growth medium used to culture the inoculum, and the density of the inoculum. It is beyond the scope of this volume to give detailed instructions for the operation of commercially produced identification systems, but most are of some use in diagnostic laboratories where cost is not an overwhelming consideration.

5. Conventional identification methods for clinically important anaerobes

The most authoritative source for identification of anaerobes is the *Anaerobe Laboratory Manual* (8, 9). The PRAS methods used are described in detail therein, but the test results tabulated therein should be augmented by reference to the relevant section of *Bergey's Manual of Systematic Bacteriology* (10, 11).

5.1 Identification using PRAS media

Roll-tube methods may be employed for performance of identification tests (8, 12), but in most clinical laboratories these are too time-consuming for routine use, unless PRAS media are readily available from a commercial supplier (e.g. PRAS II, produced by Scott Laboratories Inc). The preparation of PRAS media is described in Chapter 1. Media for identification are prepared in the same way, with the addition of the appropriate substrates (8). The range of media and substrates necessary depends upon the genus or group of genera into which the isolate is placed after the initial investigations (see *Tables 3–5*).

It is important to remember that in any anaerobic fermentation test the pH must be determined after incubation of the cultures. All pH indicators are reduced in an anaerobic atmosphere and change colour. Thus, when PRAS methods are used, it is usual to determine pH with a pH meter. A pH of 5.5 or below then indicates strong acid production (a positive fermentation test),

Table 3. Biochemical test media to be inoculated for differentiation of Gram-positive, non-sporing anaerobic rods

Actinomyces	*Bifidobacterium* and *Lactobacillus*	*Eubacterium*	*Propionibacterium*
Aesculin	Aesculin	Aesculin	Adonitol
Fructose	Amygdalin	Amygdalin	Aesculin
Galactose	Arabinose	Erythritol	Fructose
Glucose	Arginine	Fructose	Glucose
Lactose	Cellobiose	Glucose	Lactose
Maltose	Fructose	Glycogen	Maltose
Mannitol	Galactose	Inositol	Mannitol
Raffinose	Glucose	Lactose	Sorbitol
Ribose	Glycogen	Maltose	Sucrose
Starch	Lactose	Mannitol	Trehalose
Sucrose	Maltose	Mannose	Indole/nitrite
Indole/nitrite	Mannose	Raffinose	Gelatin agar
Urease	Melezitose	Rhamnose	GLC
GLC	Raffinose	Starch	
	Rhamnose	Sucrose	
	Salicin	Indole/nitrite	
	Starch	Gelatin agar	
	Sucrose	GLC	
	GLC		

Table 4. Biochemical test media to be inoculated for differentiation of Gram-negative, non-sporing anaerobic rods

Bacteroides	*Fusobacterium*	*Mobiluncus*
Aesculin	Aesculin	Arabinose
Arabinose	Fructose	Fructose
Cellobiose	Glucose	Galactose
Fructose	Lactose	Glucose
Glucose	Maltose	Glycogen
Inositol	Mannose	Inositol
Lactose	Starch	Lactose
Maltose	Sucrose	Maltose
Mannitol	Indole/nitrite	Raffinose
Raffinose	Egg yolk agar	Starch
Salicin	Bile tolerance	Sucrose
Starch	GLC	Trehalose
Sucrose		Xylose
Trehalose		Hippurate hydrolysis
Xylose		Indole/nitrite
Indole/nitrite		GLC
Gelatin agar		
Bile tolerance		
GLC		

Table 5. Biochemical test media to be inoculated for identification of spore-forming rods and anaerobic cocci

Clostridium		*Anaerobic cocci*
Aesculin	Ribose	Aesculin
Fructose	Starch	Cellobiose
Glucose	Sucrose	Glucose
Lactose	Xylose	Lactose
Maltose	Indole/nitrite	Maltose
Mannitol	Egg yolk agar	Sucrose
Mannose	Gelatin agar	Xylose
Melibiose	GLC	Indole/nitrite
		GLC

pH 5.5–6.0 a weak reaction, and pH 6.0 or above is regarded as a negative reaction. It is always necessary in addition to measure the pH, after growth, in a tube of the basal medium without additional substrates; this acts as a medium control. Some authorities regard any pH difference between control and test of 0.5 units or greater as being more reliable than a fixed definition of positive and negative results (13).

5.2 Agar plate identification method

An acceptable alternative to the use of PRAS media and methods is the use of the plate method for biochemical testing as first described by Phillips (14). This method allows the performance of carbohydrate fermentation tests in addition to tests for aesculin and starch hydrolysis. The plate method can also be used to investigate proteolytic activity, using substrates such as gelatin (15), casein, or milk.

Protocol 3. Agar plate method for fermentation reactions

1. Prepare 20% (w/v) solutions of the desired substrates in sterile distilled water, in 20 ml aliquots in Universal bottles.
2. Similarly prepare an aqueous solution containing aesculin (2% w/v) and ferric citrate (1% w/v).
3. Steam for 10 min on two successive days. After steaming store at 4°C. Discard the solutions if fungal growth becomes apparent.
4. Pour blood agar plates containing 6% sterile defibrinated blood.
5. Dry plates by inverting in a 37°C incubator.
6. Label the plates with the desired substrates. Each plate can accommodate five isolates.
7. Over the surface of each plate spread 1 ml of the appropriate substrate solution.[a] Before use, dissolve any crystals that may have formed during storage at 4°C by heating briefly in a water bath at 60°C.
8. Allow the plates to dry at 37°C.
9. Using a sterile scalpel, divide each plate into six segments.
10. Spot inoculate the isolates to be identified, placing each strain on one segment of each plate. Leave one segment uninoculated to serve as a control.
11. The choice of substrates to be tested depends upon the genus of the anaerobe to be identified (see *Tables 3–5*).
12. Also inoculate each isolate on to a segmented blood agar plate without any substrate, again to serve as a control.
13. Incubate plates at 37°C for 48 h, or until good growth has been obtained.
14. Remove plugs of agar from each agar segment using a plastic drinking straw and a small rubber teat. The plugs of agar can be placed either in the lids of the Petri dishes, or in the wells of a microtitre tray.
15. Add to each plug one drop (20 μl) of 0.04% aqueous bromothymol blue.
16. Observe for a change in colour from dark green to yellow. A blue/green coloration indicates a weak reaction. All control plugs should remain dark green.

[a] False-positive reactions will occur if maltose fermentation is tested on blood agar. Horse blood can be replaced by haemin/menadione (see *Table 1* in Chapter 2).

Protocol 4. Agar plate method for aesculin hydrolysis

1. Prepare blood agar plates flooded with aesculin/ferrric citrate solution as described in steps 2–9 of *Protocol 3*.
2. Inoculate and incubate the plates as described in *Protocol 3*.
3. After incubation inspect the plates for blackening, which indicates hydrolysis of aesculin.

Protocol 5. Agar plate method for starch fermentation and hydrolysis

1. Prepare nutrient agar containing 1% soluble starch.[a]
2. Dispense in 20-ml aliquots in Petri dishes.
3. Dry the surface of each plate before inoculating and incubating as described in *Protocol 3*.
4. After removing plugs of agar as described in *Protocol 3*, flood the plate with Lugol's iodine.
5. Zones of clearing around the areas of growth indicate hydrolysis of starch.

[a] In order to stimulate growth of some of the more fastidious non-sporing anaerobes it may be necessary to supplement the medium with haemin/menadione (see *Table 1* in Chapter 2).

Protocol 6. Agar plate method for gelatin hydrolysis[a]

1. Prepare nutrient agar plates containing 0.4% gelatin.[b]
2. Dispense in 20 ml aliquots in Petri dishes.
3. Dry the surface of each plate before inoculating and incubating as described in *Protocol 3*.
4. After incubation flood the plate with a solution of 15% mercuric chloride in 1 molar HCl.
5. After a few minutes zones of clearing develop around growth of proteolytic organisms.
6. A strongly proteolytic organism, such as *Clostridium sporogenes*, should be used as a positive control.

[a] This method may also be used to determine casein hydrolysis, using 0.4% soluble casein in place of gelatin. Milk hydrolysis requires the addition of 7% sterile skimmed milk in place of gelatin; proteolysis is indicated by zones of clearing around colonies.
[b] In order to stimulate growth of some of the more fastidious non-sporing anaerobes it may be necessary to supplement the medium with haemin/menadione (see *Table 1* in Chapter 2).

Phillips' agar plate method has been in routine use for over 15 years. Experience in the PHLS Anaerobe Reference Unit has demonstrated that results obtained by this method are directly comparable with thôse obtained using PRAS methods (8). The characteristics of some clostridia and Gram-negative anaerobic bacilli commonly encountered in clinical material are shown in *Tables* 6 and 7.

5.3 Egg yolk agar

The production of lecithinase or lipase is determined by growth on egg yolk agar (see *Protocol* 7). Production of these enzymes is an important character in the identification of clostridia and of some fusobacteria. Lecithinase C produces an intense opacity in egg yolk medium which extends away from the colony of the lecithinase-producing organism (1). In contrast, the production of lipase is marked by the development of an opacity restricted to the medium underlying the bacterial colony, accompanied by a pearly sheen overlying the bacterial growth. This superficial layer, composed of insoluble fatty acids, may be floated off the colony by allowing a drop or two of water to run over the surface of the medium.

Protocol 7. Preparation of egg yolk agar

1. Prepare and sterilize by autoclaving 500 ml of a suitable nutrient agar base (such as Columbia agar or brain–heart infusion agar).
2. Allow to cool in a waterbath to 55°C.
3. Crack a clean, fresh egg and remove the pointed end. Carefully pour off the egg white, leaving the yolk inside the shell. It is not necessary to disinfect the egg before removing the end.
4. Add the yolk to 20 ml of sterile phosphate-buffered saline in a wide-mouthed bottle and mix thoroughly. This produces an approximate 50% emulsion.
5. Add 25 ml egg yolk emulsion to the molten agar, mix well, and pour 20 ml aliquots into Petri dishes. Dry plates well before use.

Lecithinases can be neutralized by specific antitoxins, and this property can be used to identify *C. perfringens* type A (see *Protocol* 8). Some strains of *C. bifermentans* produce a lecithinase C which is partially inhibited by antitoxin to the *C. perfringens* alpha-toxin. These two organisms are differentiated by lactose fermentation. In addition, most strains of *C. bifermentans* sporulate heavily in young cultures.

Table 6. Characteristics of some clostridia and fusobacteria isolated from clinical material

Species	Spore location[a]	β-haemolysis	Lecithinase	Lipase	Indole	Acid produced from				Hydrolysis of		End products of glucose metabolism[b]
						Glucose	Lactose	Maltose	Sucrose	Gelatin	Starch	
Clostridium bifermentans	ST	+	+	–	+	+	–	–	–	+	–	A, p, ib, b, iv, v, IC
C. botulinum (types A, B, F)	ST	+	–	+	–	+	–	–	–	+	–	A, p, ib, B, iv, v, ic
C. botulinum (types B, E, F)	ST	+	–	+	–	+	–	+	+	+	+	A, B
C. botulinum (types C, D)	ST	+	–	+	–	+	–	–	–	+	–	A, P, B
C. butyricum	ST	–	–	–	–	+	+	+	+	–	+	A, B
C. difficile	ST	–	–	–	–	+	–	–	–	–	–	A, p, ib, B, iv, v, IC
C. innocuum	T	–	–	–	–	+	–	–	+	–	–	A, B
C. novyi type A	ST	+	+	+	–	+	–	–	–	+	–	A, P, B
C. novyi type B	ST	+	+	–	+	+	–	–	–	+	–	A, P, B
C. paraputrificum	T	–	–	–	–	+	–	+	+	–	–	A, B
C. perfringens	ST	+	+	–	–	+	+	+	+	+	+	A, B
C. ramosum	T	–	–	–	–	+	+	+	+	–	–	A
C. septicum	ST	+	–	–	–	+	+	+	–	+	–	A, B
C. sordellii	ST	+	+	–	+	+	–	–	–	+	–	A, p, ib, b, iv, v, ic
C. spiroforme	T	–	–	–	–	+	+	–	+	–	–	A
C. sporogenes	ST	+	–	+	–	+	–	–	–	+	–	A, p, ib, B, iv, v, IC
C. tertium	T	–	–	–	–	+	+	+	+	–	–	A, B
C. tetani	T	+	–	–	+	–	–	–	–	+	–	A, B
Fusobacterium mortiferum	—	–	–	–	–	+	+	–	+	–	–	A, B
F. necrophorum	—	+	–	+	+	–	–	–	–	+	–	A, P, B
F. nucleatum	—	–	–	–	+	–	–	–	–	–	–	A, B

[a] ST, subterminal spores; T, terminal spores.
[b] A, acetic; P, propionic; IB, isobutyric; B, butyric; IV, isovaleric; V, valeric; IC, isocaproic acids. Capitals indicate major and lower case letters indicate minor end-products.

Table 7. Characteristics of some *Bacteroides* species isolated from clinical material

Species	**Bile resistance**	**Pigmented colonies**	**Indole**	**Hydrolysis of**			**Acid produced from**[a]							
				Gelatin	**Aesculin**	**Urea**	**Ara**	**Glu**	**Lac**	**Raf**	**Sal**	**Sta**	**Suc**	**Tre**
B. asaccharolyticus	−	+	+	+	−	−	−	−	−	−	−	−	−	−
B. bivius	−	−	−	+	−	−	−	+	+	−	−	+	−	−
B. disiens	−	−	−	+	−	−	−	+	−	−	−	+	−	−
B. distasonis	+	−	−	−	+	−	−	+	+	+	+	+	+	+
B. eggerthii	+	−	+	+	+	−	+	+	+	−	−	+	−	−
B. fragilis	+	−	−	−	+	−	−	+	+	+	−	+	+	+
B. intermedius	−	+	+	+	−	−	−	+	−	−	−	+	+	−
B. levii	−	+	−	+	−	−	−	+	+	−	−	−	−	−
B. melaninogenicus	−	+	−	+	+	−	−	+	+	+	−	+	+	−
B. oralis	−	−	−	+	+	−	−	+	+	+	+	+	+	−
B. ovatus	+	−	+	+/−	+	−	+	+	+	+	+	+	+	+
B. ruminicola	−	−	−	+	+	−	+	+	+	+	+	+	+	−
B. splanchnicus	+	−	+	+	+	−	+	+	+	−	−	−	−	−
B. thetaiotaomicron	+	−	+	−	+	−	+	+	+	+	−	+	+	+
B. uniformis	+	−	+	−	+	−	+	+	+	+	+	+	+	−
B. ureolyticus[b]	−	−	−	+	−	+	−	−	−	−	−	−	−	−
B. vulgatus	+	−	−	+	+	−	+	+	+	+	−	+	+	−

Ara, arabinose; Glu, glucose; Lac, lactose; Raf, raffinose; Sal, salicin; Sta, starch; Suc, sucrose; Tre, trehalose.
[b] Produces pitting colonies on agar surfaces.

Protocol 8. Half-antitoxin plate (Nagler plate) method

1. Prepare fresh egg yolk plates as described in *Protocol 7*.
2. Prepare a 1/50 dilution of antitoxin[a] in sterile PBS.
3. With a sterile swab, spread diluted antitoxin over half the surface of an egg yolk agar plate.
4. Allow the antitoxin to dry for a few minutes, then streak the isolate to be tested across the plate. It is most important to streak from the untreated side of the plate towards the antitoxin-treated side. Several isolates can be tested on each plate.
5. Incubate anaerobically at 37°C for 24 h.
6. Inspect for lecithinase production which is inhibited by the antitoxin, forming an arrow-head appearance at the edge of the antitoxin-treated area.

[a] For the identification of *C. perfringens*, gas gangrene antitoxin is suitable.

5.4 Miniaturized methods in conventional identification schemes

Several tests carried out in broth cultures when using conventional methods can be performed by rapid, miniaturized methods (16, 17), which economize both on media and on reagents. Some of those which have been found to be reliable in comparative studies include the spot indole test (*Protocol 9*), nitrate reduction disc test (*Protocol 10*), and the bile tolerance disc test (*Protocol 11*).

Protocol 9. Procedure for spot indole test

1. Prepare a 1% (v/v) solution of *p*-dimethylaminocinnamaldehyde (DMACA; Sigma) in 10% concentrated hydrochloric acid. Store this solution in a dark bottle at 4°C until required.
2. Soak a filter paper in DMACA solution and place in the base of a Petri dish. Drain off any excess DMACA.
3. Using a 3-mm loop, pick from a colony of the organism to be tested, grown on blood agar for 48–72 h.
4. Smear the loopful of growth on to the filter paper.
5. The presence of indole is indicated by the development of a blue or green colour within a few seconds.

Protocol 10. Procedure for nitrate reduction disc test

1. Dissolve in 10 ml distilled water
   - Sodium molybdate 10 mg
   - Potassium nitrate 3 g
2. Sterilize by passing through a 0.45 μm membrane filter.
3. Aseptically dispense aliquots of 20 μl on to sterile filter paper discs. Cover and dry at 37°C. Store at 4°C.
4. Inoculate a blood agar plate with the isolate to be tested, either as described in *Protocol 1*, or by streaking.
5. Place a nitrate/molybdate disc on the inoculated plate and incubate anaerobically at 37°C for 48–72 h.
6. Remove the disc from the surface of the plate and place in the lid of the Petri dish.
7. Add to the disc one drop each of sulphanilic acid and Cleve's acid. The development of a red colour indicates the reduction of nitrate to nitrite.
8. If there has been no colour change after 3–5 min, sprinkle a little zinc dust on to the disc, from the tip of a scalpel blade.
9. The development of a red colour after addition of the zinc dust indicates the presence of unreduced nitrate and, therefore, a negative test. If, however, the solution on the disc remains colourless after the addition of the zinc dust, reduction of the nitrate beyond nitrite has occurred.

Protocol 11. Bile tolerance disc test

1. Dissolve 20 g dessicated ox-bile (Oxoid L50) in 100 ml distilled water.
2. Sterilize by autoclaving at 121°C for 15 min.
3. Aseptically dispense aliquots of 20 μl on to sterile filter paper discs. Cover and dry at 37°C. Store at 4°C.
4. Inoculate a blood agar plate with the isolate to be tested, either as described in *Protocol 1*, or by streaking.
5. Place a bile disc on the inoculated plate and incubate anaerobically at 37°C until growth is evident. Bile-tolerant organisms such as *Bacteroides fragilis* show no zone of inhibition, whereas species that are intolerant of bile invariably show a large zone of inhibition.

6. Gas-chromatographic analysis of end-products of anaerobic metabolism

Gas–liquid chromatography (GLC) is essential for the definitive identification of obligate anaerobes. Some genera are defined by their production of specific end-products, while in other genera individual species may be identified by their end-products alone. The methods described in this section are not exhaustive, but are adequate for identification of clinically important anaerobes in diagnostic laboratories by detection of volatile fatty acids (VFA) and non-volatile fatty acids (NVFA) in spent culture medium (see *Table 8*).

Table 8. Production of short-chain fatty acids by genera of clinically important anaerobes

Actinomyces	Acetic, lactic, succinic
Bacteroides	Succinic, plus others; butyric is never the major product
Clostridium	Many species produce acetic + butyric, others acetic only, others produce many acids
Eubacterium	Variable
Fusobacterium	Acetic, butyric
Lactobacillus	Lactic
Mobiluncus	Acetic, succinic
Peptostreptococcus	Variable
Propionibacterium	Acetic, propionic
Veillonella	Acetic, propionic

6.1 Media for gas-chromatographic analysis

The metabolic activities of bacteria are affected by the composition of the growth medium. This applies equally to the end-products of anaerobic metabolism. Many different basal media are used for cultivation of anaerobes for GLC analysis, including cooked meat broth (CMB), peptone–yeast extract (PY) medium (8), fastidious anaerobe broth (LabM Lab 71), and GLC broth (LabM Lab 77). Most basal media contain trace amounts of fermentable substrates; the concentration of glucose present may be enough to alter the fermentation profiles produced by anaerobes in the medium. Two solutions to this problem have been proposed (18). The first is the use of a basal medium containing minimal glucose, such as GLC broth, which is then supplemented with 0.1% (w/v) glucose; the second is the use of a less refined medium, such as CMB or PY, to which is added 1% (w/v) glucose. The first approach may be preferable, since excess glucose can lead to toxic acid accumulation and may, in addition, inhibit catabolism of amino acids, leading to decreased production of branched chain acids and alcohols. The addition of 2.5% (w/v) calcium carbonate will also help to increase the production of alcohols.

Whatever medium is used, it is important to be familiar with the fermentaiton profiles produced by a wide range of control organisms in the medium. Variation between batches must be controlled, and the presence of traces of volatile compounds in uninoculated medium must also be determined. The incubation time and conditions should be standardized within the laboratory. Finally, the extraction procedure and chromatographic conditions should be constant. Only when these precautions are taken will reproducible results be obtained, which can then be interpreted by comparison with published data (8–12, 19).

Protocol 12. Preparation of VFA standard solutions

1. Prepare standard solutions of each volatile acid by diluting the following volumes to 100 ml with distilled water.
   - Acetic acid 5.7 ml
   - Propionic acid 7.5 ml
   - Isobutyric acid 9.2 ml
   - Butyric acid 9.1 ml
   - Isovaleric acid 12.7 ml
   - Valeric acid 12.5 ml
   - Isocaproic acid 12.6 ml
   - Caproic acid 12.6 ml
2. Keep each standard solution in glass bottles with ground glass stoppers.
3. To prepare a working standard solution mix together 1 ml of each of the standards prepared in step 1; then dilute to 100 ml with distilled water. Store this standard in a stoppered glass bottle.
4. To analyse the working standard solution, follow *Protocol 13*, using the standard in place of culture medium. This should be done daily.

Protocol 13. Analysis of short-chain volatile fatty acids by gas–liquid chromatography

You will need

- 1 ml pipettes
- 15 ml conical centrifuge tubes and bungs
- 50% H_2SO_4
- Diethyl ether (special for chromatography)
- 10 μl graduated microsyringe
- GLC with appropriate column (see *Table 9*)

1. Pipette 1 ml of spent culture medium (or pus) into a centrifuge tube.

Protocol 13. *Continued*

2. Add 0.2 ml 50% aqueous H_2SO_4.
3. Add 1 ml diethyl ether.
4. Seal tube and mix for 10–15 sec on a vortex mixer.
5. Centrifuge briefly to break the ether/water emulsion.
6. Allow the aqueous and ether layers to separate.
7. Inject 1 μl of upper (ether) layer into GLC.
8. Compare retention times of peaks obtained with those of standard VFA mixture run on the same day.

Protocol 14. Preparation of NVFA standard solutions

1. Prepare standard solutions of each non-volatile acid by diluting the following volumes to 100 ml with distilled water.
    - Pyruvic acid 6.8 ml
    - Lactic acid 8.4 ml
    - Succinic acid 6.0 g
2. Keep each standard solution in glass bottles with ground glass stoppers.
3. To prepare a working standard solution mix together 1 ml of each of the standards prepared in step 1, then dilute to 100 ml with distilled water. Store this standard in a stoppered glass bottle.
4. To analyse the working standard solution, follow *Protocol 15*, using the standard in place of culture medium. This should be done daily.

Protocol 15. Analysis of non-volatile fatty acids by GLC

You will need

- 1 ml and 2 ml pipettes
- Pasteur pipettes
- 15 ml conical centrifuge tubes and bungs
- 50% H_2SO_4
- Methanol (special for chromatography)
- Chloroform (special for chromatography)
- Distilled water
- 10 μl graduated microsyringe
- GLC with appropriate column (see *Table 9*)
- Waterbath at 55°C

Protocol 15. *Continued*

1. Pipette 2 ml spent culture medium or 1 ml NVFA standard mixture into a centrifuge tube.
2. Add 2 ml methanol.
3. Add 0.5 ml 50% H_2SO_4.
4. Seal tube and place in waterbath at 55°C for 30 min. Alternatively, leave at room temperature overnight.
5. Add 1 ml distilled water.
6. Add 0.5 ml chloroform.
7. Mix for 10–15 sec on a vortex mixer.
8. Centrifuge briefly to separate chloroform and aqueous layers.
9. Using a Pasteur pipette, withdraw as much as possible of the lower (chloroform) layer and transfer to a fresh tube. Avoid transferring any of the aqueous component.
10. Inject 1 μl of the chloroform extract on to the column.
11. Compare retention times of peaks obtained with those of standard NVFA mixture run on the same day.

6.2 GLC conditions

Almost any GLC can be used for identification of anaerobes, providing an appropriate column packing is used. Most diagnostic laboratories now have access to machines equipped with packed columns and flame ionization detectors, and these are perfectly suited to work with anaerobes. More complex instruments using capillary columns and electron capture detectors can also be used but are not necessary for routine identification work. Some column packings and chromatographic conditions for detection of short chain volatile and non-volatile acids are shown in *Table 9*. Some packing materials (such as Carbowax 20M-TPA) will allow the detection of acids and alcohols under the same conditions. More detailed information for those without experience in gas chromatography will be found elsewhere (20).

6.3 Other end-products detectable by GLC

Many other products of bacterial metabolism are detectable by GLC. The use of these products as markers for the identification of anaerobes is discussed elsewhere (20). One example of the use of a specific compound for the identification of an anaerobic species is the detection of *p*-cresol production by *C. difficile* (21). Incorporation of the substrate, *p*-hydroxyphenylacetic acid, in agar plates, stimulates production of *p*-cresol. This compound gives cultures of

Table 9. Gas chromatograph conditions and column packings suitable for fatty acid end-products analysis

Column length	1.5 m × 4 mm i.d.	1.84 m × 4 mm i.d.	1.8 m × 3 mm i.d.
Packing	10% FFAP on 100–120 mesh diatomite CLQ[a]	6% Carbowax 20M-TPA on 60–80 mesh Gas Chrom Q[b]	10% SP1000 + 1% H_3PO_4 on 100–120 mesh Chromosorb 101[c]
Carrier gas	N_2	N_2	N_2
Flow rate	50 ml/min	40 ml/min	20 ml/min
Injector temperature	250°C	150°C	185°C
Detector temperature	250°C	250°C	185°C
Oven temperature	150°C isothermal	60–160°C at 8°C/min with final holding time of 8 min	160°C isothermal

[a] JJ's Chromatography Ltd.
[b] Phase Separations Ltd.
[c] Supelco. Other suitable column packings include: 15% SP1220 + 1% H_3PO_4 on 100–120 mesh Chromosorb WAW (Supelco), and 5% OV-101 on 80–100 mesh Chromosorb W (Applied Science Labs).

C. difficile a characteristic odour, sometimes described as being of elephants or rhinoceri. To confirm the growth of *C.difficile* on selective medium, *p*-cresol can be detected rapidly by GLC from agar plugs (see *Protocol 16*).

Protocol 16. Detection of *p*-cresol by GLC using agar plugs

You will need a GLC with the following column packing and conditions:

- 1.5 mm × 4 mm i.d. column packed with 3% OV1 on diatomite CLQ (J.J.'s Chromatography)
- Oven temperature 110°C, injector and detector temperatures 250°C, carrier gas N_2 at 50 ml/min

1. Prepare selective medium for *C. difficile* as described in *Protocol 5* in Chapter 2.
2. Inoculate faecal specimen on to selective medium as described in *Protocol 6* in Chapter 2. Incubate anaerobically at 37°C for 48 h.
3. Using a plastic drinking straw and a small rubber bulb, withdraw a plug of agar from an area of the plate close to growth of suspected *C. difficile.*
4. Place the plug in the lid of the Petri dish or in a microtitre tray well. Add one drop of sterile distilled water.
5. Incubate at room temperature for 10 min; then inject 1 μl of the aqueous extract on to GLC column.
6. A standard solution of *p*-cresol (10 μmol/ml) should be used as a positive control, and a plug of agar from an uninoculated plate as a negative control.

6.4 GLC of blood cultures

Anaerobes growing in blood cultures may be identified with the assistance of GLC. Detection of anaerobic metabolites in a bottle showing signs of growth indicates the presence of anaerobes and, in conjunction with a Gram stain, will often confirm the genus of the anaerobic isolate, allowing a presumptive report to be issued to the clinician. In some cases it is possible to go further; a large, square-ended, asporogenous Gram-positive rod that produced acetic and butyric acids would probably be *Clostridium perfringens.* This should be confirmed by inoculating an egg yolk plate directly from the blood culture. Similarly an anaerobic coccus producing acetic and isocaproic acids would be identified as *Peptostreptococcus anaerobius.*

6.5 GLC of clinical material

GLC can be used to provide evidence of anaerobic infection by analysis of volatile acids in specimens of pus and other exudates (see Chapter 2). Many aerobes, however, produce acetic and/or propionic acids as end-products. Thus only the presence of isobutyric acid, or longer-chain acids, can reliably indicate the presence of anaerobes. Since most anaerobic infections are polymicrobial it is not possible to identify the species involved from the GLC pattern. Even when only one anaerobic species is present, there is compelling evidence that GLC profiles produced *in vivo* and *in vitro* are often quite different, again making identification of an infecting anaerobe by GLC of an exudate impossible.

References

1. Willis, A. T. (1977). *Anaerobic Bacteriology: Clinical and Laboratory Practice* (3rd edn). Butterworths, London.
2. Mitsuoka, T. (1980). *A Colour Atlas of Anaerobic Bacteria*. Sobunsha, Tokyo.
3. Walker, P. D., Harris, E., and Moore, W. B. (1971). In *Isolation of Anaerobes* (ed. D. A. Shapton and R. G. Board), pp. 26–38. Academic Press, London.
4. Shah, H. N. and Collins, M. D. (1988). *Int. J. Syst. Bacteriol.* **38**, 128.
5. Duguid, J. P. (1989). In *Practical Medical Microbiology* (13th edn) (ed. J. G. Collee, J. P. Duguid, A. G. Fraser, and B. P. Marmion), pp. 38–63. Churchill Livingstone, Edinburgh.
6. Phillips, K. D. (1986). *Lett. Appl. Microbiol.* **3**, 77.
7. Rotimi, V. O., Faulkner, J., and Duerden, B. I. (1980). *Med. Lab. Sci.* **37**, 331.
8. Holdeman, L. V., Cato, E. P., and Moore, W. E. C. (1977). *Anaerobe Laboratory Manual* (4th edn). Virginia Polytechnic Institute and State University, Blacksburg.
9. Moore, L. V. H., Cato, E. P., and Moore, W. E. C. (1987). *Supplement to the Anaerobe Laboratory Manual* (4th edn). Virginia Polytechnic Institute and State University, Blacksburg.
10. Krieg, N. R. (1984). *Bergey's Manual of Systematic Bacteriology*, Vol. 1. Williams & Wilkins, Baltimore.
11. Sneath, P. H. A. (1986). *Bergey's Manual of Systematic Bacteriology*, Vol. 2. Williams & Wilkins, Baltimore.
12. Sutter, V. L., Citron, D. M., Edelstein, M. A. C., and Finegold, S. M. (1985). *Wadsworth Anaerobic Bacteriology Manual* (4th edn). Star Publishing Co, Belmont, CA.
13. Nakashio, S., Nakamura, S., and Nishida, S. (1982). *Microbiol. Immunol.* **26**, 877.
14. Phillips, K. D. (1976). *J. Appl. Bacteriol.* **41**, 325.
15. Frazier, W. C. (1926). *J. Infect. Dis.* **39**, 302.
16. Sutter, V. L. and Carter, W. T. (1972). *Am. J. Clin. Pathol.* **58**, 335.
17. Wideman, P. A., Citronbaum, D. M., and Sutter, V. L. (1977). *J. Clin. Microbiol.* **5**, 315.
18. Turton, L. J., Drucker, D. B., Hillier, V. F., and Ganguli, L. A. (1983). *J. Appl. Bacteriol.* **54**, 295.
19. Willis, A. T. and Phillips, K. D. (1988). *Anaerobic Infections*. Public Health Laboratory Service, London.
20. Drucker, D. B. (1981). *Microbiological Applications of Gas Chromatography*. Cambridge University Press, Cambridge.
21. Phillips, K. D. and Rogers, P. A. (1981). *J. Clin. Pathol.* **34**, 642.

4

Susceptibility testing of anaerobic bacteria

JON E. ROSENBLATT

1. Introduction

The importance of anaerobes as causes of significant infections is now widely recognized. The medical journals are replete with articles devoted to the diagnosis and treatment of these infections. In particular, studies in patients and experimental animals have documented the benefit of specific antimicrobials directed against anaerobes in the prophylaxis and therapy of infections involving these organisms (1, 2). Such benefits have been recognized even though bacteria other than anaerobes are also often isolated from these infections. A decline of 45% in the number of anaerobic bacteraemias at the Mayo Clinic between 1974 and 1988 has, at least in part, been attributed to specific antimicrobial prophylaxis and therapy (3).

Performance of susceptibility testing is generally viewed as a necessity for effective guidance of specific antimicrobial therapy. Such testing of anaerobes has not become routine in many clinical laboratories for a variety of reasons. One of these is the misconception that the susceptibility patterns of anaerobes are predictable and have not changed appreciably in recent years. In fact, a review of the activity of various antimicrobials against anaerobes illustrates the following points: development of resistance is an increasing problem; variability in susceptibility occurs from one institution to another making total reliance on someone else's published data unwise; and the introduction of many new antimicrobials creates uncertainty in physicians over their role in treating various infections.

Another reason for hesitation is the lack of a generally accepted and easy-to-perform method for susceptibility testing of anaerobes. The National Committee for Clinical Laboratory Standards (NCCLS) has recently published the second edition of *Methods for Antimicrobial Susceptibility Testing of Anaerobic Bacteria* (M11-T2) (4). This 'tentative standard' combines two previous documents, M11-A and M17-P, and represents the work of a new subcommittee on anaerobe susceptibility testing. The new document recognizes some of the shortcomings of the previously proposed standards and attempts to resolve the disagreement among experts in the area regarding the most appropriate methodology. There has also been considerable confusion arising from reports of differing results based on different methods.

Another major question concerns the extent of testing which should be performed. Most authorities agree that it would not be cost-efficient to do susceptibility tests on all anaerobes isolated from clinical specimens. The questionable significance of many isolates from mixed wound infections and the good response to standard antimicrobial regimens obviates the need for routine testing. Reliance on published surveys is not entirely acceptable, since susceptibility patterns may vary from one medical centre to another. Periodic (at least annual) surveys within an institution may be an acceptable alternative to the testing of individual clinical isolates, although rapid testing for the presence of β-lactamase in certain Gram-negative bacilli could be done (see Section 5.6).

For those laboratories testing individual isolates, there is no complete consensus on which isolates should be selected. The policy at the Mayo Clinic is to routinely test anaerobes isolated from cultures of blood, brain abscesses, bones and joints, and normally sterile body fluids as well as those isolated in pure culture from other sources; other isolates are tested when the physician has made a specific request and the isolate has been determined to be clinically significant and worthy of a full identification workup. The selection of antimicrobials to be tested should be guided by a knowledge of published *in vitro* activity and *in vivo* efficacy as well as availability on the hospital formulary.

2. Antimicrobial resistance of anaerobes

One major justification for anaerobic susceptibility testing is the development of resistance to antimicrobials which previously had been considered universally active. Although there has been little or no change in the susceptibility patterns of some agents such as chloramphenicol and metronidazole, significant resistance has developed to others such as clindamycin and penicillin. Resistance is variable among the cephalosporins and cephamycins. The most notable changes have occurred in the *Bacteroides*, especially within the *Bacteroides fragilis* group of anaerobic Gram-negative bacilli.

In an ongoing survey of susceptibility results from eight medical centres in the United States, resistance to clindamycin varied from 3% in 1982 to 9% in 1984. However, rates for individual participants ranged from 0 to 13% in 1981, from 0 to 8% in 1982, and from 6 to 13% in 1984 (5). Multicentre resistance rates of 8–12% for cefoxitin and 15–22% for moxalactam were also noted; one hospital experienced an outbreak of cefoxitin resistance with rates as high as 30%. These results illustrate the need for determining the susceptibility patterns of *B. fragilis* group isolates in individual institutions because of the significant and varying resistance to numerous antimicrobials. The results also emphasize the differing susceptibility patterns of the various species within the *B. fragilis* group,

B. fragilis being the most susceptible and *Bacteroides thetaiotaomicron* being quite resistant.

B. fragilis group organisms have long been recognized as potent producers of β-lactamase. Studies at the Mayo Clinic have pointed out that species of *Bacteroides* other than those of the *B. fragilis* group are also often potent producers of β-lactamase and suggested that they also might be considered resistant to penicillins. Antimicrobial susceptibility results were reviewed for the 6-year period 1982–87 and compared with three previous surveys at that institution (6). There appeared to be a striking increase in penicillin resistance by the 'non-fragilis' species of *Bacteroides*. Almost 80% of the 1982–87 strains (and all isolates with a penicillin minimum inhibitory concentration (MIC) >1.00 mg/l) produced β-lactamase. These 'non-fragilis' *Bacteroides* included the following species: *B. bivius*; *B. capillosus*; *B. disiens*; *B. oralis*; *B. melaninogenicus*; *B. ruminicola*; and some non-speciated strains.

Review of the Mayo Clinic survey data through 1987 not only demonstrated the development of penicillin resistance in the 'non-fragilis' *Bacteroides*, but also the appearance of four *Fusobacterium* isolates which produced β-lactamase and had penicillin MICs >64 mg/l. However, the most significant finding was that resistance of *B. fragilis* group isolates to clindamycin (MIC >8 mg/l) increased from 3% in the 1977–81 time period to 8% for 1982–87 (see *Table 1*). This increase seems more important when related to the rates of 1.4% in 1974 and 0% in 1971 from the same institution.

Table 1. Antimicrobial susceptibility for *Bacteroides fragilis* group,[a] Mayo Clinic 1982–87

Antimicrobial agent	No. of strains	MIC_{50}[b]	MIC_{70}[b]	MIC_{90}[b]	Percentage of isolates resistant at specific breakpoints[c] (Percentage of 1977–81 isolates resistant at specific breakpoints for comparison[d])
Chloramphenicol	677	4	4	8	0.2 (0)
Clindamycin	677	<0.5	<0.5	2	8 (4)
Metronidazole	677	1	1	2	0 (2)
Moxalactam	487	2	4	32	7
Penicillin	677	16	32	>64	63 (84)
Cefoxitin	555	4	8	16	5 (2)
Mezlocillin	168	8	16	64	7
Imipenem	179	<0.5	<0.5	1	1

From Musial and Rosenblatt (6), with permission.

[a] Includes *B. distasonis*, *B. fragilis*, *B. ovatus*, *B. thetaiotaomicron*, and *B. vulgatus*.

[b] Minimum inhibiting concentration for 50% (MIC_{50}), 70% (MIC_{70}), and 90% (MIC_{90}) of isolates.

[c] Breakpoints (readily achievable serum concentrations in mg/l): chloramphenicol (16), clindamycin (4), metronidazole (16), moxalactum (32), penicillin (8), cefoxitin (32), mezlocillin (64), and imipenem (8). Isolates are resistant when the MIC is greater than the breakpoint concentration.

[d] Breakpoints (readily achievable serum concentrations in mg/l): chloramphenicol (12.5), clindamycin (3.2), metronidazole (12.5), penicillin (6.25), and cefoxitin (25).

3. Mechanisms of antimicrobial resistance in anaerobes

Most isolates of the *B. fragilis* group and approximately half of other *Bacteroides* as well as occasional isolates of *Clostridium* and *Fusobacterium* produce β-lactamases. This is presumably the mechanism of their resistance to β-lactam antimicrobials. In general, the enzyme produced by the *B. fragilis* group can be described as a cephalosporinase while that of other *Bacteroides* spp. is a penicillinase. Some strains seem to be resistant by mechanisms other than β-lactamase production. Indeed, penicillin-binding proteins (PBP) have been found in *Bacteroides* and there are data indicating that the decreased susceptibility of these strains to some β-lactams may be due to low affinity between the PBP and the drug. In addition, limited outer membrane permeability to β-lactams has been shown to be a contributing factor to resistance in certain strains of *B. fragilis*. Increased negative charge of a drug and increased molecular weight have been associated with decreased uptake.

As suggested in Section 2, resistance of *Bacteroides* to clindamycin seems to be increasing slowly. Such resistance is linked genetically to erythromycin (macrolides) and streptogramin ('MLS pattern') and is similar to that seen with streptococci and staphylococci. It is caused by production of a methylase which enables methylation of adenine in 23S ribosomal RNA, the ribosome-binding site of MLS antimicrobials. Such resistance can be inducible or constitutive. Plasmids and transposons in *Bacteroides* have been shown capable of conferring clindamycin resistance to other strains of *Bacteroides* (7). The only other examples of transferable plasmids associated with resistance in *Bacteroides* involve production of the inactivating enzyme, chloramphenicol acetyltransferase, and decreased accumulation of tetracycline (demonstrated in *Escherichia coli*). Resistance of anaerobes to chloramphenicol remains rare but tetracycline resistance is so common as to make that drug clinically useless; most instances of tetracycline resistance are not plasmid-mediated.

4. Development of new antimicrobials

A myriad of new antimicrobials has been introduced in recent years, many of which are claimed by their manufacturers to have significant anti-anaerobe activity. In a number of instances, the significance of that activity is questionable. Clinicians may be interested in using some of these agents and can be guided by the results of susceptibility tests. For the newer penicillin derivations there are data suggsting that carbenicillin and ticarcillin inhibit most anaerobes, including the *B. fragilis* group, at the high concentrations usually achieved in blood after dosages needed to treat infections caused by *Pseudomonas*. The possibility of *in vivo* inactivation of these agents by *Bacteroides* β-lactamase creates some uncertainty about their clinical usefulness. Similar questions have arisen regarding the activity of the acyl-ureido penicillins (azlocillin, mezlocillin, apalcillin, and piperacillin) against anaerobes.

Clavulanic acid (CA) and sulbactam (SUL) are potent inhibitors of β-lactamases including those produced by species of *Bacteroides*. Since β-lactamase production is a major mechanism of resistance of some anaerobes to β-lactam antimicrobials, it was anticipated that CA and SUL might be able to augment the activity of these β-lactams. One or the other of these agents has been combined *in vitro* with a number of β-lactam antimicrobials including penicillin, ampicillin, amoxicillin, azlocillin, ticarcillin, cephalothin, cefazolin, cefoperazone, and several 'third-generation' cephalosporins. In general, the activity of the combination has been much greater than that of the β-lactam antimicrobial alone (with the exceptions of cefoxitin and moxalactam) against β-lactamase-producing anaerobes (8). For example, the addition of 2.0 mg/l CA to ticarcillin reduced the MIC_{90} for *B. fragilis* group isolates from 64 to 0.5 mg/l.

Recent years have witnessed the introduction of a large number of new β-lactam antimicrobials including the 'third-generation' cephalosporins and carbapenems. Rolfe and Finegold (9) performed a comprehensive comparative survey of these newer agents using 203 strains of anaerobic bacteria (see *Table 2*). An oxacephem, moxalactam, was found to be the most active of the cephalosporins. The activity of cefotaxime was not nearly so impressive. Only 54% of *B. fragilis* group isolates were considered susceptible. Other β-lactams with unimpressive activity against *Bacteroides* and other anaerobes include ceftriaxone, ceftazidime, and cefoperazone, although the latter has been combined with β-lactamase inhibitors (CA and SUL), resulting in potentiation of its activity.

Cefmetazole is a cephamycin similar in structure and activity to cefoxitin and is resistant to hydrolysis by β-lactamases. It has been used clinically in Japan for several years where its efficacy has been established. Clinical trials have been initiated in the United States. Jones *et al.* (10) using a broth microdilution method demonstrated essentially equal activity between cefmetazole and cefoxitin against 97 anaerobes.

There has been considerable debate over the anti-anaerobe activity of two other newer β-lactams with relatively long half-lives. The clinical utility of ceftizoxime and cefotetan has been promoted because of the economic benefit of less frequent dosing. Howver, marked variability in susceptibility results for ceftizoxime against the *B. fragilis* group has been noted and seems related to different inoculum sizes and types of media used. Testing of ceftizoxime is a prime example of how use of different methods, rather than one standardized one, can result in conflicting results from different laboratories.

Irrespective of these questions surrounding susceptibility-testing methodology, the activity of cefotetan seems inferior. This agent has been promoted as having good activity against '*B. fragilis*'. However, one should be quick to note that this applies only to that particular species and that cefotetan has poor activity against the other species of the *B. fragilis* group. This relatively poor activity is present in spite of its 7-α-methoxy cephalosporin structure which confers stability to β-lactamases in general.

Table 2. Percentage of strains of anaerobes susceptible to achievable serum concentrations of β-lactam antimicrobials

	Number of strains	Cefoperazone (32)[a]	Cefotaxime (32)	Cefoxitin (32)	Ceftazidime (32)	Ceftizoxime (32)	Moxalactam (32)	Thienamycin (32)
Bacteroides fragilis group	46	74	65	87	57	30	72	100
Other *Bacteroides* spp.	69	96	96	97	94	96	96	99
Clostridium perfringens	3	100	100	100	100	100	100	100
Other *Clostridium* spp.	14	64	64	86	21	50	79	100
Fusobacterium spp.	13	77	69	77	54	62	77	77
Gram-positive cocci	25	100	100	100	96	100	100	100
Gram-positive non-spore-forming bacilli	15	73	80	93	67	93	73	100

Adapted from Rolfe and Finegold (9).
[a] Achievable serum concentrations in mg/l.

The carbapenems differ from other β-lactam antimicrobics by containing a hydroxyethyl side chain on the β-lactam ring rather than an acyl amino substituent. The original compound of this group was thienamycin. Imipenem is *N*-formimidoyl thienamycin and is the first carbapenem to become available for clinical usage. This is the most active of the newer β-lactams against anaerobes. Moreover, imipenem has bactericidal activity against anaerobes. Ninety-eight to one hundred per cent of various anaerobes (including the *B. fragilis* group) are susceptible. The only exception is *Fusobacterium*, of which 23% of isolates are resistant.

The 4-quinolones are derivatives of nalidixic acid with generally enhanced activity against many bacteria. However, their activity against anaerobes is variable and somewhat limited. Ciprofloxacin is the most promising of the group although it is not yet recommended for treatment of anaerobic infections. Ciprofloxacin has little effect on the normal anaerobic bowel flora suggesting limited *in vivo* activity. A newly described fluoroquinolone (WIN 57273) has been shown to be extremely active against anaerobes, including the *B. fragilis* group; MICs were similar to those achieved with imipenem. Should further experience with this agent show it to be effective *in vivo* and free of toxicity, it could play an important role in therapy of anaerobic infections.

Metronidazole has been a mainstay of therapy of anaerobic infections for more than 10 years. Recent work has shown that its hydroxy metabolite, which is formed in humans, is almost as active against anaerobes as the parent compound. This observation suggests that the metabolite may act in concert with metronidazole during treatment. Several new nitroimidazoles also have strong anti-anaerobe activity. Tinidazole is most active having a mean MIC of 0.12 mg/l against *B. fragilis* group organisms, while ornidazole is equal in activity to metronidazole. None of the nitroimidazoles has good activity against the non-sporing Gram-positive bacilli. A small number of *Bacteroides* resistant to the nitroimidazoles have been reported but their clinical relevance is uncertain.

Metronidazole rapidly kills anaerobes. This bactericidal effect has not been observed with other antimicrobials (except for imipenem) which have good inhibitory activity. Clindamycin, for example, which acts on the 50S ribosomal subunits of bacteria to inhibit protein synthesis, is not consistently bactericidal against *B. fragilis*. Killing by clindamycin over a 24-h period is slower and less complete than that of metronidazole over only a 6-h period.

5. Methods for antimicrobial susceptibility testing of anaerobes

5.1 General

The publication by the NCCLS of Tentative Standard M11-T2 (4) provides a source for a standard method of susceptibility testing of anaerobes. Unfortunately, this agar dilution method is not entirely practical for daily testing of small numbers of clinical laboratory isolates. However, it can be used as a routine method by those laboratories which test at least seven to eight isolates daily or several times a week and does provide a standard for doing large batch surveys to determine changes in susceptibility patterns at individual institutions. Moreover, it should be the method of choice for studying the activity of new antimicrobials. Use of a standard method by different investigators will allow more valid comparisons of their results. Currently, frequent disagreements arise over the activity of new agents (especially the β-lactams) because of differing *in vitro* results achieved by different investigators using different methods.

Unfortunately, the NCCLS standard is not without problems. The medium originally specified (Wilkins–Chalgren agar) does not support the growth of a number of strains of clinically important anaerobes and is not used by some experienced and widely recognized authorities in this field. Therefore, the 'second edition' of the standard includes an alternative medium, Brucella agar supplemented with 5% laked sheep blood and Vitamin K (1.0 mg/l) as used in the 'Wadsworth agar dilution procedure'. The addition of this medium and vagueness in the recommendations for media to be used in another alternative procedure (broth-micro dilution) introduce elements of variability into the standard which may leave some readers uncertain about which media to select and, in general, makes the methodology less 'standardized'.

5.2 Antimicrobial breakpoints

Although there has been considerable debate regarding the selection of antimicrobial 'breakpoints' (the drug concentration which separates susceptible from resistant strains), the NCCLS has included a list of tentative recommendations in M11-T2 (see *Table 3*). These are based on known clinical efficacy and measured serum levels after maximum dosage therapy. There are only two categories used for reporting, either 'susceptible' or 'resistant'.

5.3 Quality control procedures

All laboratories performing susceptibility testing must have in place an adequate system of quality control to verify the accuracy of the testing procedure. The system will monitor the performance of the various components of the methodology as well as the laboratory personnel. The testing should be done using organisms of known susceptibility which are genetically stable. Three such organisms have been identified by the NCCLS and are available from them or the American Type Culture Collection (ATCC):

- *Bacteroides fragilis*, ATCC 25285
- *Clostridium perfringens*, ATCC 13124
- *Bacteroides thetaiotaomicron*, ATCC 29741

Table 3. MIC[a] breakpoints indicating susceptibility[b]

Drug	MIC
Ampicillin/sulbactam	16/8
Carbenicillin	128
Cefamandole	16
Cefoperazone	32
Cefotaxime	32
Cefoxitin	16
Chloramphenicol	16
Clindamycin	2
Imipenem	8
Metronidazole	16
Mezlocillin	128
Penicillin G	8
Piperacillin	128
Ticarcillin	128
Ticarcillin/clavulanate	128/2

Adapted from NCCLS document M11-T2 (4)

[a] Minimum inhibitory concentration in mg/l.

[b] Interpretation comprises *'susceptible', 'moderately susceptible',* and *'intermediate'* categories designated in NCCLS documents M2 and M7, Table 2. Interpretation guidelines are based on pharmacokinetic data from maximum dosage as available in drug package inserts or other sources, as well as consideration of population-distribution of MICs and studies of clinical efficacy.

The NCCLS has published mode MIC values for the above organisms and some 20 antimicrobials having activity against anaerobes which can be used for comparison in quality control testing. Such testing should be performed on each new batch of susceptibility plates and periodically thereafter. Each test run should also include organism 'growth controls', i.e. plates of media containing no antimicrobials.

An additional five anaerobes have been studied (making a total of eight NCCLS reference organisms) and published mode MIC values for selected antimicrobials are available (4). These organisms may be used to augment quality control procedures or to evaluate new methods or antimicrobials. Results with the new test or drug can be compared with those obtained by the reference agar dilution method.

5.4 Specific testing methods

Protocol 1. NCCLS Reference Agar Dilution Method[a]

1. Prepare Wilkins–Chalgren agar (WCA) containing 1:10 dilutions of pre-prepared antimicrobial solutions to achieve final concentrations ranging from 0.125 to 512 mg/l for each antibiotic to be tested (narrower ranges of concentrations may be desirable for some agents). Add 2 ml of antimicrobial diluted in distilled water to the appropriate volume (18 ml for use with round 15 mm × 100 mm Petri plates) of melted and cooled agar in a tube and invert 5–10 times.
2. Pour into 15 mm × 100 mm diameter Petri plates and allow to harden. These plates ideally should be prepared on the same day as the susceptibility tests are to be run but can be stored for 7 days at 4–10°C. It is recommended that plates containing imipenem be used on the day of preparation, although our own experience is that such plates may be stored for 2–3 days without significant loss of activity.
3. Inoculate portions of five or more colonies of a plate culture into enriched thioglycollate broth and incubate anaerobically for 6–24 h until maximum turbidity is obtained.
4. Adjust the turbidity of the actively growing broth culture to a 0.5 McFarland standard, using Brucella broth. Alternatively, make a suspension from colonies on an agar plate less than 72 h old.
4. Transfer each culture to one of the wells of a Steers replicator which is then used to transfer inoculum to the surface of plated susceptibility medium. A total of 36 separate cultures may be tested at one time including appropriate growth controls and organisms of known susceptibility. The final inoculum density should be approximately 1×10^5 c.f.u. per spot.
6. Incubate inoculated plates in a GasPak jar or a similar anaerobic environment at 35°C for 48 h.
7. Read the MIC as the lowest concentration of antibiotic yielding no growth on the agar surface, a barely visible haze, one discrete colony, or multiple tiny colonies.

[a] NCCLS document M11-T2 is available from the NCCLS, 771 E. Lancaster Ave., Villanova PA 19085, phone (215) 525-2435, and should be consulted for details concerning these procedures.

Protocol 2. Wadsworth agar dilution procedure[a]

1. Prepare Brucella agar base according to the manufacturer's recommendations.

Protocol 2. *Continued*

2. Lake (lyse) sheep blood by alternate freezing and thawing and add to the medium to a final concentration of 5%.
3. Add vitamin K_1 (menadione) to the medium (1 mg/l) and sterilize.
4. Proceed in preparation, inoculation, incubation, and reading of plates as outlined in *Protocol 1*.

[a]This method is similar to the reference agar dilution procedure described in *Protocol 1* except that Brucella blood agar rather than WCA is used. The former medium supports the growth of more anaerobes than the latter.

Protocol 3. Broth microdilution method

1. Prepare broth media according to the manufacturer's instructions. Several broth media have been used and are acceptable, including Schaedler's, WCA (without the agar—available commercially as Anaerobe Broth, MIC, Difco), and brain–heart infusion. Some of these may need supplementation (such as serum) for good growth of all anaerobes.
2. Prepare dilutions of antimicrobials as for the above methods (see *Protocol 1*) and dispense in 0.1-ml volumes per well of plastic microtitre plates. These plates can then be sealed in plastic bags and stored at −70°C for up to 4–6 months.
3. Pick portions of five or more colonies from fresh blood agar plates into enriched thioglycollate or other broth and incubate for 6–24 h. Alternatively, a turbid suspension can be achieved by scraping colonies directly from the surface of an agar plate.
4. Thaw frozen microtitre plates and reduce in an anaerobic environment for 2–4 h.
5. Adjust the turbidity of the inoculum to the 0.5 McFarland standard and dilute in broth to achieve a final inoculum in each microtitre well of 1×10^5 c.f.u. (a 1 : 10 dilution is appropriate if 0.01 ml of inoculum is to be added to each well).
6. Add 0.01 ml of inoculum to each well using a replicator or micropipette. This should be accomplished within 15 min of preparing the final dilution.
7. Incubate the plates anaerobically for 48 h at 35°C. They should be covered to prevent evaporation; stacking no more than four high is acceptable.
8. Read the MIC as the lowest concentration which inhibits growth or where the most significant reduction in growth is observed.

5.5 Additional testing methods

Two additional procedures have been used for susceptibility testing of anaerobes. The details of the methods are provided in NCCLS document M11-T2 and will not be given here. The macro broth dilution procedure is similar in principle to the broth microdilution technique but uses larger-sized test tubes and larger volumes of media and antimicrobials. This method is somewhat cumbersome and inefficient and has not been widely used. It may have some utility in those rare instances where determination of bactericidal end-points against anaerobes seems important (bacterial endocarditis, in particular). In this situation this procedure is used in conjunction with those described in NCCLS document M26-P, *Methods for Determining Bactericidal Activity of Antimicrobial Agents*.

The other additional technique is the broth disc elution procedure. This is also a macrotube method but only one or two concentrations of antimicrobial are tested and paper discs are used to deliver the drugs. This method has the advantages of simplicity and readily available materials and has been widely used, especially by smaller laboratories. However, its reliability is somewhat in doubt and problems have arisen with the testing of newer β-lactam agents. Laboratories which have been using this method should consider changing to the also practical broth microdilution technique. The availability of commercially prepared plates should make this choice easier.

5.6 Detection of β-lactamase activity using nitrocefin (chromogenic cephalosporin) (11)

While not, strictly speaking, an antimicrobial susceptibility test, detection of β-lactamase production is a simple, rapid, and inexpensive way to predict resistance to penicillins and cephalosporins which have β-lactam rings which are susceptible to hydrolysis by β-lactamases. Such agents are probably not useful in treating infections due to β-lactamase-producing anaerobes, regardless of the MIC value obtained. The test should be performed on all anaerobic Gram-negative bacilli and reported to physicians; it will provide information useful in guiding therapy even if no susceptibility testing is done. Most strains of the *B. fragilis* group and more than half of other *Bacteroides* spp. as well as a small number of *Fusobacterium* spp. produce β-lactamase. These organisms, in general, should be susceptible to combinations which include a β-lactam antimicrobial and an inhibitor of β-lactamase, such as sulbactam or clavulanic acid.

This procedure is easily performed using a commercially available disc which is impregnated with the reagent (Cefinase–BBL). Hydrolysis of the β-lactam ring in the compound by β-lactamase causes a yellow-to-red colour change which generally occurs within 15 min. Up to four organisms can be scraped on to quadrants of each disc using a loop or applicator stick. Colonies are picked from actively growing agar plates.

References

1. Bartlett, J. G. (1983). *Rev. Infect. Dis.* **5**, 235.
2. Styrt, B. and Gorbach, S. L. (1989). *New Engl. J. Med.* **321**, 298.

3. Finegold, S. M. and George, W. L. (1989). *Anaerobic Infections in Humans*. Academic Press, New York.
4. National Committee for Clinical Laboratory Standards (NCCLS) (1989). Tentative standard. *Methods for Antimicrobial Susceptibility Testing of Anaerobic Bacteria*, NCCLS Publication: Vol. 9 (no. 10 (October)), 2nd edn. Villanova, PA.
5. Cuchural, G. J., Tally, F. P., Jacobus, N. V., *et al.* (1988). *Antimicrob. Agents Chemother.* **35**, 717.
6. Musial, C. E. and Rosenblatt, J. E. (1989). *Mayo Clin. Proc.* **64**, 392.
7. Malamy, M. H. and Tally, F. P. (1981). *J. Antimicrob. Chemother.* **8** (Suppl. D), 59.
8. Wexler, H. M., Harris, B., Carter, W. T., and Finegold, S. M. (1985). *Antimicrob. Agents Chemother.* **27**, 876.
9. Rolfe, R. D. and Finegold, S. M. (1981). *Antimicrob. Agents Chemother.* **20**, 600.
10. Jones, R. N., Barry, A. L., Fuchs, P. C., and Thornsberry, C. (1986). *J. Clin. Microbiol.* **24**, 1055.
11. Lee, D. T. F. and Rosenblatt, J. E. (1983). *Diagn. Microbiol. Infect. Dis.* **1**, 173.

5

Chemotaxonomic methods

H. N. SHAH, S. E. GHARBIA, and P. A. LAWSON

1. Introduction

The taxonomy of major groups of anaerobic bacteria such as *Bifidobacterium*, *Bacteroides*, *Eubacterium*, *Fusobacterium*, *Peptostreptococcus*, and several others has until recently been unsatisfactory. Many of these taxa have been defined primarily on the basis of morphological and physiological characteristics but have undergone considerable taxonomic improvements since the introduction of acid end-products analysis. Recent studies, however, indicate that several of these genera remain extremely heterogeneous. For example, the DNA base compositions (mol % G+C) are in the range 24–55% for the genera *Clostridium*, 30–55% for *Eubacterium*, and 28–61% for the genus *Bacteroides*. Since it is now generally accepted that a difference of greater than 10 mol % G+C indicates that species are unrelated at the generic level, it is evident that new approaches are required for systematic studies of these micro-organisms.

Over the last decade biochemical/chemical methods have been applied widely to study the classification of several groups of anaerobic bacteria. This involves analysis of both free and bound components of the bacterial cell, and analyses can range from simple qualitative detection to complex analytical methods such as mass spectrometry and nuclear magnetic resonance spectroscopy. A comprehensive review of the methods used to study anaerobic bacteria is beyond the scope of this chapter. For simplicity, a practical approach to the methods described in this chapter is illustrated against a background of the large and complex family, Bacteroidaceae, where the application of biochemical/chemical methods has made considerable inroads into their classification and identification (1). In all cases, only one of several methods which have been used successfully is given, for brevity, and references are provided for alternative methods. *Table 1* gives an overview of the methods which have found wide application for studies of these and other micro-organisms and indicates some of the macromolecules which may be of diagnostic value.

2. Acid end-products

The fermentation of carbohydrates or nitrogenous substrates gives rise to a complex spectrum of metabolic end-products which characterize many anaero-

Table 1. Overview of components analysed, methods generally used, and the potential taxonomic value

Site	Component	Techniques commonly used[a]	Potential level of taxonomic discrimination
Culture supernatant	Short-chain (C_1–C_4) volatile/non-volatile fatty acids	TLC, GLC, HPLC	Genus/species
Membrane/wall	Long-chain fatty acids (C_{12}–C_{18})	GLC (using capillary columns)	Genus/species
	Polar lipids	TLC	Species
	Menaquinones	TLC, HPLC, mass spectrometry	Genus/species
	Peptidoglycan	Paper (cellulose) TLC, HPLC	Genus
Soluble cell extract	Proteins/peptides	Electrophoresis (SDS, IEF)	Species/subspecies
	Enzymes	Electrophoresis (cellulose, PAGE)	Genus/species/ subspecies
	Cytochromes	Spectrophotometry	Genus
	DNA/RNA	Spectrophotometry, electrophoresis	Family Genus/species/ subspecies

[a] TLC, thin-layer chromatography; GLC, gas–liquid chromatography; HPLC, high-performance liquid chromatography; SDS, sodium dodecyl sulphate; IEF, isoelectric focusing; PAGE, polyacrylamide gel electrophoresis.

bic bacteria. These include the volatile acids, formic, acetic, propionic, isobutyric, *n*-butyric, isovaleric, and *n*-valeric acids, and the non-volatile acids, lactic, succinic, and phenylacetic acids. In contrast, aerobic bacteria generally produce a simple pattern of mainly non-volatile acids. Consequently, acid end-product analysis is now regarded as an essential tool in both research and diagnostic laboratories. The types of acids produced by a particular species are constant for a given set of conditions but metabolic end-products vary both qualitatively and quantitively in response to variations in cultural conditions such as substrate availability, growth rate, pH, temperature, and gaseous atmosphere. Thus any attempt to use acid end-products of anaerobic bacteria as diagnostic features requires that the organisms are grown under standard conditions (see Chapter 3).

Detection of acid end-products is generally by gas–liquid chromatography (GLC) as described in Chapter 3. More recently, high-performance liquid chromatography (HPLC) has been used. However, neither of these methods can differentiate between the D(−) and L(+) isomers of lactic acid; for this purpose enzymic methods can be used.

2.1 High-performance liquid chromatography (HPLC)

HPLC also offers a simple and sensitive method of determining acid end-products and numerous commercial systems are now available. In general cation-exchange resin columns are used to separate the anions using isocratic elution with mild acid. Because organic acids and proteins both have high ultraviolet (UV) absorption, a refractive index (RI) rather than UV detector is recommended. No sample preparation is required but before injection on to the column, the supernatant should be clear and free of debris. This can be achieved by centrifugation or filtration through a 0.02 μm filter. Fatty acids can be separated conveniently by use of a cation exchange resin column fitted with a microguard column cation trap. The column is kept at room temperature and 0.024 M H_2SO_4 is used as eluent at a flow rate of 0.7 ml/min. The operating pressure is between 800 and 1000 psi.

2.2 Enzymic determination of L(+)- and D(−)-lactic acid

The method takes advantage of the high degree of specificity of lactate dehydrogenases (LDH) to oxidize only either L(+) or D(−) lactic acid in the presence of nicotinamide adenine dinucleotide (NAD). The increase in extinction at 340 nm due to NADH formation is stoichiometric with the concentration of lactic acid according to the equation

$$\text{L-lactate (or D-lactate)} + \text{NAD} \xrightleftharpoons{\text{L-LDH (or D-LDH)}} \text{pyruvate} + \text{NADH} + \text{H}^+$$

Since the equilibrium for this reaction is in favour of lactate, the pyruvate formed in the presence of glutamate is immediately catalysed to alanine and 2-ketoglutarate by the enzyme glutamate–pyruvate transaminase (GPT) as

$$\text{Pyruvate} + \text{L-glutamate} \xrightarrow{\text{GPT}} \text{L-alanine} + \text{2-ketoglutarate.}$$

A test combination kit (Boehringer Mannheim, plus D-lactate and D-lactate dehydrogenase) can be used to measure both isomers of lactic acid.

Protocol 1. Measurement of lactic acid isomers[a]

You will need

- Solution 1: 100 mM glycylglycine buffer, pH 10.0, containing L-glutamic acid, 14.667 mg/ml
- Solution 2: B-NAD, 35 mg/ml
- Solution 3: GPT, 1571 units/ml
- Solution 4: L(+) or D(−) LDH, 0.5 mg/ml
- Standard solutions containing between 50 and 300 μg/ml L(+) or D(−) lactate solutions

1. Set up assay in duplicate, one cuvette to be used as a blank. To both cuvettes add
   - 1.0 ml of solution 1

Protocol 1. *Continued*

- 0.2 ml of solution 2
- 1.0 ml water to the blank, and 0.9 ml water to the test cuvette
- 0.02 ml of solution 3
- 0.1 ml of sample to the test cuvette only

2. Mix the contents of the tubes and determine the absorbance (A_1) at 340 nm after 5 min.
3. Initiate reaction by the adding 0.02 ml of L-LDH (or 0.05 ml of D-LDH) to both cuvettes.
4. Either monitor increase in absorbance of the reaction or leave for 30 min; then take the final reading at 340 nm (A_2). Correct for any difference between blanks and test cuvette, and determine ΔA, the change in absorbance in test cuvette. Concentrations (C) can be determined from the respective calibration curves or from the general formula

$$C = \frac{V \times \text{mol. wt.}}{1000 \varepsilon d v} \times \Delta A$$

where V is the final volume, v the sample volume, mol. wt. the molecular weight, d the light path (1 cm), and ε the extinction coefficient of NADH.

Thus for L(+) lactic acid ($V = 2.24$),

$$C = \frac{2.24 \times 90.1}{6.3 \times 1.0 \times 0.1 \times 1000} \times \Delta A$$

$$= 0.320 \times \Delta A(\text{g(L(+) lactic acid)/l}).$$

For D(−) lactic acid ($V = 2.27$),

$$C = 0.325 \times \Delta A(\text{g(D(−) lactic acid)/l}).$$

[a] Although these determinations have only been carried out on about 12 species of Gram-negative anaerobic bacteria, all species tested produced both L(+) and D(−) lactic acids (2). This is potentially a very useful method for studies on lactic acid-producing anaerobic bacteria such as *Mitsuokella multiacidus* or *Leptotrichia buccalis*, in addition to being essential for definitive identification of *Lactobacillus* spp. (3).

3. Electrophoretic methods

The separation of bacterial peptides/proteins in the presence of SDS by polyacrylamide gel electrophoresis (PAGE) (4) has become the most frequently used electrophoretic procedure for studies in bacterial systematics. Gradient PAGE and isoelectric focusing (IEF) may also be used where SDS is unsuitable or unavailable. The protein bands (electropherograms) of strains of interest may be visualized and compared directly after electrophoretic separation and staining. Alternatively, quantitative comparisons of large numbers of strains can be achieved by scanning the electropherograms by use of a densitometer, which is interfaced with a computer, and then utilizing a numerical taxonomic analysis program (5) to produce a similarity matrix or dendrogram. These methods require good resolution and reproducibility of patterns and are discussed in detail elsewhere (5). Good correlation has been reported between the results of DNA–DNA hybridization and PAGE studies for a large number of genera and as rapid electrophoretic methods (e.g. Phast system) continue to develop, these techniques will undoubtedly have wider application.

Electrophoresis of cell-free extracts, followed by specific enzyme staining, provides a simple and sensitive tool for comparing the electrophoretic variation of an enzyme (zymogram) within a bacterial taxon. Unlike PAGE methods, the technique has more restricted application (e.g. at the species/subspecies level), as it often relies on single amino-acid substitutions of an enzyme. The technique has, however, been used extensively for several taxa including the family Bacteroidaceae. It should be emphasized that the mere presence or absence of an enzyme between related taxa is of considerable taxonomic value as it often implies major difference in metabolism and genetic information. The three most frequently used methods will now be discussed.

3.1 SDS-PAGE patterns

When soluble proteins are heated with an anionic detergent such as SDS, they dissociate into monomeric polypeptides and are coated with a more or less uniform distribution of negative charges. SDS-PAGE causes these SDS-polypeptide complexes to migrate towards the anode, with mobilities that are inversely related to their molecular weights. The electrophoretic system in *Protocol 2* is based on that of Laemmli (4).

Protocol 2. SDS-PAGE of cellular proteins

You will need

- Electrophoresis apparatus and powerpack
- Stock acrylamide solution: 30% (w/v) acrylamide, 0.8% (w/v) bisacrylamide
- Gel buffer: 1 M Tris–HCl (pH 8.8)
- Electrode buffer (pH 8.3): 0.303% (w/v) Tris base, 1.44% (w/v) glycine, 0.1% (w/v) SDS
- Sample buffer: 12.5% (w/v) 0.5 M Tris–HCl (pH 6.8), 12.5% (w/v) glycerol, 12.5% (w/v) aqueous 10% SDS, 1.25% 2-mercaptoethanol, 2.5% (w/v) aqueous 0.5% bromophenol, 58.75% (v/v) distilled water
- 10% (w/v) aqueous ammonium persulphate (freshly prepared)
- Aqueous solutions of 10% (w/v) and 20% (w/v) SDS
- Destain solution: 52% (v/v) distilled water, 40.5% (v/v) methanol, and 7.5% (v/v) glacial acetic acid

Protocol 2. *Continued*

- Stain solution: 0.0125% Coomassie brilliant blue R in destain solution

1. Grow the organism under standard conditions, wash three times in phosphate-buffered saline (PBS), and finally prepare a thick suspension in 1 ml of PBS.
2. Add SDS to a final concentration of 2%, then boil cell suspension for 2 min. Gram-positive bacteria may require mechanical disruption first to achieve lysis (see Section 4 in Chapter 6).
3. Centrifuge the suspension (at least 30 000 *g* for 30 min), discard the cells, and determine the protein concentration in the supernatant by standard methods.
4. Dilute in sample buffer to give a final protein concentration of 2.5 mg/ml. Samples can be stored at −20°C if necessary.
5. Assemble the slab gel cassette according to the manufacturer's instructions.
6. To a Buchner flask add the following (take care, acrylamide is a neurotoxin)
   - Stock acrylamide 33.33 ml
   - Gel buffer 37.20 ml
   - Distilled water 28.64 ml
7. Mix all the reagents and degas for about 30 min, by means of a vacuum line attached to the flask. Then add the following
   - 20% SDS 0.5 ml
   - 10% ammonium persulphate 0.33 ml
   - *N*,*N*,*N′*,*N′*-tetramethylethylenediamine (TEMED) 0.08 ml
8. Pour gel into slab cast, taking care to dislodge any bubbles trapped beneath the comb. Allow the gel to polymerize for at least 1 h (overnight may be convenient).
9. Remove the comb carefully; wash out the wells with electrode buffer using a Pasteur pipette.
10. Add samples to the wells using a microsyringe (the volume depends upon the apparatus and comb used; *ca*. 125–150 μg protein per well should give well resolved bands). Sometimes a stacking gel (same reagents as the main gel, but only 3–4% acrylamide) is cast on top of the main separation gel to facilitate better separation of the proteins as they enter the main gel. Molecular weight markers should be included in each run for comparison between gels.
11. Instructions for running gels will vary with the apparatus. It is recommended that a constant current of 15 mA per gel is used for the first hour, which can be increased to 25 mA. Electrophoresis is allowed to continue until the bromophenol blue tracking dye reaches near the bottom of the gel.

Protocol 2. *Continued*

12. Switch off the power supply, place the gel in a tray, and immerse in Coomassie blue stain for 16 h.

13. Then pour off the stain and destain the gel in several changes of destain solution until the background is clear. Gels can be stored in 7% glacial acetic acid until they can be photographed or scanned by means of a densitometer.

3.2 Isoelectric focusing (IEF)

Unlike zone electrophoretic techniques, IEF is based on moving boundary electrophoresis. In addition to the voltage between the electrodes, a pH gradient is also established. The latter is achieved by use of a complex mixture of synthetic aliphatic polyamino-polycarboxylic acids (ampholytes; mol. wt. 300–600) which migrate in an electric field to give a continuous pH gradient within a specified pH range. Proteins, being amphoteric macromolecules, will migrate in an electric field according to their net charge. The latter is influenced by the pH and the protein will cease to migrate when its net charge is zero. This pH value is termed the isoelectric point (pI) and separation based on this method is termed isoelectric focusing (IEF). Since the pI of a protein is constant, protein patterns obtained by IEF are constant, and high interlaboratory reproducibility is expected. Despite its inherent potential, IEF has so far had only limited use for systematic studies, perhaps due to the high cost of ampholytes. Gels of different

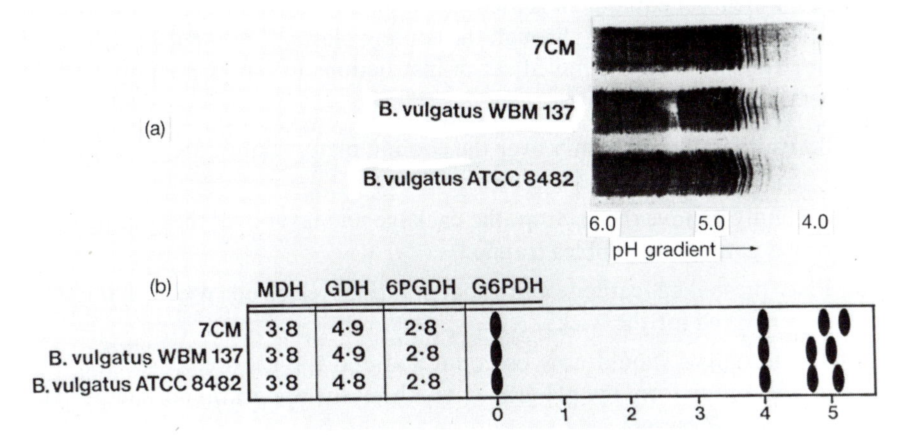

Figure 1. (a) The separation of cellular proteins of *Bacteroides vulgatus* by IEF (pH 4–6.0). The characteristic 'fingerprint' pattern was used to show that strain 7CM (the designated type strain of *Bacteroides oralis*) had almost identical bands to *B. vulgatus*. (b) This similarity was confirmed by enzyme patterns and DNA–DNA hybridization (Section 6.2) and a new type strain of *B. oralis* (now *Prevotella oralis*) was eventually proposed. MDH, malate dehydrogenase; GDH, glutamate dehydrogenase; 6PGDH, 6-phosphogluconate dehydrogenase; G6PDH, glucose-6-phosphate dehydrogenase.

pH ranges are now commercially available and are recommended because of their relatively low cost and ease of application. It is suggested that new studies begin with a wide pH range gel (e.g. pH 3–10) and then use narrower ranges. pH ranges can be selected to optimize the number and resolution of proteins under study. The proteins of *Neisseria* species, for example, focused over a wide pH range (ca. pH 3–9.5), whereas most proteins of *Bacteroides* species separate in a narrow pH gradient of 4.0–6.0 (seee *Figure 1a*).

Protocol 3. Separation of cellular proteins by IEF

You will need

- 1 M H_2PO_4 (anode solution)
- 1 M NaOH (cathode solution)
- Paraffin oil
- Fix solution. Make up by adding 17.25 g sulphosalicylic acid and 17.5 g trichloroacetic acid to a mixture of 150 ml methanol and 350 ml distilled water
- Destain solution. Mix 500 ml of ethanol and 160 ml glacial acetic acid and dilute to 2 l with distilled water
- Stain solution. Dissolve 0.23 g Coomassie Brilliant Blue R 250 and 0.115 g Coomassie Blue G in 300 ml destain solution
- IEF apparatus. Over 1200 V are generally used between 10 cm electrodes; hence, whichever equipment is used, there should be a continuous flow of cold water (6–8°C) circulating through the cooling plates. Commercially prepared plates are sold with a detailed set of instructions for specific equipment. In general these are as follows

1. Spread paraffin oil lightly over the cooling plate, taking care to stay about 1 cm from the edge of the plate.
2. Carefully remove the gel from the package and layer over the cooling plate. Avoid getting air bubbles trapped.
3. Place presoaked (cathode and anode solutions) electrode wicks on the gel in area marked for the wicks.
4. The terminals should now be connected and the gel prefocused (i.e. IEF carried out to form the pH gradient without the application of samples) for 1 h at 20 W per gel.
5. Switch off the power supply and load samples and IEF markers, either by the use of sample application strips provided with the gels, or by use of a micropipette. Unlike SDS-PAGE, sample loading is not critical as a protein will focus at its pI irrespective of where it is loaded. However, there is some distortion at the application point, so samples should be applied away from the area of interest.

Protocol 3. *Continued*

6. Reconnect the power supply and continue IEF for a further 1.5–2.0 h; then switch off the power supply.
7. Remove wicks with a pair of forceps and place the gel in the fix solution for 30–60 min.
8. Pour away fix solution and wash the gel with a little destain solution for about 5 min.
9. Stain gel at 60°C for 20 min with stain solution.
10. Destain over several days until the background is clear (commercially available destain equipment is now available).
11. Gels can be photographed or scanned with a densitometer as for SDS-PAGE (see *Protocol 2*).

3.3 Enzyme electrophoresis

Electrophoretic patterns have been performed on a variety of gel support matrices, e.g. cellulose acetate, hydrolysed starch (starch gel), polyacrylamide. SDS–PAGE, as described in *Protocol 2* cannot be used, as this will denature the enzyme. The method described below utilizes cellogel strips which are available commercially (*Figure 1b*). The electrophoretic detection of malate dehydrogenase (MDH) is used as an example.

Protocol 4. Electrophoresis of malate dehydrogenase on cellogel strips[a]

You will need:

- 50 mM Hepes buffer pH 7.5
- 40 mM barbitone–acetate buffer pH 8.6
- Sodium malate (2 mg/ml)
- NAD (0.1 mg/ml)
- Phenazine methosulphate (0.04 mg/ml) (PMS) } Store in the dark until required
- Thiazolyl blue tetrazolium (0.2 mg/ml) (MTT) } Store in the dark until required
- Agar
- Standard electrophoretic equipment with a chilled cooling plate as for IEF. However, a sample applicator (1.5 μl) is necessary to obtain reproducible results
- Cellogel strips
- Electrophoresis wicks

1. Gently float the cellogel strips on the surface of the barbitone–acetate buffer. This allows the buffer to displace air through the gel matrix. Leave for about 5 min and then totally submerge the gel.

Protocol 4. *Continued*

2. Just before use, lift the gel out with a pair of forceps, drain off excess liquid, and blot gently between sheets of filter paper.
3. Layer the gel (with its porous surface upwards) on the chilled (8–10°C) cooling plate.
4. Connect wicks (previously in barbitone–acetate buffer) on either side of the cellogel strips.
5. Using the applicator stage and holder, load 1.5 μl of cell-free extract on the surface of the gel (phenol red may be added to each sample before application to enable visual assessment of the electrophoretic progress). Switch on the power supply and electrophorese extracts at about 10 V/cm gel length.
6. When phenol red has reached some convenient distance (1 to 2 cm from the edge of the gel) switch off the power supply and note the time (as this can be used for future runs).
7. Immediately mark the origin of the sample and prepare stain for MDH by adding the following
   - Tube 1: 16 ml, 50 mM Hepes buffer + 0.3 g agar; boil and cool to 53°C
   - Tube 2: 1 ml malate
   - Tube 3: 1 ml NAD
   - Tube 4: 1 ml PMS
   - Tube 5: 1 ml MTT
8. Add the contents of tubes 2 to 5 to tube 1; mix and pour over gel.
9. Leave the gel at 37°C for 5–10 min for the enzyme to stain. The enzyme becomes visible as a bluish-purple band against a greenish background.
10. Measure migration distances using a particular strain as a reference for each run.

[a] To test for other dehydrogenases, substitute the desired substrate for malate and use NAD or NADP depending on the coenzyme specificity.

4. Peptidoglycan composition

Peptidoglycan, common to cell walls of both Gram-positive and Gram-negative bacteria, is a heteropolymer which consists of polysaccharide (glycan) strands cross-linked through short peptides (*Figure 2*). The glycan moiety is made up of β(1,4)-glycosidically linked *N*-acetylglucosamine residues but each alternate residue contains a D-lactic acid ether at its C-3 hydroxyl group (referred to as muramic acid). The carboxyl group of *N*-acetyl muramic acid is linked to an oligopeptide (stem peptide) which contains both L- and D-amino acids.

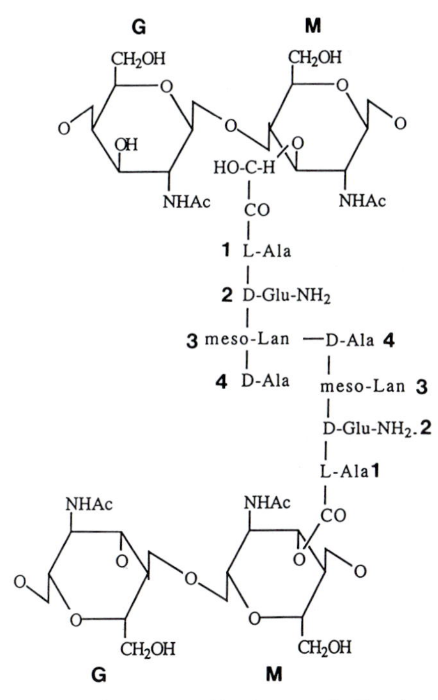

Figure 2. Primary structure of the peptidoglycan of *Fusobacterium nucleatum.* (M, *N*-acetylmuramic acid; G, *N*-acetylglucosamine; 1–4, stem peptide amino acids; ala, alanine; glu, glutamate; meso-Lan, meso-lanthionine.

Position 3 or (less commonly 2) of the stem peptide may be cross-linked to position 4 of another stem peptide by means of different interpeptide bridges.

While the glycan structure is highly conserved in Eubacteria, the amino-acid sequence of the stem peptide shows some variation, with position 3 having the greatest variation (11 different amino acids are known). These variations give rise to the nomenclature introduced by Schleifer and Kandler (6; see *Table 2*). Simple qualitative amino-acid analysis is often sufficient to provide the taxonomic information required. This is particularly true of Gram-negative bacteria where there is considerably less diversity than in Gram-positive organisms. If it is necessary to determine the nature of the cross-linkage, the sequence of amino acids or their configuration, reviews by Schleifer and Kandler (6) and Schleifer and Seidl (7) are recommended.

In this section, attention will be given to sample preparation and qualitative detection. A standard amino-acid autoanalyser or HPLC can be used to obtain quantitative information on the purified cell walls.

4.1 Preparation of cell walls

Several methods exist for the preparation of cell walls. In general, bacterial cells

Table 2. Summary of the peptidoglycan chemotypes proposed by Schleifer and Kandler (6)

Two groups, A and B

A-stem peptides cross-linked between amino acids 3 and 4 (see *Figure 3*)
B-stem peptides cross-linked between amino acids 2 and 4.

Group A subdivisions (based on the nature of the cross-linkage)

A1: direct cross-link (no peptide bridge, see *Figure 3*).
 A1*α*, A1*β*, A1*γ*, A1*δ* if the amino acid in position 3 of the stem peptide is L-lysine, L-ornithine, *meso*-diaminopimelic acid or *meso*-lanthionine, respectively.

A2: cross-linked by polymerized peptide subunits.

A3: cross-linked by monocarboxylic L-amino acids or glycine or both.
 (A3*α*, A3*β*, or A3*γ* as for A1 (A3*γ*=LL, isomers)).

A4: cross-linked by dicarboxylic acids.
 A4*α*, A4*β*, A4*γ* (as in A1), A4*γ* (diaminobutyric acid).

A5: cross-linked by D-dicarboxylic acid dibasic amino acids.
 A5*α*, A5*β* (as in A1).

Group B subdivision

B1: interpeptide bridge composed of L-diamino acids.
 B1*α*, B1*β*, B1*γ*, B1*δ*, if the amino acid 2 of the stem peptide is L-lysine, L-histidine, D-glutamic acid, or D-alanine, respectively.

B2: interpeptide bridge composed of D-diamino acids.
 B2*α*, B2*β*, or B2*γ* if D-ornithine, D-histidine, or diaminobutyric acid, respectively.

are first mechanically broken (Vibrogen cell mill, Biox pressure cell or Mickle tissue disintegrator) then digested with proteases or boiled with 2–4% SDS (see Section 4 in Chapter 6). Wall polysaccharides may be removed by several methods such as trichloroacetic acid treatment or boiling with formamide at 170°C for 20 min. Purified preparations are then hydrolysed with 4 M HCl at 100°C for 16 h. Simplified procedures for routine preparation of cell walls and their chemical composition are given in *Protocols 5–8*.

Protocol 5. Extraction of wall polysaccharides using trichloracetic acid

You will need

- 10% (w/v) trichloracetic acid (TCA)
- 4% (w/v) SDS
- 0.2 M phosphate buffer (pH 8.0)
- Pronase or proteinase K (Sigma)
- Chloroform

1. Suspend 500–1000 mg bacterial cells in 20 ml of TCA in a boiling tube.
2. Boil for 20 min.

Protocol 5. *Continued*

3. Transfer the boiled suspension to centrifuge tubes, centrifuge at 7000 *g* for 10 min. Discard the supernatant.
4. Using a Pasteur pipette, wash the deposit with distilled water and repeat centrifugation step.
5. Discard the supernatant; repeat step 4 twice.
6. Resuspend the deposit in 20 ml of 4% SDS, transfer to a fresh boiling tube, and boil for 20 min.
7. Centrifuge at 20 000 *g* for 20 min.
8. Discard the supernatant and wash three times with distilled water.
9. Resuspend the deposit in 25 ml 0.2 M phosphate buffer and add pronase to give a final concentration of 50 μg/ml (e.g. 1.25 mg).
10. Add 0.5 ml chloroform (to prevent bacterial growth) and incubate at 37°C for 18 h.
11. Centrifuge at 20 000 *g* for 30 min and wash the deposit three times in distilled water.
12. Freeze-dry the sample and store in a dry atmosphere at room temperature until required.

Protocol 6. Mechanical disruption of cells using a Mickle tissue disintegrator

1. Resuspend cells in a small volume of distilled water in 'Mickle pot'.
2. Add Ballotini beads (grade 12) to about one-third the volume of the cell suspension so that the volume of beads to air is equivalent.
3. Add 1–2 drops of octan-2-ol and allow cell disruption to take place. The integrity of the cells may be checked by microscopy.
4. After cell disruption, allow the beads to settle; then transfer the supernatant to a centrifuge tube.
5. Centrifuge at 3000 *g* for 10 min to deposit any remaining Ballotini beads or whole cells.
6. Transfer supernatant to a fresh tube and centrifuge at 38 000 *g* for 30 min. Discard the supernatant.
7. Repeat step 6 and freeze dry the sample.

Protocol 7. Hydrolysis of cell walls for detection of amino acids

1. Weigh directly into a glass ampoule, 2 mg of dried cell wall material from *Protocol* 5 or 6.

Protocol 7. *Continued*

2. Add 0.5 ml 4 M HCl and carefully seal the ampoule in a hot flame.
3. Hydrolyze overnight at 10°C.
4. Carefully break open the ampoule (after etching the glass with a diamond) and pour contents into a glass vial.
5. Place vial into a dessicator and carefully dry in a vacuum over phosphorus pentoxide and sodium hydroxide pellets (both can cause severe burns).
6. When dry, wash with distilled water and dessicate to dryness. Repeat 2–3 times until the pH is neutral. Store dry until needed for chromatography.

Protocol 8. Hydrolysis of cell walls for detection of carbohydrates

1. Prepare cell walls using *Protocol 6*.
2. Weigh directly into a glass ampoule, 2 mg of cell wall material.
3. Add 0.5 ml 1 M H_2SO_4 and seal the ampoule in a hot flame.
4. Hydrolyze at 100°C for up to 2 h. Examine at 15-min intervals; if the suspension starts to go brown/yellow, remove from the oven and allow to cool.
5. Neutralize the hydrolyzate using an ion-exchange resin (Zerolit FF (IP) in the CO_3 form).
6. Filter and freeze-dry the filtrate.

4.2 Chromatographic methods

Samples can be examined directly by HPLC or TLC, or by using an amino-acid autoanalyser. Where this is not available, simple paper chromatographic solvent systems can be used as described in *Protocols 9–12*.

Protocol 9. Detection of amino acids by two-dimensional chromatography

1. Resuspend freeze-dried sample (from *Protocol* 7) in 100–200 μl 10% (v/v) isopropanol.
2. Spot 20–40 μl of the hydrolysate at the corner of 20 cm × 20 cm Whatman no. 1 chromatography paper.
3. Run overnight in a solvent system of ethanol/butan-1-ol/water/propionic acid (10:10:5:2 by volume); then dry chromatograms thoroughly.
4. Run chromatograms overnight again in the second dimension in a solvent system of butan-1-ol acetone/water/dicyclohexylamine (10:10:5:2 by vol), and dry.

Protocol 9. *Continued*

5. Develop papers by dipping in ninhydrin (0.2 g) in glacial acetic acid (7 ml) in acetone (100 ml).
6. Dry in air and heat at 100°C for 3–5 min. Compare coloured spot with authentic standards.

Protocol 10. Detection of diaminopimelic acid by descending chromatography

1. Spot 40–60 μl of hydrolysate (from *Protocol* 7) on a pre-labelled Whatman no. 1 chromatography paper for descending chromatography.
2. Spot standards, e.g. *meso* and *LL*-diaminopimelic acids, lysine, etc.
3. Run for 16–18 h in a solvent system of methanol/pyridine/10 M HCl/water (80:10:2.5:17.5 by vol.).
4. Dry chromatogram and develop in a solution of 0.1% ninhydrin in acetone.
5. Dry and heat at 100°C for 3–5 min.
6. Diaminopimelic acid spots are yellowish-green while other amino acids are purple. *Meso*-diaminopimelic acid migrates slower than the *LL*-isomer.

Dibasic amino acids, such as lysine, ornithine, diaminobutyric acid, or aspartic and glutamic acids, which are often difficult to resolve by the methods in *Protocols 9* and *10*, can be separated conveniently by high-voltage electrophoresis (1000–1500 V). The method requires a thicker chromatography paper (Whatman no. 3MM).

Protocol 11. High-voltage electrophoresis for separating amino acids with very similar pK_a values

1. Mark with a pencil the origin line, to aid spotting of the samples.
2. Soak the paper in 0.25 M pyridine in 0.158 M acetate buffer (pH 5.2).
3. Blot excess buffer, then place the paper on the cooling plate (8–10°C).
4. Avoid getting air bubbles trapped between the paper and the cooled plate.
5. Load sample at the origin line.
6. Fix the electrode to the wicks, set voltage at 100 V, and run electrophoresis for 1.5–2 h.
7. Turn off the power and dry the paper in air for several hours. Stain as described in *Protocol 10*.
8. Identify unknown amino acids by reference to standards.

Protocol 12. Detection of reducing sugars by descending chromatography

1. Spot 50–100 μl of sample, prepared using *Protocol 8*, on to Whatman no. 1 chromatography paper.
2. Spot on a range of standards such as glucose, galactose, rhamnose, arabinose, mannose, 6-deoxytalose, etc.[a]
3. Equilibrate in the chromatography tank and run overnight in a solvent system of butan-1-ol/pyridine/water (6:4:3 by vol).
4. Remove paper from the tank and dry for 1–2 h.
5. Dip paper in a developing solution containing phthalic acid (1.6 g), aniline (1.5 ml), water (1.0 ml) in 100 ml acetone.
6. Hang the paper to dry for 5 min, then heat at 100°C for up to 10 min.

[a] The following R_g values (movement relative to glucose standard) in addition to their characteristic colours should facilitate identification of unknown spots: glucose, brown, $R_g=1.0$; galactose, brown, $R_g=0.88$; rhamnose, red, $R_g=1.63$; arabinose, red, $R_g=1.15$; fucose, red, $R_g=1.34$; mannose, brown, $R_g=1.14$; 6-deoxytalose, brown, $R_g=1.9$; ribose, red, $R_g=1.42$.

5. Lipids as major chemotaxonomic markers

The term lipid is used loosely to denote a chemically heterogeneous group of natural products which are soluble in non-polar solvents such as chloroform but not in water. They may either be 'free', that is easily extracted from the bacterial cell, or 'bound' to macromolecules such as lipopolysaccharide. The latter requires acid or alkaline hydrolysis for extraction. Because of their immense structural variation, lipids have an enormous potential for use as chemotaxonomic markers. Among Gram-negative anaerobic bacteria, three classes of lipids (phospholipids, long-chain fatty acids (C_{12}–C_{18}), and menaquinones) have been used extensively in classification and identification. Comprehensive reviews on the use of other lipids in bacterial systematics have been published elsewhere (8, 9).

5.1 Polar lipids

The most common polar lipids are phospholipids whose structures are based on the parent compound of the series glycerol-3-phosphoric acid (stereo-specific numbering). In phosphoglycerides, one of the primary hydroxyl groups is esterified to phosphoric acid (see *Figure 3*), which can be substituted with a variety of compounds to form the corresponding phospholipid. They are readily extracted and analysed by simple two-dimensional TLC to produce diagnostic patterns (see *Figure 4*).

```
                           OH
                           |
                CH2—O—C—R1
                 |         ||
                 |         O
R2—C—O—C—H
     ||          |         OH
     O           |          |
                CH2—O—P—O—X
                           ||
                           O
```

Figure 3. General structure of polar lipids commonly found in bacteria. 'X' can have a variety of substitutions (e.g. if 'X' is substituted with ethanolamine, the polar lipid is referred to as phosphatidylethanolamine). R_1 and R_2, long-chain acyl groups.

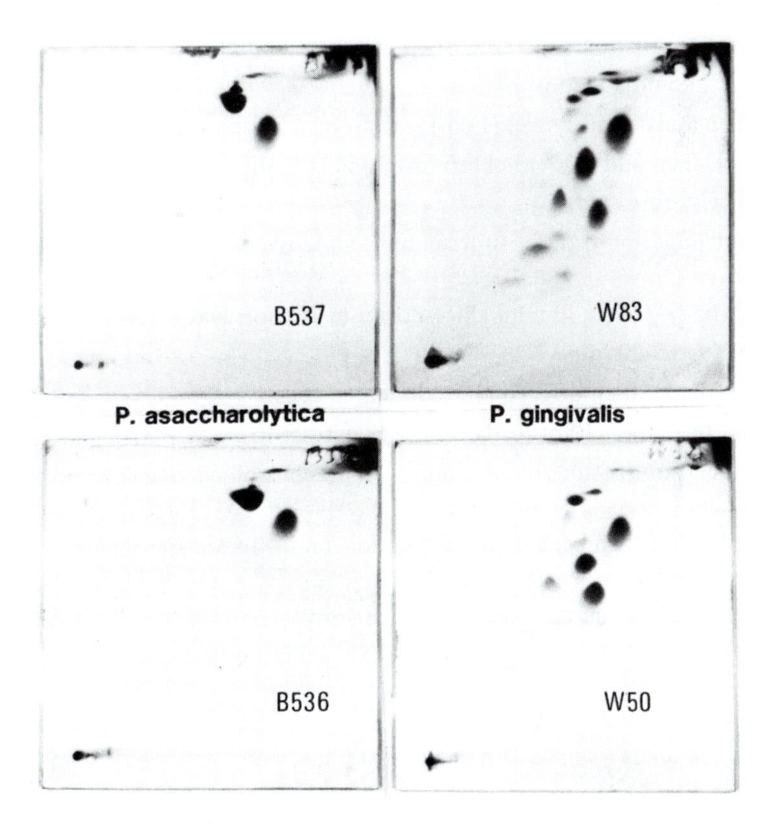

Figure 4. Two-dimensional TLC of the polar lipids of *Porphyromonas gingivalis* (W50, W83) and *Porphyromonas asaccharolytica* (B536, B537). These strains were originally classified as *B. melaninogenicus* subspecies *asaccharolyticus* (see ref. 1). Chemotaxonomic analyses such as that in *Protocol 14* showed that two centres of variation existed within this subspecies. These were later given separate species status.

Protocol 13. Extraction of polar lipids

You will need

- Aqueous methanol (10 ml of 0.3% aqueous NaCl in 100 ml methanol)
- Chloroform
- Chloroform: methanol: 0.3% aqueous NaCl (50:100:40 by vol)
- 0.3% (w/v) aqueous NaCl
- Chloroform: methanol (2:1 v/v)
- Phosphate-buffered saline (PBS)

1. Grow the culture in broth or on agar plates using standardized conditions. Suspend the cells in PBS and wash three times.
2. Freeze-dry the suspension overnight.
3. Place approximately 50 mg dried cells in a bijou bottle.
4. Add 2.75 ml aqueous methanol.
5. Add a magnetic follower (1 cm) and stir on a heated stirring block for 2 min.
6. Cool, then add 1.25 ml chloroform and 0.75 ml aqueous NaCl.
7. Stir vigorously for 2 h.
8. Centrifuge and transfer supernatant to a Universal bottle with a 'Tuf-bond' seal.
9. To the residue, add 2 ml chloroform: methanol: NaCl.
10. Stir for 15–20 min.
11. Centrifuge, then discard the sedimented cells.
12. Add the supernatant to the previous extract (from step 8).
13. To the pooled supernatant, add 1.75 ml chloroform and 1.75 ml aqueous NaCl.
14. Shake and then centrifuge (8000 *g* for 5 min) to aid the separation of the layers.
15. The polar lipids are located in the lower chloroform layer. The upper aqueous layer is removed with a Pasteur pipette and the lower organic layer is evaporated to dryness in a stream of high purity nitrogen.
16. Resuspend in a few drops of chloroform/methanol (2:1) before analysis.

Protocol 14. Analysis of polar lipids by two-dimensional TLC

You will need:

- Solvent 1 for TLC: chloroform:methanol:water (65:25:4 by vol)

Protocol 14. *Continued*

- Solvent 2 for TLC: chloroform : methanol : acetic acid : water (80 : 12 : 15 : 4 by vol)
- General spray reagent; molybdophosphoric acid reveals all lipids (see *Figure 4*): 10% (w/v) dodecamolybdophosphoric acid in absolute ethanol
- Ninhydrin detects aminolipids: 0.2% (w/v) ninhydrin in water-saturated butan-1-ol
- Zinzadze phospholipid reagent
  (a) Solution A: add 40.1 g molybdenum trioxide to 1 l of 12.5 M H_2SO_4 and boil until the residue dissolves;
  (b) Solution B: add 1.5 g powdered molybdenum to 500 ml of solution A and boil gently for 15 min, then allow to cool; mix equal volumes of solutions A and B, then dilute with 2 volumes of distilled water
- Periodate reagent for α-glycols: fresh 1% (w/v) aqueous sodium metaperiodate
- Schiff reagent for α-glycols: reduce a 1% (w/v) rosaniline hydrochloride solution by passing SO_2 through it
- α-naphthol reveals lipids containing sugar residues (e.g. glycolipids): mix together 6.5 ml conc. H_2SO_4, 4.5 ml 95% ethanol, 4 ml distilled water, and 10.5 ml 15% (w/v) α-naphthol in 95% ethanol
- 10 cm × 10 cm aluminium backed plates (Merck Kieselgel $60F_{254}$)
- SO_2 gas

1. Spot the sample in one corner of the plate.
2. Develop in one dimension, using solvent 1.
3. Dry for 30–60 min; then develop in the second dimension in solvent 2.
4. Dry the plate thoroughly; then spray with the appropriate reagent to detect specific lipid types. Spray with molybdophosphoric acid until the plate is wet; then heat at 150°C for 15 min. Lipids appear as blue-black spots on a yellow background.
5. To detect aminolipids, spray the plate lightly with ninhydrin solution; then heat the plate at 100°C for 5 min. Lipids containing free α-amino groups appear as red/violet spots on a white background. Mark these components with a pencil.
6. Spray Zinzadze reagent very lightly over the same ninhydrin-covered plate. Phospholipids appear as 1–2 mm blue spots on a white background within 5 min at room temperature.
7. To detect α-glycols, spray the plate with periodate solution until saturated; then leave at room temperature for 5 min. Pass a stream of SO_2 gas over the plate until it has been decolorized. Spray very lightly with Schiff's reagent; then expose again to reduction by SO_2 (until the brown coloration disappears). α-glycols appear as purple spots on a white background.

Protocol 14. *Continued*

8. Spray α-naphthol lightly on to the plate and heat at 120°C for 10 min. Glycoplipids appear as purple/red spots on a pink background.

5.2 Fatty acids

Fatty acids are the most widely studied class of bacterial lipids for systematic studies. Bacteria contain straight-chain ($C_{n:o}$), mono-unsaturated ($C_{n:i}$), iso-methyl branched (*iso*-C_n), *anteiso*-methyl branched (*ai*-C), and hydroxylated (OH–C_n) fatty acids, in addition to other acids (see *Figure 5*). Although these acids are commonly found in very diverse species, the pattern and concentration of specific acids are important criteria for defining many taxa. For example, it was shown previously (1) that many '*Bacteroides*' species contain only straight-chain and mono-unsaturated acids, whereas members of the genus *Bacteroides sensu stricto* possess in addition methyl-branched acids. Analysis of cellular fatty acids necessitates a preparative step (see *Protocol 15*) followed by gas chromatography.

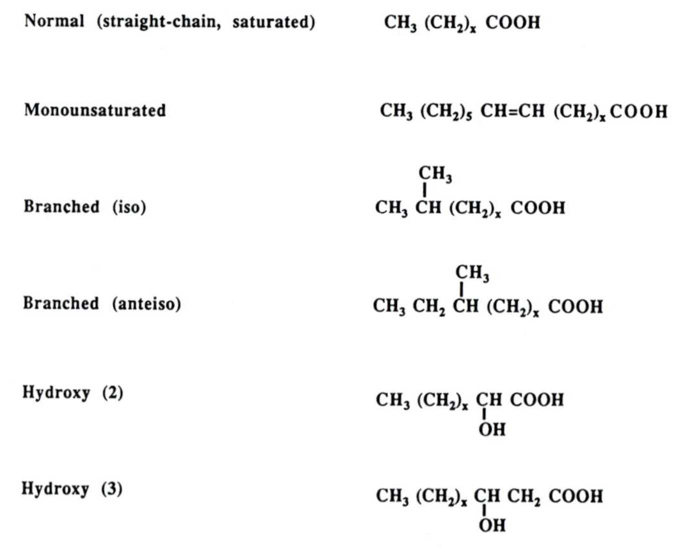

Figure 5. Long-chain fatty acid structures commonly found in Gram-negative anaerobic bacteria.

Protocol 15. Preparation of fatty acid methyl esters (FAME)

You will need

- Organic solvents: diethyl ether, *n*-hexane, methanol, petroleum ether (b.p. 60–80°C), and toluene

Protocol 15. *Continued*

- Sulphuric acid
- Molybdophosphoric acid (general spray reagent, from *Protocol 14*)
- 0.25 mm thickness preparative TLC plates (Merck art. no. 573)

1. Place about 50 mg of dried cells (see step 2 in *Protocol 13*) in a 10-ml tube with a Teflon cap.
2. Add the following reagents
   - 2.5 ml methanol
   - 2.5 ml toluene
   - 0.1 ml H_2SO_4
3. Incubate overnight at 50°C, for methanolysis to take place.
4. Extract with approximately 1.5 ml of *n*-hexane. Two layers result; use the upper hexane layer for TLC.
5. Spot the hexane extract in a line on to a preparative TLC plate.
6. Develop in petroleum ether (b.p. 60–80°C)/diethyl ether (85:15 v/v).
7. Cover the central part of the plate with foil and spray only the margins of the plate with molybdophosphoric acid, to detect FAME.
8. Heat at 150°C for 15 min.
9. FAME have an R_f (movement relative to solvent front) value of approximately 0.8. Scrape FAME from the TLC plate and elute from the silica gel with chloroform, using a sintered glass funnel.
10. Dry under a stream of N_2.
11. Take up the sample in a few drops of *n*-hexane immediately prior to GLC.
12. Hydrogenate half the sample for 5 min in the presence of H_2 gas and palladium catalyst or charcoal.

A large number of gas chromatographs can be used for the analysis of cellular FAME. A suitable system should have flame ionization detectors, a stream-splitter injection system, and twin wide-bore, fused-silica capillary columns, at least 25 m in length (0.32 mm diameter). Two stationary phases are necessary, such as FFAP (polar) and OV101 (non-polar). Both columns can be operated isothermally at 220°C with helium (2 ml/min) as the carrier gas. An alternative is a column of SE64 operated at 180°C. FAME are identified by comparison of retention times with those of standard mixtures obtained from Sigma or Supelco. The relative proportions of FAME present are determined by integration of peak areas. Examples of the separation of cellular FAME by this method are shown in *Figure 6*.

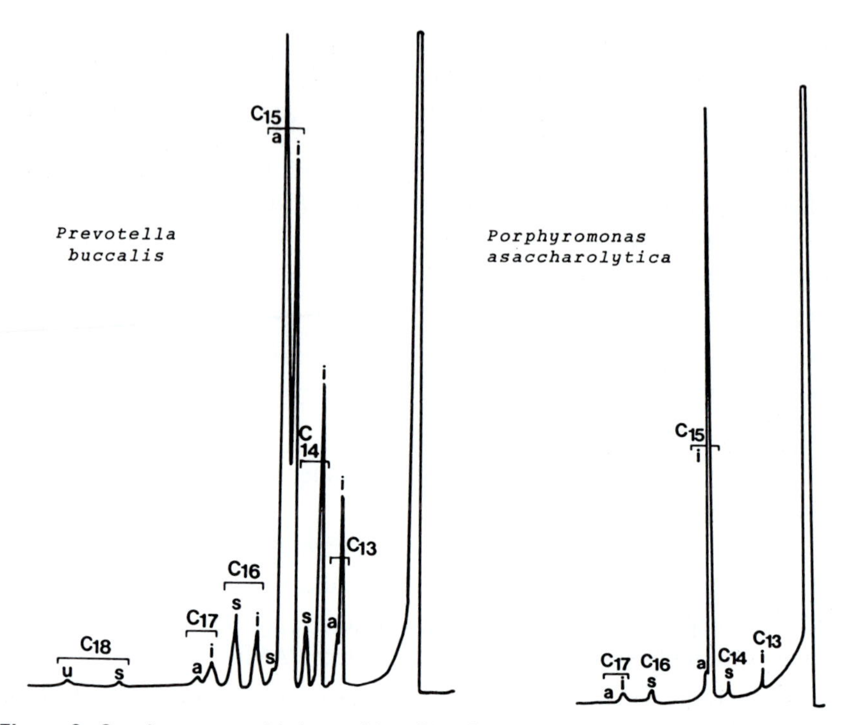

Figure 6. Gas chromatographic fatty acid profiles of two species which until recently were both classified in the same genus.

5.3 Menaquinones

Many members of the family Bacteroidaceae have a growth requirement for menaquinones. Also referred to as menadione, or vitamin K, menaquinones which are present in many anaerobic bacteria are 2-methyl-3-polyprenyl-1,4 naphthoquinones. They form a large class of molecules, in which the length of the C-3 polyprenyl side-chain may vary both in the number of isoprene units ($n = 1$–16; see *Figure 7*) and in the degree of hydrogenation. These compounds are of major taxonomic importance and are now used in the definition of several species and genera (9).

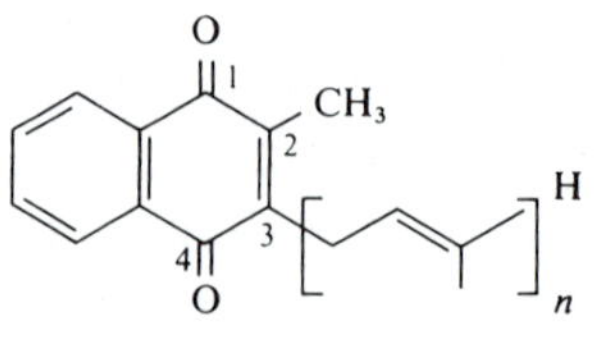

Menaquinone-*n*
(MK-*n*)

Figure 7. Structure of menaquinones found in *Bacteroides* and related genera ($n = 1$–16).

Menaquinones are 'free' lipids and can be readily extracted with organic solvents. They are, however, susceptible to degradation by strong acid or alkaline conditions and photo-oxidize rapidly in the presence of oxygen and UV radiation. Precautions are therefore necessary during purification. Menaquinones were initially difficult to analyse and required mass spectrometry. However, recent developments in analytical TLC and the availability of reverse-phase TLC plates have made this detection routine. Quantitative determinations can be achieved by reverse-phase partition HPLC.

Protocol 16. Extraction of menaquinones

You will need

- 10 cm × 10 cm Kieselgel $60F_{254}$ preparative TLC plates (Merck)
- chloroform/methanol (2:1, v/v)
- Petroleum ether (60–80°C)/diethyl ether (85:15 v/v)
- Sodium borohydride
- Rotary evaporator
- Oxygen-free nitrogen

1. Add 30 ml of chloroform/methanol (2:1 v/v) to 100–200 mg dried cells (from *Protocol 13*) and stir gently in the dark for 2–3 h.
2. Filter the suspension using filter paper and discard the cells.
3. Using a rotary evaporator, dry the filtrate under reduced pressure.
4. Resuspend lipids in a small volume of chloroform/methanol (2:1 v/v) and apply to a preparative TLC plate.
5. Develop in petroleum ether (60–80°C)/diethyl ether by ascending chromatography and dry.
6. Briefly expose the plate to UV light (254 nm) and mark the band of menaquinones (R_f approximately 0.8) with a pencil.
7. Transfer the silica gel containing the menaquinones to a sintered glass filter and elute with chloroform.
8. Evaporate to dryness under a stream of oxygen-free nitrogen.
9. The presence of menaquinones can be confirmed by an absorption spectrum in the UV range, and by reduction on the addition of a few crystals of sodium borohydride. Maxima at 242, 248, 260, 269, and 326 nm indicate the presence of menaquinones.

The number of isoprenologues can be determined readily by reverse-phase partition chromatography.

Protocol 17. Analysis of menaquinones by reverse-phase partition TLC

You will need

- 10 cm × 10 cm Merck HPTLC $RP18F_{254}$ plates

1. Spot samples prepared using *Protocol 16* on to the plate and dry.
2. Develop in acetone/water (99:1, v/v) until the solvent reaches the top of the TLC plate.
3. Locate menaquinones by examination under UV light (254 nm) and locate with authentic standards. Score their intensity on a scale of + to + + + +, as this gives a good indication of their relative proportions (see *Figure 8*).

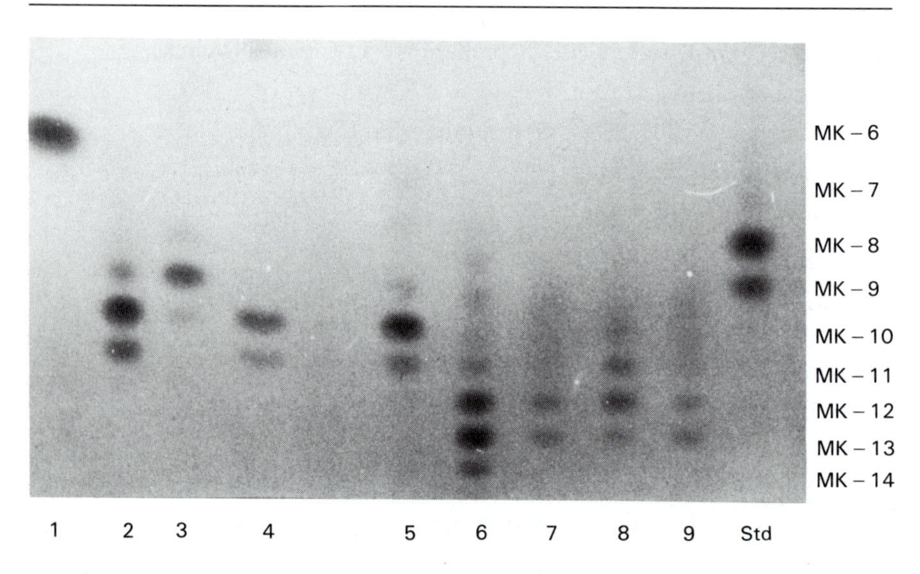

Figure 8. Separation of menaquinones from Gram-negative anaerobic bacteria by reverse-phase partition TLC. 1: *Capnocytophaga* sp. (formerly *Bacteroides*); 2,5–9: *Prevotella* spp.; 3,4: *Porphyromonas* spp.; Std: standard.

Confirmation of the precise nature of the menaquinones present can be achieved by mass spectrometry. *Figure 9* shows the mass spectrum of *Porphyromonas gingivalis* W83 done on an AE1 MS9 instrument using a direct insertion probe, an ionizing voltage of 70 eV and a temperature range of 200–220°C.

Excellent separation and quantitative information on menaquinones can also be achieved by HPLC. The separation of MK-7 through to MK-14 (present in *Bacteroides*) is well resolved using a spherisorb ODS (5 μm) LDC column (250 × 4.6 mm i.d.) with methanol-1-chlorobutane (100:10) as the mobile phase (1.5 ml/mm). For the separation of small unsaturated (e.g. MK-1–MK-6) or small hydrogenated (e.g. MK-6(H_2), MK-5(H_2)) menaquinones, other conditions are required. An excellent review of this topic is given by Collins (10).

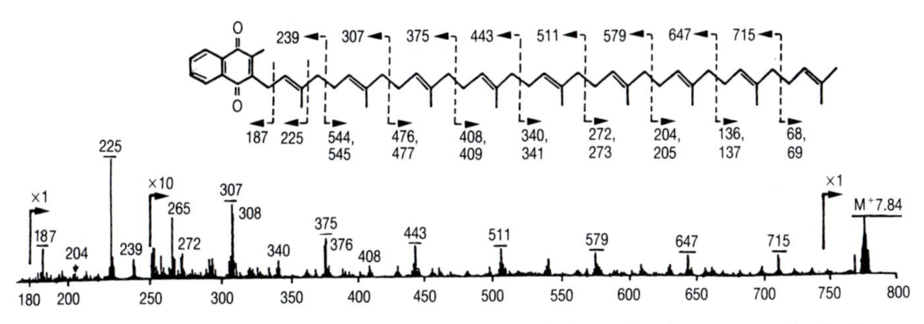

Figure 9. Mass-spectrum of menaquinone-9 (MK-9) from *Porphyromonas gingivalis.*

6. DNA base composition, DNA–DNA hybridization, and ribosomal RNA gene restriction patterns

Nucleic acid analyses are now an essential part of bacterial systematics. Distant phylogenetic relatedness can be estimated by techniques such as ribosomal RNA sequence analysis. This method is beyond the scope of this chapter but several excellent reviews have been published (11). DNA base composition helps define a genus (*ca.* 10–12 mol % G+C) while DNA–DNA reassociation provides evidence of relatedness within a species. Subspecific differences can be assessed by nucleic acid restriction fragment length polymorphism (RFLP). All of these techniques require the preparation of highly purified DNA or RNA; the methods described in the following subsections are used by the authors for studies of the Bacteroidaceae.

6.1 Isolation and purification of DNA

Several methods are available for the isolation and purification of DNA. Lysis of Gram-positive cells is generally more difficult than of Gram-negative bacteria, but the latter may contain a wider spectrum of nucleases which rapidly degrade the nucleic acids. The method described in *Protocol 18* has been used successfully for the preparation of DNA from a wide range of Gram-negative, non-spore-forming anaerobes.

Protocol 18. Extraction and purification of chromosomal DNA

You will need

- Tris–saline–EDTA (TES) buffer: 0.05 M Tris, 0.05 M NaCl, 0.005 M EDTA, pH 8.0
- 25% (w/v) sodium dodecyl sulphate (SDS)
- 5 M sodium perchlorate
- Chloroform: octanol (24:1 v/v)

Protocol 18. *Continued*

- Ethanol 95% (also 70%, 80%, and 90%)
- Saline–citrate (SSC): 0.15 M NaCl, 0.015 M-trisodium citrate, pH 7.0
- Concentrated saline–citrate (CSC) = × 10 SSC
- Dilute saline–citrate (DSC) = 10^{-1} SSC
- Ribonuclease in 0.15 M NaCl, pH 5.0 (heated to 80°C for 10 min to inactivate any DNase present) added at 50 μg/ml final concentration
- Isopropanol
- Proteinase K, 50 μg/ml final concentration
- Acetate–EDTA: 3.0 M sodium acetate plus 0.001 M EDTA, pH 7.0

1. Harvest *ca.* 1–2 g wet weight of cells from either solid or liquid medium and wash at least once in 50 ml of TES buffer. Harvest cells by centrifugation and then resuspend in 18 ml of the same buffer. (Aim to get a well dispersed suspension.)
2. Add 2 ml of SDS and place in a 60°C water-bath for 10 min. The temperature can vary between 30 and 60°C depending on the ease of lysis of the cells. Nuclease inhibitors such as diethylpyrocarbonate (10 μg/ml) or proteinase K (50 μg/ml) may be added to prevent degradation.
3. Add 5 ml of 5 M perchlorate.
4. Add an equal volume of chloroform/octanol mixture, cap the vial securely, and shake for 30 min.
5. Centrifuge the emulsion at 20 000 *g* for 10 min. Three layers will form. Carefully remove the top aqueous layer with a wide bore pipette.
6. Add *ca.* 5–6 ml of aqueous mixture from step 5 to a fresh tube and carefully layer on 2 vol of cold ethanol. Use a glass rod to 'spool' the nucleic acid fibres by stirring the rod at the interface. Remove the rod and drain off excess alcohol.
7. Redissolve in 18 ml of DSC. Repeat the procedure with a fresh rod until all the nucleic acids have been collected into one 18-ml tube. At this stage the ethanol and aqueous layers become more or less homogeneous.
8. Leave to dissolve overnight at 4°C in an equal volume of chloroform/octanol.
9. Continue deproteinization/centrifugation procedures (steps 4–7) without precipitation of nucleic acids, until most of the protein interface disappears.
10. Precipitate with ethanol as in step 6 and redissolve in 10 ml of SSC.
11. Add RNase (50 μg/ml final concentration) and incubate at 37°C for 30 min.
12. Repeat deproteinization/centrifugation procedure (2–3 times), depending on the amount of protein seen at the interface.
13. Precipitate DNA (step 6) and redissolve in 9 ml DSC.

Protocol 18. *Continued*

14. Add 1.0 ml acetate–EDTA and precipitate DNA with 0.54 vol of cold iso-propanol.
15. Wash the DNA fibres in 70, 80, 90, and 95% alcohol, then drain off the excess alcohol.
16. Redissolve in 2 ml of SSC and dialyse against 500 vols of SSC. Three changes, for 24 h each at 5°C, are recommended to remove contaminating divalent cations, polyamines, and traces of phenol (if phenol is used instead of chloroform).

6.1.1 Determination of DNA base composition (mol % G+C)

Thermal denaturation is rapid, reliable, and has been widely used. In practice, the DNA is heated in DSC or SSC (0.5–1.0°C/min) and its absorbance continually monitored at 260 nm until the observed sigmoid curve plateaus. The temperature which corresponds to half its absorbance value is its T_m. From this temperature, the mol % G+C may be calculated from published equations (12).
For DNA melted in SSC,

$$\text{Mol } \% \text{ G} \times \text{C} = 2.44\ T_m - 169.0.$$

For DNA melted in DSC,

$$\text{Mol } \% \text{ G} \times \text{C} = 2.08\ T_m - 106.4.$$

6.2 DNA–DNA hybridization

DNA–DNA hybridization for taxonomic studies has been carried out by two main techniques: in free solution or by using membrane support systems. Both approaches work well and have merits and drawbacks. A free solution method, often referred to as the S_1 nuclease method, is described here. Details of other methods were reviewed by Owen and Pitcher (13). The S_1 nuclease method can be divided conveniently into five steps:

(a) preparation of unlabelled DNA from test strains;
(b) nick translation of probe DNA;
(c) hybridization;
(d) termination of hybridization;
(e) calculation of percentage reassociation.

6.2.1 Preparation of DNA for hybridization

DNA prepared using *Protocol 18* can be subjected to electrophoresis using 0.8% agarose and viewed in the presence of ethidium bromide under UV radiation to determine its quality. DNA sheared or denatured during purification runs ahead of the main band. The DNA is dialysed for at least 16 h in 0.42 M NaCl and sheared by use of a sonicator (15 sec, 3–4 amperes) and stored at −20°C until required.

Protocol 19. Preparation of [^{3}H] dCTP labelled DNA by nick translation (e.g. based on an Amersham kit)

You will need

- 2–3 μg of DNA to be labelled (probe). Shear DNA by sonication (3–4 amperes) for 15 sec (ca. 400–800 nucleotides)
- Nucleotide/buffer solution: 100 μM dATP, 100 μM dGTP, 100 μM dTTP in 0.5 M Tris–HCl pH 7.8 containing 0.1 M $MgCl_2$ and 100 mM dithiothreitol or 2-mercaptoethanol
- 20 μM [5-^{3}H] dCTP
- *E. coli* DNA polymerase 1, and 100 pg DNase 1 in 0.5 M Tris–HCl pH 7.5, containing $MgCl_2$, 50% glycerol, and 500 μg/ml bovine serum albumin.
- 0.5 M EDTA, pH 8.0
- 10 mM Tris–100 mM EDTA, pH 8.0 buffer
- 7.5 M ammonium acetate

1. To a polypropylene tube (1.5 ml capacity) containing 2–3 μg DNA (in distilled water, volume will vary) to be labelled, add the following
   - 2 μl of nucleotide/buffer solution
   - 10 μl of [^{3}H] dCTP
   - Water to give a total volume of 90 μl, and mix
   - 10 μl of enzyme solution (D)
2. Incubate at 15°C for up to 60 min.
3. To monitor the reaction, remove 1- to 2-μl aliquots at 15-min intervals, precipitate with TCA as described in *Protocol 20*, and place in a chilled scintillation vial on ice.
4. Add 2 μl of 0.5 M EDTA to stop the reaction.
5. Add 10 ml scintillation fluid (Dioxan-based) and determine the percentage uptake of [^{3}H]dCTP. When the c.p.m. have reached a plateau (*ca.* 30–60 min), terminate the reaction and separate the nick-translated DNA from unincorporated dNTPs.
6. Pour a Sephadex G-50 column (10 × 1 cm wide) and determine its void volume with blue dextran (12–15 drops).
7. Add 200 μl of ice-cold 0.5 M EDTA to the incubation tube and leave on ice.

Protocol 19. *Continued*

8. Apply the mixture to the sephadex G-50 column and elute with Tris–EDTA pH 8.0 buffer.
9. Collect 100-μl fractions (*ca.* 2 drops) and determine the c.p.m. radioactivity of 1 μl of each fraction. (All the c.p.m. are usually recovered in the first six fractions.)
10. To all tubes with high c.p.m. add 0.5 volume of 7.5 M ammonium acetate and 2 volumes of cold ethanol, mix, and freeze at -20°C to precipitate DNA.
11. Centrifuge the tubes at 10 000 g for 10 min, decant liquid, and redissolve DNA in DSC.
12. Pool the DNA solutions from all tubes and add to a microdialysis tube. Dialyse against DSC overnight.
13. Determine the radioactivity and the DNA concentration of an aliquot, and calculate its specific activity.

6.2.2 Hybridization

Britten and Kohne (14) showed that the initial reassociation of DNA in solution follows second-order kinetics and that hybrid formation is a function of the initial concentration of each DNA species and its incubation time. They introduced the acronym Cot, for the product of the concentration (moles of nucleotides per litre) (Co) and time (t) in seconds. When the temperature, salt concentration, and fragment size are fixed, Cot values can be used to predict the extent of DNA–DNA reassociation. Denatured DNA in a free solution is generally reassociated for k_{Cot} (k, rate constant) values of 100, to enable maximum reassociation. The lower the concentration of test DNA, the longer the incubation period ($k_{Cot} = 100$). Thus for an 83-h incubation, 100 μg unlabelled DNA per millilitre is required. In general experiments are set up with a high ratio of unlabelled to labelled DNA of 1500:1, to ensure a minimum chance of self-reassociation.

Experiments can be carried out conveniently in 1.5-ml polypropylene tubes using 200-μl volumes. For an 83-h incubation, 20 μg of unlabelled DNA is therefore required. For a 1:1500 ratio, 0.013 μg [^{3}H] DNA: 20 μg unlabelled DNA is required. Maintain this ratio against the DNA of all test strains, and carry out the experiment in triplicate. To avoid evaporation, add a few drops of liquid paraffin to cover the surface of all hybridization mixtures. Keep the tubes gently agitated during incubation. Incubation can be either at 35°C below its T_m (non-stringent) or 15°C less than T_m (stringent). The optimum temperature (T_{OR}) is calculated from the equation $T_{OR} = [0.51 \times (G+C)] + 47$.

Protocol 20. Termination of hybridization

You will need

- S_1 nuclease mix
  - (i) 1 mM $ZnSO_4$/0.3 M sodium acetate: 12.02 μl
  - (ii) denatured and sheared calf thymus DNA[a]: 1.49 μl
  - (iii) S_1 nuclease: 119 μl
  - (iv) Sterile distilled water: 64.27 μl
- Trichloroacetic acid (TCA), 25% (w/v) and 5% (w/v)

1. Transfer hybridization reaction tubes to an ice bath and leave for 20 min.
2. Add 54 μl of S_1 nuclease mix and incubate tubes at 50°C for 20 min.
3. Stop the reaction by the addition of 0.7 ml ice-cold 25% TCA and leave for 1 h.
4. Filter through a 0.45-μm microfilter and wash three times with 3 ml 5% TCA.
5. Place the filter in a scintillation vial and dry at 70°C for 1 h.
6. Add 10 ml scintillation fluid and determine c.p.m.

[a]Stock calf-thymus DNA (1 mg/ml in 0.42 M NaCl) sheared by sonication (4–6 amperes for 15 sec) and denatured by boiling for 10 min.

6.2.3 Calculation of percentage reassociation

Some c.p.m. are always lost during various manipulations and recovery is never 100%. Thus the average c.p.m. of the homologous reaction (the c.p.m. obtained by reassociation of the unlabelled DNA with its labelled probe) is taken as 100% homology. All of the test samples are expressed as a percentage of the homologous reaction. Correction should also be made for any non-specific hybridization from control experiments using denatured, or non-denatured, calf thymus DNA.

6.3 16S ribosomal RNA gene restriction patterns

DNA can be digested with restriction enzymes to give a range of fragments, which can then be separated by electrophoresis, providing a 'fingerprint' pattern for a strain. This is termed a restriction fragment length polymorphism (RFLP). This widely used technique is very simple but has many drawbacks, one of which is the difficulty of comparing RFLPs between isolates. The method described in *Protocol 21* involves the use of 16S rRNA genes only, which, after digestion and electrophoresis, are hybridized with digested DNA of the test strains. The patterns produced are simple, reproducible, and easily recorded for interstrain comparisons. *Figure 10* shows the patterns derived by this method to separate two of the three subspecies of *Fusobacterium nucleatum*.

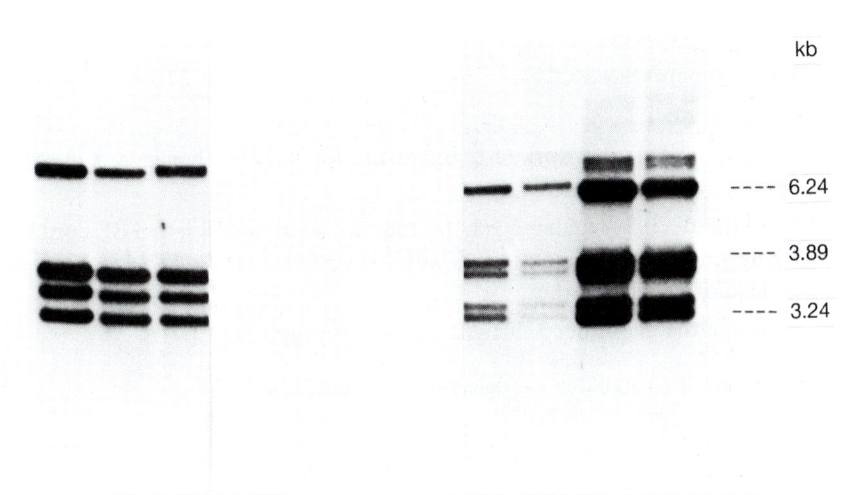

Figure 10. RFLPs of *Fusobacterium* species using *Hind*III.

Protocol 21. DNA restriction enzyme digestions

You will need

- Tris–boric acid–EDTA buffer (TBE): 0.089 M Tris, 0.089 M boric acid, 0.002 M EDTA, pH 8.0
- Enzyme mix: in practice, the DNA is usually digested with several enzymes before specific ones are selected for further use; *Eco*RI, *Hind*III, and *Taq*1 generally provide good diagnostic patterns
- Stopper mix

Final concentration	*per 100 ml*
10 mM Tris–HCl (pH 7.5)	1 ml of 1 M stock solution
20 mM EDTA	4 ml of 0.5 M stock solution
10% glycerol	10 ml
0.025% bromothymol blue	25 mg
0.025% bromophenol green	25 mg

- Lambda DNA digested with *Hind*III is used as a molecular weight marker; this is available commercially and requires incubation at 65°C for 3 min and chilling on ice before use

1. For each DNA sample set up a digest containing
   - DNA (5 μg) ×

Protocol 21. *Continued*

• 10 × assay buffer[a]	5
• Enzyme mix + water[b]	×
Total volume	50 μl

2. Incubate at the appropriate temperature for 4–16 h; then add 5 μl stopper mix.
3. Load on to a 0.8% agarose gel and run at 1 V/cm for 18 h in TBE containing 0.5 μg/ml ethidium bromide. *Hin*dIII-digested lambda DNA is used as a molecular weight marker.
4. Examine the gel under UV light and photograph.

[a] Assay buffer is supplied with each buffer by the manufacturer.
[b] Add 5 enzyme units per μg of DNA.

Protocol 22. Southern blotting

You will need

- Denaturing solution: 1.5 M NaCl, 0.5 M NaOH
- Neutralizing solution: 1.5 M NaCl, 0.5 M Tris–HCl (pH 7.2), 0.001 M EDTA
- Whatman no. 3 MM filter paper
- Saline–citrate (SSC) 20 × : 3.0 M NaCl, 0.3 M tri-sodium citrate, pH 7.0

1. After photographing the gel (step 4 in *Protocol 21*), cover with denaturing solution and agitate gently for 40 min.
2. Remove the denaturing solution, rinse with water, and then cover with neutralizing solution and agitate for 40 min.
3. Set up the blot ensuring there are no air bubbles in the wells or between the gel and the nylon membrane:
   (i) 0.5–1 kg weight;
   (ii) 15 cm depth of paper towels;
   (iii) two sheets of Whatman no. 3 MM, soaked in 2 × SSC;
   (iv) gel (face up), surrounded by parafilm;
   (v) nylon membrane;
   (vi) Whatman 3 MM wick, overlying a support, with the ends within a reservoir of 20 × SSC
4. Blot for approximately 24 h.
5. Remove the towels carefully, mark the wells, and label the nylon membrane with the date and enzyme used, with a marker pen.

Protocol 22. *Continued*

6. Place the membrane in 2 × SSC and gently shake for 2 min to remove agarose.
7. Air-dry for 30 min. Bake in an oven at 80°C for 2 h to fix the DNA to the nylon membrane.

Protocol 23. Isolation and labelling of Probe DNA

1. Digest 6 μg plasmid pKK3535 DNA with *Bcl* I and *Bst* EII as described in *Protocol 21*.
2. Load on to a 1% low gelling agarose gel with the molecular weight marker lambda cleaved with *Hin*dIII. Electrophorese at 50 mA until the blue tracking dye has migrated to 5 cm.
3. Examine under UV light and, using a scalpel blade, excise the 1.5-kb fragment containing the 16S rRNA gene and transfer it to a preweighed 1.5-ml microcentrifuge tube.
4. Determine the weight of the gel slice, add 3 vol water, incubate at 100°C for 7 min, vortex-mix, and store at −20°C.
5. α-^{32}P-dCTP is then incorporated using a Random Primer Extension Labelling System (DuPont) following the manufacturer's instructions, using 28 μl of the agarose probe mix. Probe DNA is used at a concentration of 1×10^6 c.p.m. per ml per membrane, or, alternatively, a quarter of the labelled probe can be used.

Protocol 24. Hybridization

You will need

- Saline–citrate (SSC) 20 × : 3.0 M NaCl, 0.3 M tri-sodium citrate, pH 7.0
- 100 × Denhardt's stock: 2% polyvinylpyrrolidone (Sigma P-5288), 2% Ficoll (Sigma F-9378), 2% BSA Pentax fraction V (Sigma A-4503). Store in 50-ml aliquots at −20°C
- Hybridization and pre-hybridization mix:

Final concentration	*per litre*
6 × SSC	300 ml of 20 × stock solution
0.5% SDS	25 ml of 20% SDS
5 × Denhardt's	50 ml of 100 × Denhardt's
herring sperm DNA (25 μg/ml)[a]	2.5 ml of 10 mg/ml solution
dextran sulphate (in hybridization mix only)	200 ml of 50% solution

Protocol 24. *Continued*

- Wash solution 1: 2 × SSC, 0.1% SDS
- Wash solution 2: 0.2 × SSC, 0.1% SDS

1. Wet the filter prepared using *Protocol 22* with water and place into a polythene bag. Heat-seal the bag along three sides and examine for possible leaks.
2. Add 25 ml pre-hybridization mix heated to 65°C and seal the bag completely. Snip off one corner and remove all air bubbles by laying on a flat surface and running a ruler over the bag. Reseal the bag.
3. Incubate in a shaking water-bath at 65°C for 2–4 h.
4. Denature the double-stranded probe (see *Protocol 23*) by heating at 100°C for 3 min; then chill quickly on ice for 5 min.
5. Add the probe to 25 ml hybridization mix at 65°C.
6. Open the bag and drain the pre-hybridization mix. Replace with the hybridization mix and reseal the bag, ensuring no air bubbles remain. Incubate at 65°C for 18 h.
7. Open the bag and discard the hybridization mix into a sink designated for radioactive disposal.
8. Place the filter in a plastic sandwich box and cover with an excess of wash solution 1 at 65°C. Agitate for 15 min at 65°C.
9. Replace the wash solution with fresh wash solution 1 at 65°C. Repeat incubation as described in step 8.
10. Replace the wash solution with an excess of wash solution 2 at 65°C. Incubate as described in step 8.
11. Repeat step 10.
12. Remove the filter and wrap in Saran wrap. Expose to X-ray film for 12–24 h using intensifying screens at −70°C; then develop according to the manufacturer's instructions. A longer period of exposure may be necessary, based on experience.

[a] Shear herring sperm DNA by sonication to 1×10^6 base pairs (generally 2 × 15 sec bursts). Boil for 10 min, chill quickly on ice, and then add to the mix.

References

1. Shah, H. N. and Collins, M. D. (1983). *J. Appl. Bacteriol.* **55**, 403.
2. Shah, H. N., Nash, R. A., Hardie, J. M., Weetman, D. A., Geddes, D. A., and McFarlane, T. W. (1984). In *Chemical Methods in Bacterial Systematics* (ed. M. Goodfellow and D. E. Minnikin), pp. 317–40. Academic Press, London.
3. Holdeman, L. V., Cato, E. P., and Moore, W. E. C. (1977). *Anaerobe Laboratory Manual* (4th edn). Virginia Polytechnic Institute and State University, Blacksburg.

4. Laemmli, U. K. (1970). *Nature* **224**, 680.
5. Jackman, P. J. H. (1984). In *New Methods for the Detection and Characterisation of Micro-Organisms* (ed. C. S. Gutteridge and S. Chichester), pp. 117–28. John Wiley, Chichester.
6. Schleifer, K. H. and Kandler, O. (1979). *Bacteriol Rev.* **36**, 407.
7. Schleifer, K. H. and Seidl, H. P. (1977). In *Microbiology—1977* (ed. D. Schlessinger), pp. 339–51. American Society for Microbiology, Washington, DC.
8. Minnikin, D. E., Goodfellow, M., and Collins, M. D. (1978). In *Coryneform Bacteria* (ed. I. J. Bousfield and A. G. Callely), pp. 85–160. Academic Press, London.
9. Collins, M. D. and Jones, D. (1981). *Microbiol. Rev.* **45**, 316.
10. Collins, M. D. (1984). In *Chemical Methods in Bacterial Systematics* (ed. M. Goodfellow and D. E. Minnikin), pp. 267–87. Academic Press, London.
11. Woese, C. R. (1987). *Microbiol. Rev.* **51**, 221.
12. Owen, R. J. and Hill, L. R. (1988). In *Identification Methods for Microbiologists* (ed. F. A. Skinner and D. W. Lovelock), pp. 277–96. Academic Press, London.
13. Owen, R. J. and Pitcher, D. (1984). In *Chemical Methods in Bacterial Systematics* (ed. M. Goodfellow and D. E. Minnikin), pp. 67–93. Academic Press, London.
14. Britten, R. J. and Kohne, D. E. (1966). *Carnegie Inst. Yearbook* **65**, 78.
15. Lawson, P. A., Gharbia, S. E., Shah, H. N., and Clark, D. R. (1989). *FEMS Microbiol. Lett.* **15**, 41.

6

Immunochemical methods

I. R. POXTON

1. Introduction

As in any branch of modern bacteriology, immunological techniques are used for a vast and increasing number of applications. The applications in the past were mainly in medical microbiology, especially in the diagnosis of disease, in epidemiological typing, and in the fundamental investigation of pathogenic mechanisms. In the future this trend is likely to continue, but the exploitation of immunological methods for the detection and quantitation of any bacterial component or product is likely to increase. In anaerobic bacteriology, however, it is by no means an area of great activity. Many workers have been disillusioned by the complexity of the subject, and it is true to say that much of the past work has been greatly compromised by a lack of appreciation of the complexity of the antigenic composition of anaerobes. They are no more complex than aerobes but, especially for the *Bacteroides* spp., there are no traditional serotyping schemes available as there were for the enterobacteria, and attempts at extrapolating from enterobacterial serology have been largely unsuccessful. Apart from certain clostridial infections there is little evidence of a one organism/one disease relationship and again this has confused the unwary.

This chapter will give details of the immunological methods that will be of use in the investigation of most problems likely to be encountered in anaerobic bacteriology. Some examples of applications will be given where relevant and the many problems and pitfalls highlighted.

2. Historical applications of immunological methods

Before beginning to describe the techniques and their applications, it is perhaps worthwhile reviewing, and reminding the reader, of the immunological methods used in the past. It is in clostridial identification that they have been most useful. From the universally used Nagler reaction for the specific identification of *C. perfringens* on egg-yolk agar (see Chapter 3) to the definitive identification of *C. tetani* and *C. botulinum* by toxin neutralization tests in mice (see Chapter 9), the use of specific antibody neutralization tests has been wide. Immunofluorescent antibody methods have been developed for certain clostridia. Capsular typing based on agglutination tests has been used for investigation of *C. perfringens* food poisoning strains. Several attempts have been made to develop a somatic typing scheme for the *Bacteroides fragilis* group, but these have been largely unsuccessful and their relevance has been questioned.

3. Antigens of anaerobic bacteria

Anaerobic bacteria can be considered as typical Gram-positive and Gram-negative bacteria (*Figure 1*). As far as we know there are no fundamental differences in structure between aerobes and anaerobes. The range of antigens that can be exploited include both cell-surface components and extracellular products (exotoxins and enzymes).

These antigens include

in Gram-positive bacteria

- secondary wall carbohydrate (teichoic acid)
- membrane carbohydrate (lipoteichoic acid)
- surface proteins (regular arrays and fibrillar proteins)

in Gram-negative bacteria

- lipopolysaccharide
- 'common antigens' (equivalent to the enterobacterial common antigen)
- outer membrane proteins
- lipoproteins
- fimbriae/pili

in both

- capsular and other exopolysaccharides
- peptidoglycan
- flagella
- extracellular products (exotoxins and exoenzymes)

4. Preparation of some relevant antigens

The preparation of all the antigens just listed cannot be covered in this chapter. Only those considered most important and relevant to anaerobes will be described here. The investigator who requires more detailed information on the preparation and characterization of bacterial antigens is referred to the book by Hancock and Poxton (1). The extraction of antigens from the bacterial cell usually begins with either the whole cell where extractants such as detergents or solvents are employed or, alternatively, the cell may first be fractionated into

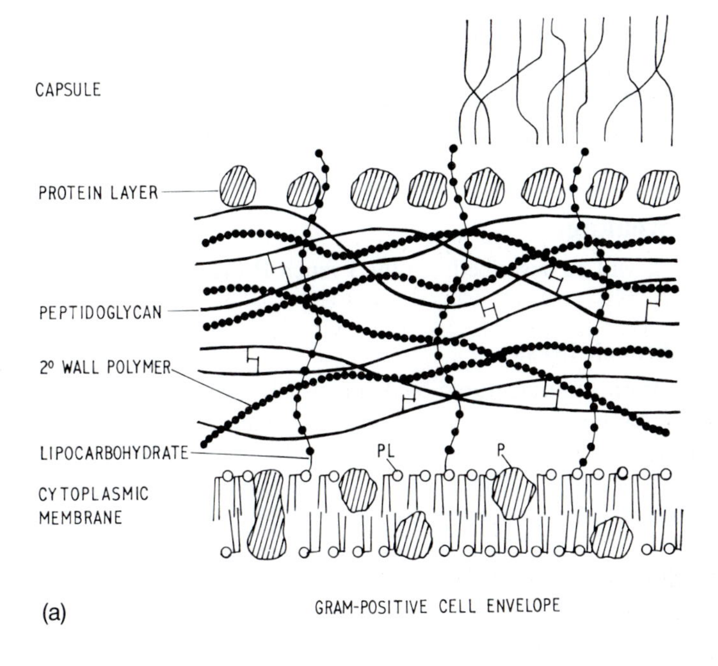

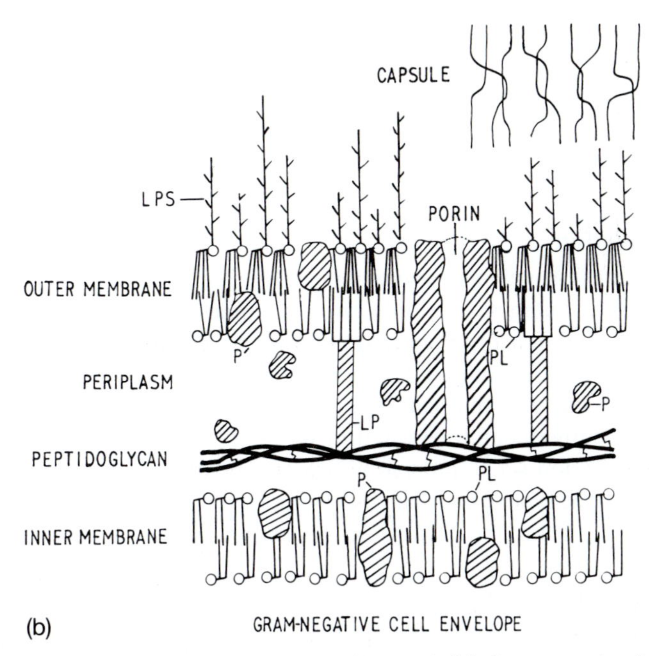

Figure 1. The cell envelopes of (a) Gram-positive and (b) Gram-negative bacteria. PL, phospholipid; LPS, lipopolysaccharide; P, protein; LP, Braun's lipoprotein.

subcellular components such as cell walls, cell envelopes, or membranes, and then subsequent extraction methods are employed on these fractions. For this latter approach, cell breakage is necessary and is most important for Gram-positive bacteria.

4.1 Cell breakage

Two methods will be described: (a) the French Press; and (b) sonication.

4.1.1 The French Press

The French Press (manufactured by Aminco, Silver Springs, Maryland, USA) is a convenient method for breaking all types of bacteria. It has the advantage that the wall/envelope remains largely intact and is not highly fragmented. The bacteria are made into a suspension in a physiological buffer containing 1 mM magnesium, and the cell density can range widely in concentration from very dilute to 50% (wet w/v). The suspension (5–40 ml) is poured into the pre-cooled pressure cell and is assembled in the hydraulic press. Pressures in the range of 10^7 Pa are required for breakage. Generally, Gram-negative rods break most easily, while Gram-positive cocci are the most difficult, and may require several passes through the cell for complete breakage. The efficiency of breakage must be monitored by microscopy and any unbroken cells are removed by two cycles of low-speed centrifugation (5000 *g* for 10 min). In very dense suspensions the viscosity may rise alarmingly after breakage because of the release of nucleic acids. If this is a problem 10–20 μg of DNase and RNase can be added prior to passage through the cell, and incubation at room temperature for 15 min or so after breakage is usually sufficient to reduce viscosity.

4.1.2 Sonication

This technique is usually more widely available than the French Press, but suffers from the disadvantages that only small volumes (usually 10 ml or less) are handled. Heating is a problem and cooling in ice/ethanol is necessary. Some bacteria, especially Gram-positive cocci, can be difficult or impossible to break, and, for those bacteria that can be broken, tiny fragments are often produced which make subsequent fractionation difficult. A suspension is made up as described in Section 4.1.1, and a 5–10% (wet w/v) concentration is usually ideal. Depending on the type of bacteria, five 1-min bursts (with 30-sec cooling periods) of peak power from the ultrasonic generator are often sufficient. Monitor breakage by microscopy.

Once cells are broken, it must be recognized that most bacteria possess autolytic enzymes which can greatly modify cellular components in a short time. Subsequent steps must take account of this and procedures must involve inactivation steps, or speed and low temperatures must be employed.

Protocol 1. Wall preparation in Gram-positive bacteria

1. After cell breakage and removal of any unbroken cells, centrifuge the suspension at 40 000 *g* for 20 min. The cell wall/envelope fraction forms a dense layer at the bottom of the tube. For many anaerobic bacteria this layer is often black in colour because of the presence of insoluble sulphides produced in the highly reduced growth conditions.
2. Decant the supernate containing the cell membranes and cell contents and retain for later. Resuspend the wall pellet in water and wash once.
3. Resuspend the pellet in water and add an equal volume of 4% (w/v) sodium dodecyl sulphate (SDS) which has just boiled.
4. Stir the suspension for several hours (conveniently overnight) at room temperature.
5. Wash the suspension free of SDS by at least six cycles of centrifugation (40 000 *g*) with water at *room temperature* (SDS precipitates in the cold). SDS removal can be confirmed by the water not frothing. The cell walls should now appear white.
6. Resuspend in a small volume of water and lyophilize.

After the breakage of Gram-negative bacteria, the envelope fraction, i.e. inner membrane, outer membrane, and associated peptidoglycan layer, is pelleted by centrifugation at forces of 50 000 *g* or greater.

4.2 Preparation of the outer membrane of Gram-negative bacteria

Traditionally, the outer membrane (OM) was prepared from a whole envelope fraction followed by density gradient centrifugation. Although this is probably the most authentic method, an approximation to the method can be made using the detergent sodium *N*-lauroyl sarcosinate (Sarkosyl) which selectively solubilizes the inner membrane from a broken cell suspension, leaving the OM as an insoluble pellet.

Protocol 2. Outer membrane preparation

1. Break cells by either of the methods described in Section 4.1 and remove any unbroken cells.
2. Add Sarkosyl (1 vol of 24% (v/v) solution of Sarkosyl (Sigma: 30% (w/v) solution) to 9 vol. supernate) to give a final concentration of 0.7% by weight.
3. Centrifuge at 50 000 *g* for 1 h to pellet insoluble outer membranes.
4. Resuspend pellet in distilled water and wash as in step 3.

4.3 EDTA extraction of Gram-positive surface antigen

A method that has been used for the extraction of a surface antigen from various clostridia makes use of the chelating agent ethylenediamine tetraacetic acid (EDTA). It appears that a range of surface proteins together with the lipocarbohydrate membrane antigen can be released from clostridia without the lysis of the bacterial cell. This soluble antigen is representative of the whole surface of the bacterium. It has been used in a variety of applications such as in the immunotaxonomy of clostridia (2), especially *C. botulinum* and related species (3), and in the immunological fingerprinting of *C. difficile* (4).

Protocol 3. EDTA extraction of whole bacteria

1. Harvest bacteria from 100 ml broth and wash three times in phosphate-buffered saline (PBS: 50 mM phosphate buffer containing 0.15 M NaCl, pH 7.4) by centrifugation at 10 000 *g* for 10 min.
2. Resuspend pellet in 4 ml PBS containing 10 mM EDTA.
3. Incubate at 45°C for 30 min.
4. Remove cells by two cycles of centrifugation as in step 1.
5. The supernate, which will contain in the order of 1 mg/ml protein, can be used without further treatment in ELISA and immunoelectrophoresis assays.

4.4 Secondary cell-wall carbohydrate antigens

The secondary cell-wall carbohydrate antigen or teichoic acid analogue is probably, as in the aerobic Gram-positives, an important grouping antigen in clostridia. It has been investigated only in *C. difficile* (5).

Protocol 4. Preparation of secondary cell-wall carbohydrate

1. Prepare pure cell walls (see *Protocol 1*).
2. Resuspend 1 g in 40 ml of either[a] 0.5 M NaOH or 0.1 M HCl and stir at room temperature for 1 h.
3. Neutralize with the equivalent concentration of HCl or NaOH and remove extracted walls by centrifugation (10 000 *g*, 15 min).
4. Dialyse supernate against at least two changes of 5 l of distilled water over 18 h.
5. Concentrate by rotary evaporation and then freeze-dry.

[a] Although both have been used successfully with *C. difficile*, it is well known for aerobes that, depending on the nature of the glycosidic linkages, it is possible that either may depolymerize such polymers. It will be necessary to test each new species individually.

4.5 Lipocarbohydrate antigens

This polymer is analogous to the lipo- or membrane teichoic acid of aerobic Gram-positives, where it is often an important antigen. In aerobes it has been investigated only in *C. difficile* and close relatives (6) and most recently in *C. tyrobutyricum* (7).

Protocol 5. Preparation of lipocarbohydrate

1. The supernate prepared in step 2 of *Protocol 1* is freeze-dried and used as the starting point of the antigen extraction.
2. Dissolve the lyophilized cell supernate in chloroform/methanol (2:1) at a concentration of 20 mg/ml and stir for several hours at room temperature. After filtration through Whatman no. 1 paper, re-extract with another volume of chloroform/methanol by stirring overnight.
3. Air-dry the filtered, defatted material and then suspend to a concentration of approximately 5% (w/v) in distilled water. Add an equal volume of 80% (w/w) aqueous phenol[a] and stir at room temperature for 1 h.
4. Centrifuge in phenol-resistant tubes at 10000 *g* for 20 min.
5. Remove the upper aqueous phase and dialyse against running water overnight.
6. Remove any insoluble material by centrifugation (10000 *g*, 10 min).
7. For many purposes it may be convenient to interrrupt the preparation at this point and test the material for antigenicity. If a highly purified product is required, it will be necessary to fractionate the material on a column of Sepharose 6B where it will elute in the void volume. Final purification may require immunoadsorbent chromatography (see references 5 and 6 for details).

[a] *Caution*! Phenol is extremely caustic and toxic; handle with care.

4.6 Lipopolysaccharide

As in all Gram-negative bacteria, lipopolysaccharide (LPS) is present in Gram-negative anaerobes. It is usually considered less endotoxic than its aerobic (*Escherichia coli*) equivalent, but only the clinically important *Bacteroides* and *Fusobacterium* spp. have been investigated. Various serotyping schemes have been developed for the *B. fragilis* group, but their relevance is uncertain. There is confusion as to the relationship between capsular polysaccharide and LPS. There is some debate as to the method for the extraction of LPS (8). The best starting approach when investigating an unknown LPS is to use the proteinase K method developed by Hitchcock and Brown (9).

Protocol 6. Proteinase K method for LPS preparation

1. Harvest bacteria from broth or solid medium, wash once in PBS, and resuspend in PBS to an A_{525} (absorbance at 525 nm) of between 0.5 and 0.6.
2. Centrifuge 1.5-ml volumes in a microcentrifuge at 10 000 g for 3 min.
3. Suspend pellet in SDS-PAGE sample buffer and heat to 100°C for 5 min.
4. When cool add 25 μg proteinase K (Protease type XI, Sigma) in 10 μl SDS-PAGE sample buffer and incubate at 60°C for 60 min.
5. Analyse 10-μl samples on 14% polyacrylamide gels, omitting the SDS from the stacking and separating gel buffers.
6. Stain gels with silver or transfer to nitrocellulose for immunoblotting (see *Protocol 14*).

Once the chemotype of the LPS (i.e. rough or smooth) is established by the proteinase K method, it is possible to select the classical aqueous phenol method for the preparation of smooth LPS or the phenol/chloroform/petroleum method for rough LPS (see reference 1 for methods). A combination of these methods has been suggested by Weintraub *et al.* (10), but it should be noted that it is possible that the smooth component of the LPS could inadvertently be discarded (8), especially if the LPS only appeared rough because of the silver stain failing to reveal the 'O' polysaccharide chains: these are sometimes periodate-resistant and only revealed by specific antibody in immunoblots.

4.7 Exopolysaccharides

Capsules are found on a wide range of anaerobes, both Gram-positive and Gram-negative. As described in Section 4.6 there is a certain amount of confusion between the exo- or capsular polysaccharide of *Bacteroides* species and the O antigen of the LPS. It is probably true to say that no definitive method exists for the preparation of exopolysaccharide. In attempting to begin the preparation of the capsule, mechanical methods of shearing off the capsule should be the first approach.

Protocol 7. General method for capsule preparation

1. Harvest bacteria from solid medium by gently scraping with a glass rod and suspend in PBS containing 0.5% formaldehyde.
2. Shear off capsule: gentle stirring to violent agitation in a blender or homogenizer may be required. Monitor by microscopy with indian ink staining.
3. Remove bacteria by centrifugation with forces $>10\,000\ g$ because of high viscosity. Monitor removal of cells by microscopy.

Protocol 7. *Continued*

4. Add four volumes of ice-cold acetone to the supernate and allow the capsular material to precipitate overnight at 4°C.
5. Wash precipitate several times in acetone, dissolve in water, and lyophilize.
6. At this stage the material should be checked for antigenicity. It will almost certainly be contaminated with proteins and other carbohydrates, especially LPS in Gram-negative bacteria. The full purification will require often difficult biochemical and immunological separation techniques which are compounded by the high viscosity and high molecular mass of the material.

4.8 Preparation of appendages

Bacterial appendages are well known antigens, but in anaerobes very little is known of fimbriae or pili. Only the preparation of flagella will be covered here. They have been investigated in several clostridia (11, 12).

Protocol 8. Preparation of flagella

1. Harvest motile bacteria from a log phase broth culture (1 l) by centrifugation at 10 000 *g* for 15 min. *Gently* resuspend in 1 l of PBS and recentrifuge.
2. Resuspend in 20 ml PBS and homogenize in a rotating blade blender at maximum speed for 1–2 min.
3. Remove bacteria by at least two cycles of centrifugation at 10 000 *g* for 15 min.
4. Pellet crude flagella by centrifugation at 100 000 *g* for 90 min. The pellet may be extremely difficult to see, but it should be resuspended in a small volume of PBS and monitored by electron microscopy (by shadowing or negative staining).
5. Prepare a caesium chloride solution (1.3 g/ml) by dissolving optical grade CsCl in 1 ml PBS to give a refractive index of 1.3630 (*ca.* 2.125 g is required). Add this to the pellet of flagella and mix well. Centrifuge at 180 000 *g* for 20 h in a swinging bucket rotor.
6. The flagella form a band in the central region of the tube and are removed with a fine-gauged syringe needle.
7. Recover the flagella by diluting the CsCl solution in the removed band in about 10 ml distilled water and centrifuge at 100 000 *g* for 1 h.

5. Production of antisera

Only the preparation of conventional antiserum in rabbits will be described here. The production of monoclonal antibodies is a much more specialized procedure and requires special facilities. Nevertheless the immunization procedures described here for rabbits can be scaled down 10- or 100-fold for the immunization of mice prior to hybridoma production.

Only two methods are given here for raising conventional polyclonal antiserum—one for whole bacteria and the other for subcellular or soluble antigens. The doses are based on New Zealand White rabbits of 2–2.5 kg. For the smaller Dutch rabbits, which are becoming more common because of cage size regulations, the dose should be scaled down on a weight for weight basis. As the response to antigens can vary greatly from one animal to another it is usual to use more than one animal for raising antisera.

5.1 Whole-bacteria vaccine

For most anaerobes, a washed suspension of live bacteria in PBS is all that is required. All *Bacteroides* spp., and many of the clostridia, permit doses of 10^9 in 1 ml to be injected from day 1. For the more endotoxic *Fusobacterium* spp. and the clostridia which produce potent exotoxins, it is necessary to reduce the dose to 10^6 organisms for the first three injections and irradiate the bacteria (in a thin film in a glass Petri dish) with a lethal dose of ultraviolet light. It is convenient to prepare one batch of antigen for the whole immunization series, storing volumes for each injection in individual tubes and deep freezing.

The procedure is as follows.

(a) Inject 1-ml doses of bacteria into the marginal ear vein on days 1, 2, 3, 8, 9, 10, and 22.

(b) Test bleed from the ear on day 29 and, if antibody is of sufficient strength, exsanguinate the rabbit by cardiac puncture under terminal anaesthesia.

Boosting with more vaccine after day 29 is unlikely to result in a more than twofold increase of titre.

5.2 Subcellular or soluble vaccine

Toxic antigens, such as powerful exotoxins, should be toxoided prior to injection by treatment with 0.5% formaldehyde for 18 h at room temperature.

(a) Prepare vaccines containing 0.1–2.0 mg antigen in 1 ml amounts by mixing equal volumes of the antigen in aqueous solution/suspension and Freund's complete adjuvant. Thorough mixing in a small homogenizer or by passing from one syringe to another through a fine-bore tube is recommended to produce a water-in-oil emulsion. This can be confirmed by placing a small drop on water where it should remain without dispersing quickly.

(b) Inject the vaccine subcutaneously in several sites in the scapular region of the back and massage firmly. Repeat after about 4 weeks using Freund's incomplete adjuvant.

(c) Test bleed after 2 weeks and boost if necessary; then exsanguinate.

The preparation of serum is conveniently accomplished by allowing the blood to clot in a glass vessel overnight at 4°C. If more than one rabbit was used, the serum should only be pooled after checking for an adequate response in each rabbit. After pooling, sera should be divided up into convenient volumes and stored deep frozen. Freezing and thawing should be kept to an absolute minimum.

For some applications, such as crossed and rocket immunoelectrophoresis, a much cleaner result can be obtained if the IgG fraction is prepared. Also, if an antibody is to be labelled, it can be done to a much greater specific activity if it is pure. Several methods exist for the preparation of IgG, but perhaps the simplest, and the most suitable for most purposes, is by ammonium sulphate precipitation.

Protocol 9. Preparation of IgG

1. Prepare a saturated solution of ammonium sulphate in water and add 0.67 ml dropwise at the rate of one drop per second to 1.0 ml serum which is stirred gently in an ice bath.
2. Continue stirring for 15 min; then centrifuge the precipitated IgG at 3000 *g* for 15 min.
3. Resuspend the pellet in 1 ml or less of 0.05 M Tris buffer, pH 8.0 and dialyse against the same buffer. Alternatively a 10 kd cut-off ultrafilter may be used.
4. If the IgG is not to be labelled, it is recommended that bovine serum albumin is added to a final concentration of 1%.

---

6. Immunological methods

The last two decades or so have seen great advances in immunological techniques. Before this, double gel diffusion, complement fixation tests, and haemagglutination assays were the main techniques available. These have now largely been superseded. In the early 1970s improvements in gel diffusion techniques resulted in the widespread use of two-dimensional or crossed immunoelectrophoresis. This was the first technique to permit resolution of complex mixtures of antigens. It is a relatively difficult technique requiring a certain degree of dexterity; it can only be done with a few samples at a time and was only useful for precipitating antigens. The related rocket immunoelectrophoresis allowed quantitation of multiple samples but did not give good resolution. For most purposes immunoblotting has taken over these techniques. The development of radioimmunoassays opened up the field to automation of immunoassays and multiple specimen handling; however, enzyme immunoassays, especially enzyme-linked immunosorbent assays (ELISA), have superseded this technique. For this chapter, precipitation in gel techniques will only be covered briefly and most emphasis will be placed on ELISA, immunoblotting, and immunogold electron microscopy.

6.1 Precipitation in gels

Simple Ouchterlony-type gel diffusion is now largely superseded and will not be described here. Rocket immunoelectrophoresis is probably a more sensitive substitute and is recommended for screening multiple samples containing precipitating antigens. It is quantitative but it has poor resolution for complex antigen mixtures. Crossed immunoelectrophoresis is recommended for complex mixtures where resolution is required. It is only applicable to a relatively small number of samples, but is useful for investigating large-molecular-mass antigens which are impossible to analyse by the more recent immunoblotting technique (see Section 6.3).

Protocol 10. Preparation of electrophoresis buffer and agarose gel

1. Prepare electrophoresis buffer solution 1 by dissolving in 2 l of distilled water
   - Barbitone sodium 26 g
   - Barbitone 4.14 g
2. Prepare solution 2 by dissolving in 2 l distilled water
   - Glycine 112.5 g
   - Tris 90.4 g
3. Mix equal volumes of solutions 1 and 2 and confirm pH to be 8.8. Use this buffer undiluted in buffer reservoirs.
4. Mix together
   - Agarose (low EEO) 1 g
   - Electrophoresis buffer (from step 3) 25 ml
   - Distilled water 75 ml

Heat with constant stirring in a boiling bath until agarose has dissolved. Triton X100 or other detergents can be added to a final concentration of 1% at this stage, if membranous or poorly soluble antigens are being investigated.

5. Cool to 55°C before pouring gels.[a]

[a] It is convenient to prepare in bulk and dispense in suitable volumes for storage.

Protocol 11. Rocket immunoelectrophoresis

Up to 10 samples can be analysed on a 50 mm square of agarose. For more samples a longer length of 50 mm width is required.

1. Cut a 50-mm square of Gelbond® and place it hydrophilic side up on to a supporting glass plate on an absolutely level surface.

Protocol 11. *Continued*

2. Blank off two-thirds of the sheet with a glass plate and cast a layer of agarose (see *Protocol 10*; *ca.* 1.3 ml required).
3. After setting, cut up to 10 3-mm diameter wells in a linear or staggered pattern.
4. For many purposes it is convenient to show identity between antigens and, if this is the case, the wells should be loaded at this stage with 10-μl volumes of antigen and then left for 30 min in the cold to allow antigens to diffuse into the agarose. This technique is then referred to as fused rocket immunoelectrophoresis. If showing identity is not a concern then the wells may be loaded after the next step.
5. Gently mix 2.2 ml molten agarose held at about 55°C with 0.5 ml antiserum which may be diluted in 1 : 4-diluted electrophoresis buffer, and pour on to the remaining two-thirds of the Gelbond®.
6. After setting, place the gel on to a horizontal electrophoresis tank with the wells towards the cathode and connect the ends of the agarose to the electrophoresis buffer (see *Protocol 10*) with presoaked filter paper wicks.
7. Apply 12 V/cm for 16 h at 4°C.
8. Press, wash, and stain with Coomassie blue as described in *Protocol 13*.

Protocol 12. Crossed immunoelectrophoresis

1. Prepare the first-dimension gel by pouring 15 ml molten agarose containing 1% Triton X100 (see *Protocol 10*) on to an 80-mm square glass plate.
2. After setting, place plate on the template (see *Figure 2a*) and cut out wells. Fill with 10–15 μl antigen.
3. Place in a horizontal electrophoresis tank with wells towards the cathode. Connect to buffer with filter paper as described in step 6 of *Protocol 11*.
4. Apply 100 V/cm for 1–1.5 h at 4°C.
5. Remove gel from tank and replace on template. Cut out strips and transfer to the edge of a 50-mm square of Gelbond®, making sure the hydrophilic side is uppermost, as in *Figure 2b*.
6. Pour the second-dimension antiserum-containing agarose (0.5 ml of antiserum, suitably diluted in 1:4-diluted electrophoresis buffer, plus 3 ml agarose at 55°C) against the first-dimension strip.
7. After setting, replace in the electrophoresis tank with the first-dimension strip towards the cathode and connect the edges to the buffer with wicks.
8. Apply 12 V/cm for 16 h at 4°C.
9. Stain as described in *Protocol 13*.

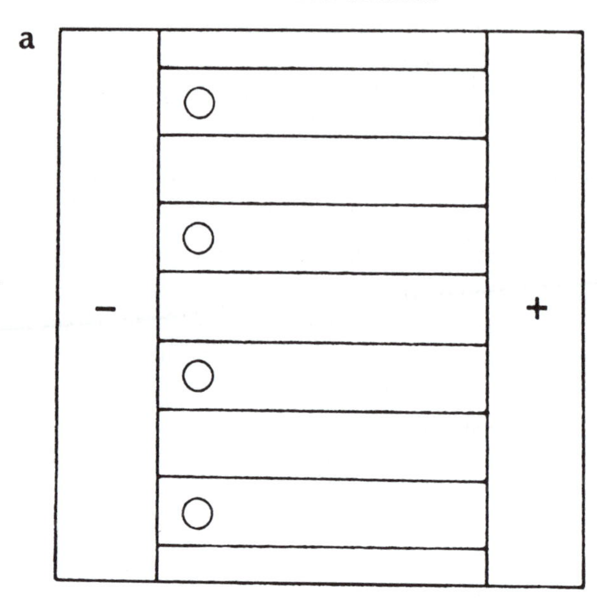

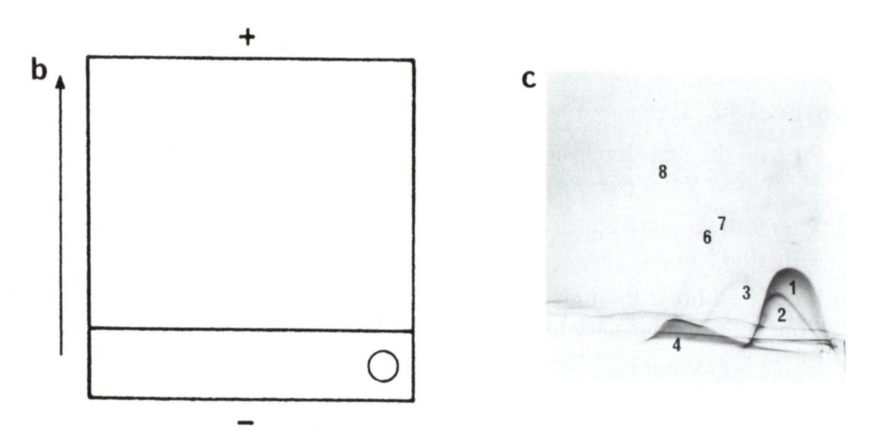

Figure 2. Crossed immunoelectrophoresis (CIE). (a) Template for first dimension—use an 80-mm square of glass; (b) template for second dimension—use a 50-mm square of Gelbond®; (c) an example of CIE: an EDTA extract (see *Protocol 3*) of *Clostridium difficile* (10 μl) run against whole cell antiserum (diluted 1 in 2).

Protocol 13. Staining of gels

1. Place gels on a flat surface and carefully cover with a sheet of Whatman no. 1 paper, avoiding air bubbles.
2. Cover with a stack of blotting paper and apply an even pressure with a weight of about 1 kg (e.g. a glass plate and several books) for 15 min.
3. Remove gels, discard filter paper, and wash in two changes of 0.01 M NaCl for 15 min, followed by a 15-min water wash.
4. Repeat the pressing as in step 2, and then dry the gels in air or with a hot air blower.
5. Place in Coomassie blue stain for 10 min
   - Coomassie blue R250 5 g
   - Ethanol 45 ml
   - Acetic acid 10 ml
   - Water 45 ml
6. Destain with two or three changes of the solution in step 5 from which the dye is omitted.

An example of the use of CIE is shown in *Figure 2c*.

6.2 Enzyme-linked immunosorbent assay (ELISA)

Since its development by Engvall and Perlmann in 1972 (13), this has become the most useful and widely used of all immunological techniques. It has a vast number of applications and many modifications and variations have been made. Both antigen and antibody can be detected and it is extremely sensitive, quantitative, and can readily be automated. The major disadvantage of the technique is that, when it is used for complex mixtures of either antigens or antibodies, it must be remembered that the final colour is proportional to the sum of perhaps many individual antibody/antigen reactions and results must be interpreted with caution. For many complex applications it is perhaps worth combining the technique with immunoblotting, which will resolve the individual reactions (see *Protocol 15*). The technique described here is an example of the most commonly used system where antigen is bound to the solid phase (plastic) and specific antibody can be detected. After reaction of antigen and antibody, the complex is detected with a second, antispecies antibody–enzyme conjugate. Its applications range from monoclonal antibody screening to diagnostic serology.

In the latter case it must be remembered that the antigen is all important. If it is impure, minor side-reactions can greatly distort the true picture. This is especially true for anaerobes where, generally, the antigenic makeup of an individual organism is uncertain and the degree of reaction and cross-reaction is extremely uncertain. Perhaps its safest use is the detection of antibodies to purified toxins. For the detection of antigen the system must be inverted and the antibody solid-phased for use as a capture for the antigen. A second antibody which is enzyme-labelled is then used to create a 'sandwich'. One of the antibodies should be monospecific polyclonal or monoclonal.

Protocol 14. A standard ELISA

1. Make dilutions of antigen in coating buffer (0.05 M sodium carbonate buffer pH 9.6 containing 0.02% sodium azide) and add 100-μl volumes containing 10–100 μg antigen to wells of microtitre plates. Cover and allow to incubate at 37°C for 4 h and then at 4°C overnight.
2. Wash three or four times with 0.9% NaCl containing 0.05% Tween 20.
3. Make dilutions of antibody in antibody/conjugate buffer (0.05 M phosphate buffer, pH 7.4, containing 0.85% NaCl, 0.05% Tween 20, and 0.02% sodium azide) and add to wells. Incubate for up to 4 h at room temperature or 1 h at 37°C.
4. Wash as in step 2.
5. Add 100-μl volumes of suitably diluted anti-first species antibody–enzyme conjugate (dilutions of 1 in 500, to 1 in several thousand are typical) to the wells and continue incubating for several hours (conveniently overnight) at room temperature or an hour or so at 37°C. The most popular enzyme-conjugates are alkaline phosphatase and horseradish peroxidase.
6. Wash as in step 2.
7. Add enzyme substrate diluted according to the manufacturer's instructions and incubate at room temperature for 1 h. The results are conveniently read in a purpose-made spectrophotometer.

6.3 Immunoblotting

Immunoblotting or Western blotting is the technique by which antigens, usually proteins, are separated by polyacrylamide gel electrophoresis and are then electrophoretically transferred to a nitrocellulose or similar membrane, where they are subsequently probed with antibody to locate antigens. Since the transfer method was first developed about 10 years ago by Towbin and colleagues (14) it has become one of the most widely used of immunological techniques with an ever-increasing range of applications and modifications. It has largely superseded the immunoelectrophoresis techniques such as crossed and rocket immunoelectrophoresis. It is used in all areas of immunological research, in diagnostic serology, and is becoming the standard method for the high-resolution detection of any molecule against which an antibody is available. In anaerobic bacteriology it has been used for immunological fingerprinting of *Clostridium difficile* (6) as well as in numerous other studies for investigating the antigenic composition of anaerobic bacteria.

Protocol 15. Immunoblotting

1. Separate antigens on a conventional polyacrylamide slab gel, and place the unfixed gel on the cathodic side of the transfer apparatus cassette.
2. Carefully overlay with a sheet of nitrocellulose which has been soaked in transfer buffer, avoiding any air bubbles.
3. Assemble transfer cassette and place in a buffer tank containing a high pH buffer (e.g. Tris/glycine/methanol, pH 8.3: Tris 12 g; glycine 57.68 g; methanol 1 l; water 4 l, adjusting final pH with 1 M NaOH if necessary).
4. Connect to power supply and allow transfer to proceed. An overnight run of 10–12 V, 40 mA is convenient, but faster runs can be made if higher voltages are possible.
5. Remove the membrane and wash for 10 min in Tris-buffered saline (TBS: Tris 4.84 g; NaCl 58.48 g; water 2 l, pH 7.5.
6. Block in 3% gelatin in TBS for 30–45 min.
7. Place in first antibody, suitably diluted in 1% gelatin in TBS, for an hour or more. Dilutions range from as low as 1 in 2 for human serum or monoclonal antibody supernates to 1 in several hundred for hyperimmune rabbit antisera.
8. Wash briefly in distilled water followed by two 10-min washes in TBS with 0.05% Tween 20.
9. Place in second antibody (enzyme-conjugate) diluted in 1% gelatin buffer for at least 1 h. Dilutions of 1 in 500 to 1 in several thousand are usual for most commercial conjugates.
10. Wash as in step 8.
11. Add colour development reagent (which must produce an insoluble coloured product), made up according to the manufacturer's instructions, and allow desired colour to develop.
12. Stop reaction by washing thoroughly in distilled water.

The recently developed semi-dry transfer apparatus with two flat graphite electrodes is gaining in popularity. There are savings in buffer costs and the transfer is more rapid because of the much steeper voltage gradient. An extension to the original concept, where the electrophoretic transfer step is circumvented, is by applying the antigen directly to the membrane as a dot (dot blotting). A further modification to this technique is the recently described 'line blot' (15) where the lines of antigen are applied with an ink pen point.

The main problems encountered with immunoblotting include: (a) a difficulty in transfer or a lack of binding of the antigen; (b) the antigen losing its antibody-binding capacity after separation and transfer, presumably by denaturation, and the subsequent need for renaturation of the antigen; (c) the problems of high background staining or non-specific binding, and the requirement of blocking agents; (d) a lack of sensitivity when compared to ELISA; (e) the quantitation of the technique; and (f) the inherent problem that the technique is difficult to miniaturize for use in large-scale screening. Many of these problems have been addressed recently with a limited amount of success, and the reader is encouraged to see the recent review by Stott (16).

The final immunological method to be described here (immunogold electron microscopy) is not commonly used, probably because of its complexity, but it is proving invaluable for the detection and location of antigens on intact bacteria. It has been used to great effect with *Bacteroides* by Patrick and colleagues (17).

Protocol 16. Immunogold microscopy[a]

1. Harvest and wash bacteria and resuspend in 0.01 M sodium cacodylate buffer, pH 7.2.
2. Fix in cacodylate buffer containing 2% (w/v) paraformaldehyde and 0.1% glutaraldehyde for 1 h at 4°C.
3. Wash in cacodylate buffer and dehydrate twice in graded alcohols and finally twice in 100% ethanol which has been dried over anhydrous sodium sulphate.
4. Embed in LR (London resin) white resin as follows: mix in 50% resin, 50% ethanol for 1 h at room temperature. Next mix in 100% resin in open containers in a fume cupboard overnight. After two further changes of resin at 2–3 h intervals, transfer the bacteria to gelatin capsules which have been dried at 60°C for 3 h, fill the remaining space with 100% resin, seal, and polymerize the resin in a 60°C oven for 18 h.
5. Cut thin sections, in an ultramicrotome and place on nickel grids.
6. Perform immunoassay at room temperature by placing sequentially in: (a) 1% bovine serum albumin (BSA) in PBS, pH 7.2 for 15 min; (b) antibody suitably diluted in 0.1% BSA in TBS (see *Protocol 15* but adjust pH to 8.2); (c) hold in TBS/BSA prior to washing in a stream of 8 ml TBS coming from a burette; (d) further block in 1% BSA in TBS for 15 min, followed by (e) addition of an appropriate gold conjugate (15–20 nm particles from Janssen Pharmaceuticals); (f) hold in BSA/TBS prior to washing as in (c); and (g) finally rinse in distilled water.
7. Stain with saturated aqueous uranyl acetate in the dark for 30 min.
8. Wash once in distilled water, then in Reynold's lead citrate, and desiccate for

Protocol 16. *Continued*

several minutes before viewing in the electron microscope. To avoid the film of resin disintegrating, anneal the section by playing the electron beam over it at low power before going to high power.

[a] For negatively stained specimens, bacteria are suspended in distilled water and placed on Formvar-coated nickel grids. The labelling is performed as in step 6 and negatively stained with 1% aqueous ammonium molybdate.

7. Conclusions

The selected methods given in this chapter will be a useful initiation to those wishing to investigate anaerobes by immunological methods. The techniques are now well established in many laboratories, but a thorough awareness of the complexity of the organism must always be borne in mind. A final point, which has not been covered elsewhere in this chapter, cannot be stressed too much; that is the need to recognize that the antigenic composition of any bacterium may be strongly under the influence of its growth environment. This is as true for anaerobes as for any other bacterium. Attempts must always be made to culture the organism in a medium approximating as closely as possible its natural habitat.

References

1. Hancock, I. C. and Poxton, I. R. (1988). *Bacterial Cell Surface Techniques*. Wiley, Chichester.
2. Poxton, I. R. and Byrne, M. D. (1984). *J. Med. Microbiol.* **17**, 171.
3. Poxton, I. R. (1984). *J. Gen. Microbiol.* **130**, 975.
4. Sharp, J. and Poxton, I. R. (1985). *J. Immunol. Meth.* **83**, 241.
5. Poxton, I. R. and Cartmill, T. D. I. (1982). *J. Gen. Microbiol.* **128**, 1365.
6. Sharp, J. and Poxton, I. R. (1986). *FEMS Microbiol. Letts.* **34**, 97.
7. Gueguen, F., Robreau, G., Talbot, F., and Malcoste, R. (1990). *Microbiol. Immunol.* **34**, 55.
8. Poxton, I. R. and Brown, R. (1986). *J. Gen. Microbiol.* **132**, 2475.
9. Hitchcock, P. J. and Brown, T. M. (1983). *J. Bacteriol.* **154**, 269.
10. Weintraub, A., Larsson, B. E., and Lindberg, A. A. (1985). *Infect. Immun.* **49**, 197.
11. Chandler, H. M. and Gulaskharam, J. (1974). *J. Gen. Microbiol.* **84**, 128.
12. Sharp, J. (1988). In *Anaerobes Today* (ed. J. M. Hardie and S. P. Borriello), pp. 169–76. Wiley, Chichester.
13. Engvall, E. and Perlmann, P. (1972). *J. Immunol.* **109**, 129.
14. Towbin, H., Staehelin, T., and Gordon, J. (1979). *Proc. Nat. Acad. Sci. USA* **76**, 4350.
15. Raoult, D. and Dasch, G. A. (1989). *J. Immunol. Meth.* **125**, 57.
16. Stott, D. I. (1989). *J. Immunol. Meth.* **119**, 153.
17. Reid, J. H., Patrick, S., and Tabaqchali, S. (1987). *J. Gen. Microbiol.* **133**, 171.

7

Methods for biochemical studies

S. J. FORSYTHE

1. Introduction

Numerous methods of quantitative and qualitative biochemical analysis are now available for the study of anaerobic micro-organisms. Fortunately, most biochemical studies use standard protocols and thus the major problem for the anaerobic microbiologist is being able to grow sufficient quantities of the organism. Certain procedures, however, such as reductase enzymes, require strictly anaerobic assay conditions, but even this problem can be overcome through the use of anaerobic cabinets, anaerobic cuvettes, and artificial electron carriers. This chapter will concentrate, therefore, on the analysis of oxygen-sensitive components from anaerobic micro-organisms.

2. Cell culture

2.1 Preparation of media for anaerobes

Anaerobic organisms, by definition, must be grown under low or non-detectable levels of oxygen. Various procedures are available for producing anaerobic media and these are described in other chapters in this volume. For assaying oxygen-sensitive cell components it is often necessary to prepare the growth medium in centrifuge pots (50 ml to 1 l). The pots can be inoculated under anaerobic conditions and incubated under the required conditions. For small volumes (5–20 ml) Universal bottles or serum vials sealed with butyl rubber septa and aluminium caps can be used.

2.2 Formulation of media

Maximum expression of certain enzymes requires particular nutrients to be present in the growth medium. For example, some *Bacteroides* spp. can only synthesize fumarate reductase when haemin (a cytochrome precursor) is available. Commercially available preparations of Schaedler broth contain sufficient haemin (10 mg/l) for cytochrome synthesis. However, if an alternative medium is used it must be supplemented with haemin at a final concentration of 10 mg/l.

The addition of the reducing agent cysteine must be avoided when using nitrate-containing media, since cysteine acts as an electron donor to the nitrate anion. It therefore leads to a false positive result when testing for nitrate reductase. Media must therefore be reduced using evacuation (see Section 3.2) or by prior incubation in an oxygen-free atmosphere.

3. Harvesting anaerobic cultures

In order to assay oxygen-sensitive cell components one must minimize the exposure of the cells to air during all stages prior to the analysis, including the step of centrifugation to concentrate the culture. To minimize exposure of the culture to air the cells can be grown anaerobically in centrifuge pots and serum vials (see Section 2.1). For cultures which produce large amounts of gas, such as *Clostridium* spp. and *Veillonella alcalescens*, the excess head-space pressure must be vented through an air filter (0.45 μm pore size) prior to centrifugation. Alternatively, the culture can be poured into the centrifuge pots in an anaerobic cabinet or under a stream of oxygen-free nitrogen in a manner similar to that of the Hungate technique for roll tubes (see Section 3.1).

3.1 Manipulation of cell suspensions under anaerobic conditions on open bench tops

The most versatile and cheap way of manipulating anaerobic cultures in a laboratory is to use a football bladder as an oxygen-free nitrogen reservoir (see *Figure 1*). This method enables continuous gassing of the cell suspension with nitrogen during all manipulations prior to the assay. Nitrogen is the most economic inert gas for this purpose and can be replaced with hydrogen for manipulation of methanogens.

After centrifugation the cell pellet is resuspended in the appropriate buffer using a glass homogenizer. Care is necessary during this step as vortexing can easily aerate the cell suspension if the nitrogen gas flow is insufficient to keep an anaerobic head space over the homogenizer.

3.2 Preparation of anaerobic buffers

Standard buffers are used in anaerobic assays, but dissolved oxygen must be removed by degassing. This can be achieved either by evaculation or by autoclaving; both methods are preferable to the boiling method used to produce pre-reduced anaerobically sterile (PRAS) media.

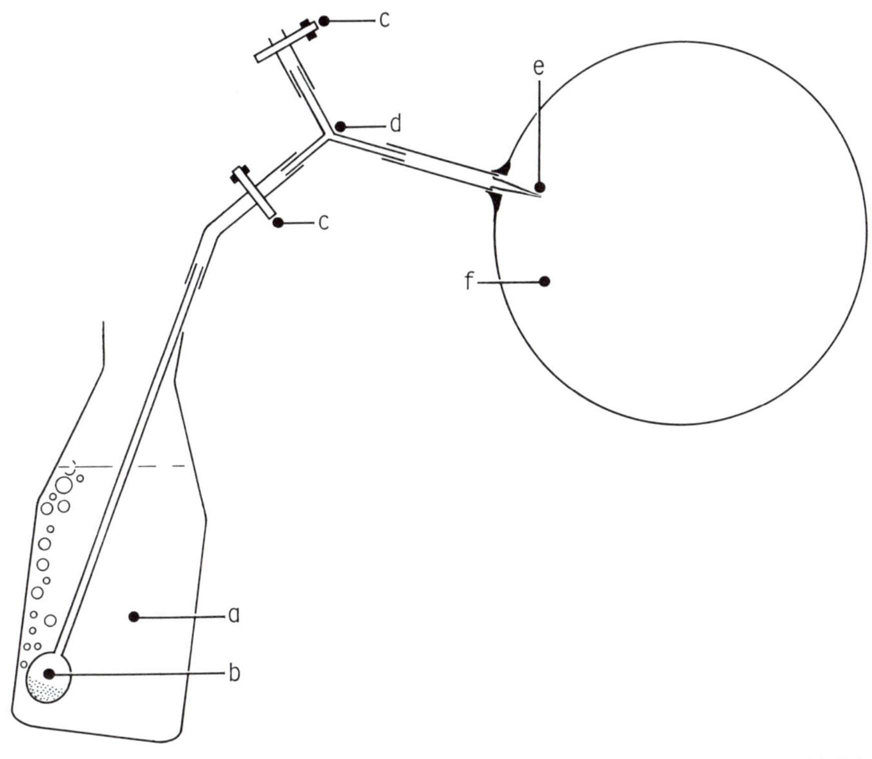

Figure 1. Sparging buffers with oxygen-free nitrogen gas: (a) screw-capped bottle; (b) fish tank aspirator; (c) spring clips; (d) Y-piece for additional gassing jets; (e) rubber tubing to short wide-gauge needle inserted through bladder valve; (f) football bladder (diameter <25 cm).

Protocol 1. Evacuation of anaerobic buffers

1. Pour the buffer into a large side-armed conical flask containing a magnetic flea and place on a magnetic stirrer.
2. Using thick-walled rubber tubing, connect the side-arm to a liquid trap which is connected either to a vacuum pump or, if there is sufficient water pressure, to a water-tap vacuum connector (see *Figure 2*).
3. Evacuate until the buffer ceases to bubble, approximately 15 min for 1 l of buffer.
4. Dispense the degassed buffer under a stream of nitrogen using two gassing jets, one for the conical flask and the other for the buffer receptacle.

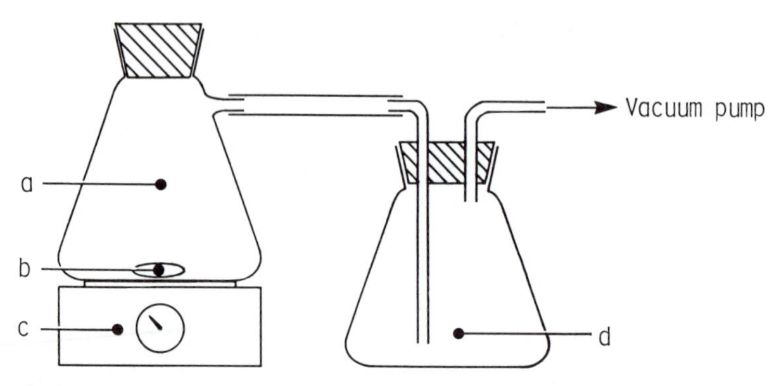

Figure 2. Equipment for degassing buffers and media: (a) stoppered side-arm conical flask, containing buffer or medium; (b) magnetic flea; (c) magnetic stirrer; (d) side-arm conical flask acting as a liquid trap.

Protocol 2. Anaerobic buffer preparation by autoclaving

1. Pour the buffer into medical flasks or Universal bottles according to the volume required.
2. Autoclave in a short cycle autoclave or a pressure cooker at 121°C for 15 min.
3. Purge with oxygen-free nitrogen as soon as the temperature has dropped to about 80°C. Take case when inserting the gassing jet as the buffer may spit and scald the skin. Ideally, the user should wear protective asbestos gloves.

4. Cell fractionation

4.1 Methods of cell fractionation

Certain enzymes cannot be assayed in whole cells, either because the substrates are impermeable, or because the enzyme system needs to be studied in a more compartmentalized fashion. The procedure for preparing cell fractions for anaerobic assays is more exacting than that for whole cells. This is because there is a much greater risk of exposing the cell components to oxygen. Fortunately, various methods are available to break open bacterial cells and the method of choice will depend upon the facilities available (see *Table 1*). All the methods in

Table 1. Methods of cell fractionation

Principle	Equipment
Disintegration	Mickle disintegrator
Pressure cell	Hughes Press and French Press
Chemical lysis	SDS, lysozyme, and toluene
Ultrasonic	Sonicator

Table 1 can be used provided an anaerobic headspace is vigorously maintained throughout the procedure (see Section 3.1).

Centrifugation is used to pellet any remaining whole cells and to remove cell membranes if desired. The cell-free extract is protected from the air by using screw-capped centrifuge tubes which are thoroughly flushed with an inert gas, usually nitrogen, as they are prepared. It is easier to balance an anaerobic centrifuge tube against a water-filled tube than against another anaerobic tube because of the need to maintain the anaerobic headspace.

4.2 Separation of oxygen-sensitive cell components

Because of the extreme oxygen sensitivity of certain enzymes and cell components it may be necessary to construct an anaerobic chromatography system (1). This system has the anaerobic elution buffer and fraction collector housed inside an anaerobic cabinet (see *Figure 3*). The fraction collector and peristaltic pump need to be small in order to pass through the interchange into the cabinet. The buffer is pumped via ports in the cabinet to the separating column, and the eluted fractions are subsequently collected inside the cabinet, protected from exposure to air. The molecular weight of the cell component can be determined by the inclusion of molecular weight markers (for example cytochrome *c* and catalase) in the sample. In addition, the eluted cell components can be purified by dialysis and ammonium sulphate precipitation inside the cabinet.

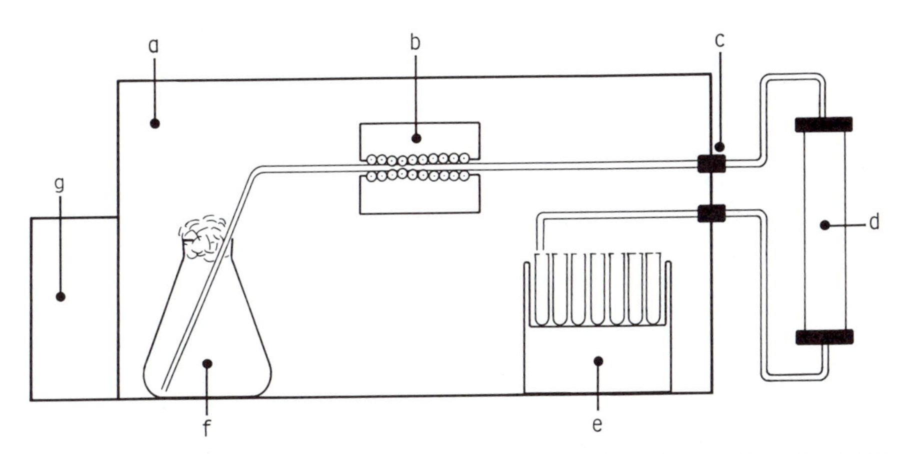

Figure 3. Diagrammatic scheme for anaerobic chromatography: (a) anaerobic cabinet; (b) peristaltic pump; (c) needle port connections; (d) separating column; (e) fraction collector loaded with screw-capped test tubes; (f) anaerobic buffer; (g) air lock.

Protocol 3. Anaerobic column chromatography

1. Place all test tubes, stoppers, and buffer in the anaerobic cabinet 1 day before required, in order to pre-reduce.

Protocol 3. *Continued*

2. Pharmacia columns and end adapters are most suitable for the construction of an anaerobic separating column, as the connectors have a tight fit with neoprene O-rings, which help maintain anaerobic conditions. The column inlet and outlet are connected to thick-walled polyethylene tubing.
3. Pack the column (aerobically) with appropriate material and equilibrate by pumping through it four column volumes of anaerobic buffer (see Section 3.2). This is sufficient to remove oxygen from the column.
4. Apply cell-free extract to the column by pumping it from the anaerobic cabinet on to the column, followed by the elution buffer.
5. Collect eluted fractions from the column in small test tubes, stoppered with neoprene bungs. The tubes can then be transferred to ice if required. The sample tubes will have a positive pressure on removal from the anaerobic cabinet since the cabinet is maintained at a higher pressure than the atmosphere. This will help in preventing air from entering the sample.

5. Biochemical studies of whole and perforated cells

Fortunately, much information concerning anaerobic organisms can be obtained without resorting to cell fractionation. This includes the detection of a diverse range of enzymes using commercial kits, and estimation of other enzymes whose functions range from energy generation to cellulose degradation.

5.1 Dehydrated enzyme kits

Recent years have seen great advancement in the production of simple enzyme kits in which the substrate and colorimetric indicator are dehydrated in small ampoules. API-bioMerieux produce a wide range of such kits, three of which are pertinent to the study of anaerobic organisms: API 20A; API Zym; and ATB 32A (see *Table 2*). Although these kits will not quantify the rate of any enzyme reaction they can be very useful for rapid determination of the presence of the enzyme of interest, without having to prepare a large number of biochemical reagents. It should be noted that the nitrate reductase test can give false negative results, as no electron donor is supplied in the assay. Instructions for using these kits are provided by the manufacturer and are sufficiently clear as to not warrant any further comment.

5.2 Fluorogenic substrates

A wide range of enzyme activities, particularly substrate specificity for proteolytic enzymes, can be determined very simply using fluorogenic substrates (see *Table 3*). These substrates are incubated with the cell suspension in the wells of a microtitre tray inside an anaerobe cabinet or an anaerobe jar. The tray is then removed and the results recorded.

Table 2. Biochemical tests used in API strips

API ZYM	API 20 A	ATB 32 A
Phosphatase alkaline	Indole	Urease
Esterase (C4)	Urease	Arginine dihydrolase
Esterase lipase (C8)	Glucose fermentation	α-galactosidase
Lipase (C14)	Mannitol fermentation	β-galactosidase
Leucine arylamidase	Lactose fementation	β-galactosidase-6-P
Valine arylamidase	Sucrose fermentation	α-glucosidase
Cystine arylamidase	Maltose fermentation	β-glucosidase
Chymotrypsin	Salicin fermentation	α-arabinosidase
Trypsin	Xylose fermentation	β-glucuronidase
Phosphatase acid	Arabinose fermentation	β-N-acetyl-glucosaminidase
Phosphoamidase	Glycerol fermentation	Mannose fermentation
α-galactosidase	Cellobiose fermentation	Raffinose fermentation
β-galactosidase	Mannose fermentation	Glutamate decarboxylase
β-glucuronidase	Melezitose fermentation	α-fucosidase
α-glucosidase	Raffinose fermentation	Nitrate reduction
β-glucosidase	Sorbitol fermentation	Indole production
N-acetyl-β-glucosaminidase	Rhamnose fermentation	Alkaline phosphatase
α-mannosidase	Trehalose fermentation	Leucine arylamidase
α-fucosidase	Gelatinase	Arginine arylamidase
Lipase (C10)	Aesculin hydrolysis	Proline arylamidase
L-arabinosidase	Catalase	Leucyl glycine arylamidase
Phospho-β-galactosidase		Phenylalanine arylamidase
		Pyroglutamate arylamidase
		Tyrosine arylamidase
		Alanine arylamidase
		Glycine arylamidase
		Histidine arylamidase
		Glutamylglutamate arylamidase
		Serine arylamidase

Table 3. Examples of fluorogenic substrates

4-Methylumbelliferyl (4MU) based	7-Amido-4-methylcoumarin (7AMC) based
N-acetyl-β-galactosaminide 4MU	Ala-leu-lys 7AMC
N-acetyl-β-glucosaminide 4MU	Gly-pro 7AMC
α-arabinoside 4MU	Pro-7AMC
β-cellobiopyranoside 4MU	*N*-succinyl-leu-tyr 7AMC
β-galactoside 4MU	*N*-t-BOC-leu-ser-thr-agr 7AMC
β-glucoside 4MU	*N*-E-CBZ-lysine 7AMC

Protocol 4. Detection of cellulase activity (2) using methylumbelliferyl -β-D-cellobioside (MUC)[a]

1. Dispense 0.1 ml of the reaction mixture (0.5 mg/ml MUC in 25 mM sodium phosphate buffer, pH 7.0) into microtitre tray wells.

Protocol 4. *Continued*

2. Add 0.1 ml of sample (cell suspension or cell-free extract) and incubate at 37°C for up to 18 h. Observe at hourly intervals initially.
3. Record positive cellulase activity when a well fluoresces under UV illumination.

[a] Further cellulase assays are described in *Protocols 16–18*.

5.3 Enzyme specific activity

It is standard procedure to express enzyme activity per mg cell biomass on a dry weight basis, or per mg protein. However, it is plausible when studying intestinal organisms to extrapolate from the *in vitro* assay to the *in vivo* activity by using the units per mg cell wet weight of 10^{10} bacteria/g intestinal contents, etc.

5.3.1 Protein determination

Use the protein-dye binding method of Bradford (3) for protein determination as this technique is not susceptible to interference by thiol groups as is the standard Folin–Lowry method (4).

5.4 Thunberg anaerobic tubes and cuvettes

Enzyme assays which require anaerobic conditions require specially constructed glassware. The simplest is the Thunberg tube, which is the size of a boiling tube, has a side arm, and has a ground glass neck into which fits a single chamber stopper (see *Figure 4*).

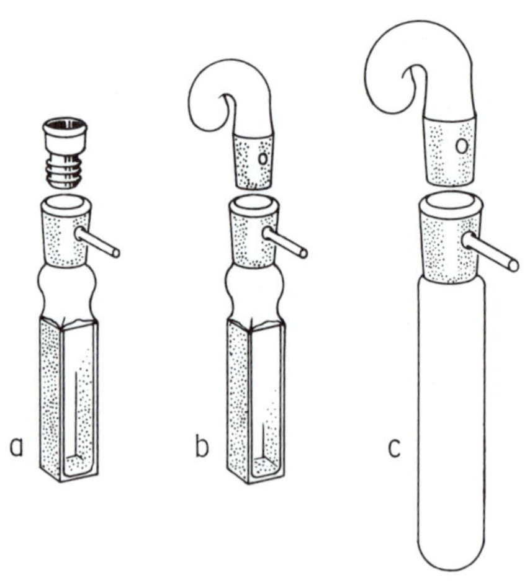

Figure 4. Anaerobic cuvettes: (a) anaerobic cuvette with rubber septum; (b) anaerobic cuvette with single substrate bulb; (c) Thunberg tube.

Protocol 5. Use of the Thunberg tube for enzyme assays

1. Put a very fine smear of silicone grease around the Thunberg tube neck. Excess grease will cause the accumulation of dust and dirt.
2. Pipette the assay mix, except for the cell suspension and substrate, into the Thunberg tube, under a stream of nitrogen gas as described in Section 3.1.
3. Pipette the enzyme substrate into the stopper chamber.
4. Pipette the cell suspension into the Thunberg tube, giving a final volume no more than 10 ml, and fit the stopper so that the side-arm is in the open position.
5. Evacuate the tube as described in *Protocol 1*. The contents should bubble for about 3 min.
6. Close the side-arm and disconnect from the vacuum.
7. Connect the tube to nitrogen supply and open the side-arm for 1 min.
8. Repeat steps 5–7.
9. Repeat for replicate tubes.
10. Place tubes in waterbath at either 25°C or 37°C for 5 min to allow for temperature equilibration.
11. Start reaction by mixing the tube contents with the substrate.
12. Remove the stopper from each tube at times intervals and analyse for appropriate metabolite.

5.4.1 Using Thunberg cuvettes

Enzyme assays involving intermediates such as NADH and benzyl viologen can be determined spectrophotometrically using Thunberg (anaerobic) cuvettes. There are two types of Thunberg cuvette. The first is sealed with a septum through which a hypodermic needle can be inserted to add a substrate or cell suspension. The second type has a ground glass neck and a stopper which may have one or two chambers for substrate addition (see *Figure 4*). These cuvettes are fragile and very expensive.

Protocol 6. Use of Thunberg cuvettes

1. Ensure that the cuvette faces are clean.
2. Add the assay mix and cell suspension to the cuvette using a syringe needle gassing jet to maintain a nitrogen atmosphere.
3. Carefully seal the cuvette with the rubber stopper whilst withdrawing the gassing jet.
4. Incubate if necessary in the thermostatically controlled cell chamber of the spectrophotometer for 5 min.

Protocol 6. *Continued*

5. Add the substrate through the septum using a syringe with a narrow-gauge needle.
6. Measure the rate of absorbance change at the appropriate wavelength.
7. Determine the enzyme activity using the coefficient of extinction for the appropriate metabolite.

Assays using the Thunberg cuvettes with one or two chamber stoppers are prepared essentially as outlined for the Thunberg tube (see *Protocol 5*) except that the cuvette is subsequently placed in a spectrophotometer rather than water bath.

An alternative method for preparation of Thunberg cuvettes is to flush the contents in the cuvette with nitrogen for up to 15 min prior to starting the reaction. This can be tiresome but can ensure reproducible results for assays of enzymes which are extremely oxygen-sensitive.

Sometimes it is possible to assay an oxygen-sensitive enzyme in open tubes provided they are narrow (100 mm × 10 mm), giving a small surface area, and all manipulations are made under an inert atmosphere. This is the case with the nitrite reductase of *V. alcalescens*. For preliminary studies it is advisable to compare results obtained using open tubes and Thunberg tubes since the former technique is considerably less laborious.

5.5 Artificial electron carriers

Artificial electron carriers are frequently used in assays for anaerobic enzymes involved in redox reactions. The most common electron mediators used are benzyl viologen (λ_{max} 600 nm, Eh = −350 mV) and methyl viologen (λ 600 nm, Eh = −446 mV). (Eh is the redox potential.) These are a deep blue colour in the reduced state and turn colourless upon oxidation. Other mediators include phenazine methosulphonate and 2,6-dichlorophenolindophenol. These compounds must be pre-reduced in order to donate electrons into the redox enzyme system. Sodium dithionite is often used to reduce the viologen dye. Care must be taken since sodium dithionite can inhibit the activity of enzymes such as nitrous oxide reductase. Alternatively, viologen dyes can be photochemically reduced using proflavin by exposure to a blue fluorescent light.

5.6 Electrophoresis of enzymes

Determining the electrophoretic mobility of enzymes is a very useful chemotaxonomic method which has not been fully investigated (5). This topic is more fully covered in Chapter 5. The molecular weight and subunit composition of enzymes are determined using SDS-PAGE, which is also described in Chapter 5.

6. Oxidoreductase enzymes

Enzymes which catalyse redox reactions are important in many aspects of cell biochemistry. Because anaerobes are unable to fully oxidize their energy substrate they have a low ATP yield per mole substrate metabolized. However, many anaerobes are able to generate additional ATP via non-substrate level phosphorylations using the phosphoclastic reaction and fumarate reductase. This section covers a wide range of oxidoreductase enzymes (commonly referred to as reductases) which enable strict anaerobes to metabolize various substrates and thus to out-compete facultative anaerobes in their microniche.

6.1 Fumarate reductase

Many strict anaerobes are able to generate ATP from the reduction of fumarate to succinate via fumarate reductase. This reaction is the reverse of that catalysed by succinate dehydrogenase, yet uses a separate enzyme system. As mentioned in Section 2.2, certain *Bacteroides* spp. can only synthesize an active fumarate reductase system provided haemin is provided in the growth medium. This is because haemin is required for cytochrome b_{FR} synthesis, which is involved in the electron transfer to fumarate reductase.

The method for assaying fumarate reductase in *B. fragilis* cells (6) is described in *Protocol 7*. Procedures for anaerobic cell culture, preparation of anaerobic buffer, and use of Thunberg cuvettes are described in *Protocols 1*, *2*, and *6*. The volume of cells required will vary according to the species studied.

Protocol 7. Assay for fumarate reductase of *Bacteroides fragilis*

1. Centrifuge 20 ml culture at 6000 *g* for 10 min in a bench centrifuge.
2. Resuspend the cell pellet in 2 ml of anaerobic sodium phosphate buffer (pH 7.0, 10 mM).
3. To a 3 ml Thunberg cuvette flushed with nitrogen gas add
   - Sodium phosphate buffer (pH 7.0, 10 mM) 2.8 ml
   - Cell suspension 0.1 ml
   - Benzyl viologen (20 mg/ml) 10 μl
   - Sodium dithionite (40 mg/ml) 20 μl
4. Incubate the cuvette at 37°C in the spectrophotometer.
5. Set up replicate cuvettes.
6. Measure the absorbance change at 600 nm for 5 to 10 min to determine the endogenous rate of benzyl viologen reduction. Absorbance should be approximately 1.
7. Add 50 μl sodium fumarate (116 mg/ml) through the septum using a syringe.

Protocol 7. *Continued*

8. Continue to measure the absorbance change until the cuvette contents are colourless.
9. Determine the fumarate reductase activity using the molar coefficient of extinction for benzyl viologen of 7.8×10^3 M^{-1} cm^{-1}.

6.2 Pyruvate dehydrogenase

Pyruvate dehydrogenase catalyses the production of acetyl-CoA from pyruvate. The acetyl-CoA is subsequently converted into acetyl-phosphate which is used to generate ATP via the phosphoclastic reaction. This enzyme is inhibited by the anti-anaerobic nitroimidazole drugs and is therefore important in maintaining cell viability. A procedure for assaying pyruvate dehydrogenase in *B. fragilis* is reproduced in *Protocol 8*. This method uses Triton X-100 to permeabilize the cells and avoids the need for cell fractionation (7, 8). The assay is extremely oxygen-sensitive since any trace oxygen will catalyse the reverse reaction of the assay, the conversion of reduced benzyl viologen to the oxidized form.

Protocol 8. Assay of pyruvate dehydrogenase in *Bacteroides fragilis*

1. Grow the cells overnight in 20 ml Schaedler broth at 37°C in sterile Universal bottles.
2. Harvest the cells by centrifugation at 6000 *g* for 10 min in a bench centrifuge at room temperature.
3. Resuspend the cell pellet in 5 ml anaerobic sodium phosphate buffer (pH 7.0, 50 mM) under a stream of nitrogen gas and repeat step 2.
4. Perforate the cells by resuspending the resulting pellet in 0.5 ml of anaerobic buffer containing 0.1% Triton X-100.
5. Add a few crystals of sodium dithionite to help maintain anaerobic conditions.
6. To a 3 ml anaerobic cuvette, under a stream of nitrogen, add the following
   - Sodium phosphate buffer (pH 7.0, 50 mM) 2.7 ml
   - Cell suspension 100 μl
   - Benzyl viologen (20 mg/ml) 50 μl
   - Coenzyme A (4 mg/ml) 50 μl
7. Incubate the cuvette in the spectrophotometer at 37°C for 5 min and then add 0.1 ml sodium pyruvate (8.25 mg/ml) through the septum.
8. Monitor the increase in absorbance at 600 nm.
9. Calculate the rate of benzyl viologen reduction from the coefficient of extinction of 7.8×10^3 M^{-1} cm^{-1}.

6.3 Nitrogen oxide reductases

Many organisms use nitrate, nitrite, and N_xO as terminal electron acceptors. Since oxygen is the preferred electron acceptor to nitrate and nitrite, due to their redox potentials, assaying nitrogen oxide reductases requires strictly anaerobic techniques. This applies equally to nitrite reductase assay in the aerobe *Staphylococcus aureus* (9). The following sections cover protocols for nitrogen oxide reductases involved in assimilatory and dissimilatory nitrate reduction and denitrification. Why some organisms do not possess a complete chain of reduction from nitrate to ammonia or dinitrogen (denitrification) has not yet been explained. It is known that some organisms use nitrite reduction for ATP generation (*V. alcalescens*) whilst others reduce nitrite as a detoxification mechanism, as in *Propionibacterium* spp. (10, 11).

6.3.1 Nitrate reductase

Nitrate reductase can be determined by a number of methods: nitrate disappearance rate; nitrite appearance rate; NADH oxidation; and benzyl or methyl viologen dye oxidation. Nitrite appearance rate is more commonly used for whole assays where the complete glycolytic pathway is present. It should be used in conjunction with the nitrite reductase assay as some organisms reduce both nitrate and nitrite at the same time, rather than sequentially as would be predicted from redox potentials. The use of ion-specific electrodes for nitrate and nitrate reductase assays is not recommended.

Protocol 9. Benzyl viologen—nitrite reductase assay

1. Using a Thunberg cuvette flushed with nitrogen gas add
   - Sodium phosphate buffer (50 mM, pH 7.5) 2.7 ml
   - Cell-free extract (0.2–0.5 mg protein) 0.1 ml
   - Benzyl viologen (2 mg/ml) 0.1 ml

   Seal with a rubber septum (see *Protocol 6*).
2. Incubate cuvette at 37°C in the spectrophotometer for 5 min.
3. Inject 0.1 ml potassium nitrate (60 mM) through the septum and monitor the absorbance change at 600 nm. Molar coefficient of extinction = $7.8 \times 10^3\ M^{-1}\ cm^{-1}$.

6.3.2 Nitrite reductase

Nitrite reductase may produce ammonium ions which can be further assimilated or excreted. Alternatively, nitrite can be reduced to nitric oxide or nitrous oxide, which is subsequently reduced to dinitrogen gas (denitrification); whether nitric oxide is an enzyme-free intermediate is uncertain.

Most studies have used variations of *Protocol 10* for measuring nitrite reductase of whole cells. *Protocol 9* can be adopted to measure NADH-linked nitrite reductase in cell-free extracts by using 0.1 ml NADH (6 mM) as the electron donor, sodium nitrite (60 mM) as the electron acceptor, and monitoring the absorbance change at 340 nm. Some organisms such as *V. alcalescens* will not accept electrons from NADH but can use benzyl viologen (10).

Protocol 10. Nitrite reductase of *Veillonella alcalescens*

1. Harvest the cells by centrifugation at 5000 *g* for 15 min at 5°C.
2. Resuspend the cells in 1% (w/v) Bactopeptone (Difco) in sodium phosphate buffer (pH 7.5, 50 mM).[a]
3. Recentrifuge the cells as described in step 2.
4. Resuspend the washed cell pellet in 20 ml of Bactopeptone buffer.
5. Keep the cell suspension on ice under nitrogen until required.
6. Using a narrow test tube (100 mm × 10 mm) under a stream of nitrogen add
   - Anaerobic buffer 4.3 ml
   - Electron donor (lactate, 1 M) 0.1 ml[b]
   - Cell suspension 0.5 ml
7. Use parafilm to seal the test tube; then mix by inverting gently.
8. Place in a waterbath at 37°C for 3 min.
9. Start the reaction by adding 0.1 ml sodium nitrite (60 mM).
10. At 5-min intervals remove 50 μl aliquots into 3.7 ml acidified sulphanilamide (20 g sulphanilamide/l of 2 M HCl).
11. Add 0.3 ml 0.1% α-naphthylenediamine.
12. Allow colour to develop for 30 min.
13. Read the absorbance at 540 nm in a spectrophotometer.
14. Determine the rate of nitrite removal from a standard curve for sodium nitrite in the range 0 to 1 mM.

[a] Bactopeptone is only required for *V. alcalescens* lactate–nitrite reductase.
[b] Alternative electron donors: glucose, formate, and pyruvate.

End-products from nitrite reduction can be determined by the indophenol reaction for ammonia production and gas production (presumed to be dinitrogen) by placing a small upturned glass tube (Durham tube) in the growth medium prior to inoculation. Identification of the intermediates of nitrite reduction to dinitrogen gas requires gas chromatography with Poropak Q as the solid phase (11).

6.4 Nitroreductase activity of strict anaerobes

The nitroimidazole antibiotics specifically inhibit the growth of strictly anaerobic organisms. This selectivity is possibly due to the activation of the nitro group of nitroimidazoles to certain bactericidal radicals by the enzyme nitroreductase. There are four nitroreductases in *B. fragilis*, but the usual method for assaying the enzyme is to use the structurally similar substrate *p*-nitrobenzoic acid which is converted into *p*-aminobenzoic acid and is subsequently quantified by a deazoation reaction (12, 13).

Protocol 11. Determination of nitroreductase activity

1. Harvest 20 ml culture cells by centrifuging at 6000 *g* for 10 min in a bench centrifuge.
2. Resuspend the cells in 20 ml anaerobic buffer and recentrifuge as described in step 1.
3. Resuspend the cell pellet in 2 ml of anaerobic potassium phosphate buffer (pH 7.5, 0.1 M) under a stream of oxygen-free nitrogen.
4. To a Thunberg tube, whilst still under nitrogen, add
   - Cell suspension 1.0 ml
   - Potassium phosphate buffer (pH 7.5, 0.1 M) 3.9 ml
5. Incubate the Thunberg tube for 10 min at 37°C in a water bath.
6. Start the reaction by adding 0.1 ml of *p*-nitrobenzoic acid (PNBA, 1.16 mg/ml).
7. At timed intervals remove 0.5 ml aliquots and mix with 0.5 ml 10% trichloroacetic acid, in a 1.5 ml microcentrifuge tube previously chilled on ice.
8. Centrifuge the acidified sample for 5 min in a microcentrifuge at high speed (13 000 r.p.m.).
9. To 0.5 ml of the supernatant add
   - 0.1% sodium nitrite, 0.125 ml. Allow 3 min to react.
   - 0.5% ammonium sulphamate, 0.125 ml. Allow 3 min to react.
   - 0.1% naphthylenediamine hydrochloride, 0.125 ml. Allow 3 min to react.
   - Distilled water, 1.625 ml. Leave for 30 min.
10. Read the absorbance at 545 nm.
11. Calculate the amount of *p*-aminobenzoic acid (PABA) produced from a standard curve, prepared as above using volumes of PABA (0.5 mM) in the range 12.5 to 100 μl.

6.5 Methanogens

The methanogens pose one of the most challenging yet rewarding areas of anaerobic microbiology. They possess biochemical pathways and cell structures unique to Archaebacteria, which have yet to be fully elucidated (see Chapter 12 for a fuller account of these organisms).

6.5.1 F_{420}-dependent hydrogenase of methanogens

Although the coenzyme F_{420} is not directly involved in methanogenesis it is very important in the electron transfer system of methanogens. It may be analogous to ferredoxin (a low redox potential electron carrier) in other anaerobes in that it acts as an electron acceptor for hydrogenase, NADPH, formate, pyruvate, and 2-oxoglutarate dehydrogenase. *Protocol 12* describes the assay of F_{420}-dependent hydrogenase for *Methanobacterium thermoautotrophicum* (14). Methods for growing methanogens are described in Chapter 12 and the preparation of cell-free extracts is described in *Protocol 14* of Chapter 12.

Protocol 12. F_{420}-dependent hydrogenase

1. To a 1-ml Thunberg cuvette under a hydrogen gas phase add
   - F_{420} sufficient to give an absorbance (420 nm) equal to 1 (approximately 0.2 ml)
   - Tris-HCl buffer (pH 8.6, 50 mM) to give a final volume of 1 ml (*ca.* 0.8 ml)
2. Seal the cuvette and place in the spectrophotometer at 35°C for 5 min.[a]
3. Add 1 μl cell-free extract (0.02 mg) through the septum. Mix well.
4. Monitor the absorbance decrease at 420 nm.
5. Determine the F_{420}-dependent hydrogenase activity using the molar coefficient of extinction, 42.5×10^3 M^{-1} cm^{-1}.

[a] Since *M. thermoautotrophicum* is a thermophile this reaction can be monitored at 60°C.

6.5.2 Hydrogenase activity

The hydrogenase activity of methanogens can be quantified (12) using *Protocol 12* by replacing F_{420} with 25 μl methyl viologen (10 mM) and monitoring the absorbance at 578 nm; the molar coefficient of extinction is 9.7×10^3 M^{-1} cm^{-1}.

6.5.3 Methyl-coenzyme M reductase

The final stage in methanogenesis requires a membrane-bound hydrogenase, methyl reductase, and a low-molecular-weight coenzyme, possibly coenzyme M

(2-mercaptoethanesulphonic acid). Coenzyme M carries a methyl group, which is the substrate for the methylreductase reaction. Methyl-coenzyme M can be measured in cell-free extracts of *Methanosarcina barkeri* from methanol and methylamine (15). The reaction requires ATP, which has a catalytic role rather than being a substrate.

Protocol 13. Methanol: 2-mercaptoethanesulphonic acid methyltransferase

1. Using 10-ml serum vials with hydrogen : nitrogen (50 : 50) gas phase add
   - ATP to final concentration 9.38 mM
   - Magnesium chloride to final concentration 6.25 mM
   - Bromoethanesulphonic acid (50 μm) in 10 mM *N*-tris(hydroxylmethyl) methyl-2-aminoethane sulphonic acid buffer (pH 7.2), *ca.* 90 μl
   - Cell-free extract (0.4 to 1 mg protein)

   Final volume 100 μl.
2. Seal the vial with a butyl rubber stopper and aluminium cap.
3. Prepare replicate vials.
4. Place the vials in a water bath at 37°C for up to 30 min.[a]
5. Replace gas phase with nitrogen.
6. Add 7 μl methanol (12.5 mM) and 7 μl propanol (7 mM) through the septum to start the reaction.
7. To stop the reaction place the vial on ice and remove the cap.
8. Determine the methanol concentration by gas chromatography. Use a column packed with 0.2% Carbowax 1500 on Carbopak 80/100 mesh (Pharmacia) with a column temperature 130°C, injection temperature 170°C, and a flame ionization detector at 170°C. The carrier gas is nitrogen (25 ml/min flow rate). Sample size injected on to column is 0.1 μl. As an internal standard 2-propanol (7 mM final concentration) is added to the incubation mixture.

[a] This is to activate the methyltransferase enzyme.

---

7. Membrane potential generation

7.1 Production of proton motive force by methanogens

Methanogens can generate ATP by conserving the energy released by the reduction of carbon dioxide to methane. This energy conservation may be due to the generation of a proton motive force across the cell membrane which subsquently drives ATP synthesis. In order to measure the membrane potential of methanogens the lipophilic cation tetraphenylphosphonium chloride (TPP^+) concentration is monitored under strictly anaerobic conditions. pH, Eh, pH_2, and pCO_2 must be carefully controlled in these experiments. The construction of the TPP^+ sensitive electrode and reaction vessel were described by Butsch and Bachofen for *Methanobacterium thermoautotrophicum* (16).

7.2 Sodium-transport decarboxylases

Oxaloacetate and methylmalonyl-CoA decarboxylases convert the energy from decarboxylation into sodium ion gradients across the cell membrane, generating a proton motive force of approximately 114 mV. This electrochemical gradient can be used for the uptake of sugars or amino acids. The methylmalonyl-CoA decarboxylase of *V. alcalescens*, in addition to being essential to lactate degradation, generates an electrochemical sodium gradient (17). A method for measuring methylmalonyl-CoA decarboxylase is described in *Protocol 14*.

Protocol 14. Methylmalonyl-CoA decarboxylase activity of *V. alcalescens*

1. To an anaerobic cuvette add

• Potassium phosphate buffer (pH 7.0, 250 mM)	0.36 ml
• Sodium chloride (400 mM)	50 μl
• Potassium hydrogen carbonate (4 M)	50 μl
• Magnesium chloride (100 mM)	50 μl
• Dithiothreitol (40 mM)	50 μl
• Phosphoenolpyruvate (20 mM)	50 μl
• ATP (40 mM)	50 μl
• NADH (6 mM)	50 μl
• Pyruvate kinase (1000 U/ml)	10 μl
• Lactate dehydrogenase (4000 U/ml)	10 μl
• Propionyl-CoA carboxylase (500 U/ml)	10 μl
• (R, S)-methylmalonyl-CoA (8 mM)	10 μl

2. Incubate the cuvette at 25°C in a spectrophotometer.
3. Prepare replicate cuvettes.
4. Start the reaction by anaerobically adding 100 μl cell-free extract.
5. Monitor absorbance change at 340 nm.
6. Determine methylmalonyl-CoA decarboxylase activity using the molar coefficient of extinction of $6.22 \times 10^3 \ M^{-1} \ cm^{-1}$.

8. Proteolytic enzymes

Proteolysis is a very important enzyme activity as it can be a virulence factor and enables anaerobes to assimilate complex nitrogenous compounds in low ammonia-N environments. Non-specific protease is assayed using the substrate azocaesin (18, 19) and more specific activities are studied using fluorogenic substrates (see *Table 3*).

Protocol 15. Protease assay using azocaesin

1. To a narrow test tube under a nitrogen headspace add
   - Potassium phosphate buffer (100 mM, pH 7.5) containing 2 mM dithiothreitol — 2 ml
   - Cell suspension — 1 ml
2. Incubate at 37°C in a water bath.
3. Prepare replicate tubes.
4. Start the reaction by adding 1 ml azocaesin (4 mg/ml).[a]
5. At timed intervals stop the reaction by adding 1 ml 25% (w/v) trichloroacetic acid.[b]
6. Prepare control tubes in which the azocaesin is added after enzyme activity has been inactivated by the trichloroacetic acid.
7. Centrifuge at 30 000 *g* for 10 min at 4°C.
8. Mix 1 ml of supernatant with 1 ml 0.5 M NaOH.
9. Measure absorbance of hydrolysed caesin at 450 nm.
10. Determine the rate of azocaesin hydrolysis using a standard curve prepared with azocaesin diluted as described for the treated sample.

[a] Protease inhibitors can be added.
[b] Avoid intervals greater than 8 h as azopeptides are destroyed during prolonged incubation.

---

9. Cellulose degradation

Anaerobic degradation of lignocellulose waste can provide an alternative carbon source for the production of commercially important solvents, acetone and ethanol (20). Current interest has focused on the use of the thermophilic anaerobes, *C. thermocellum* and *C. thermohydrosulfurium*, to degrade cellulose, because of the increased enzyme stability associated with thermophilicity.

In 1984 the Commission on Biotechnology (IUPAC) published a series of procedures for measuring cellulase activity. These assays recently have been reviewed by Wood and Bhat (21) who included a number of alternative assays for consideration.

9.1 Cellulase activity detected using carboxymethylcellulose

A straightforward procedure for screening bacterial colonies for cellulase activity uses the substrate carboxymethylcellulose (2; J. Oldham, personal communication). This method is useful for screening *Escherichia coli* recombinants containing cellulase genes from *Clostridium thermocellum*.

Protocol 16. Screening colonies for cellulase activity

1. Inoculate agar plates containing 0.4% carboxymethylcellulose (CMC) with various cell dilutions and incubate to obtain single colonies.
2. Flood the plate with 1% Congo red and wash with 1 M sodium chloride.
3. Lyse the colonies by exposure to chloroform vapour for 15 min.
4. Cellulolytic colonies are surrounded by a yellow halo on a red background.

Protocol 17. Quantitative carboxymethylcellulase assay (2)

1. Pipette 4 ml 1% (w/v) CMC in 0.025 M sodium phosphate buffer (pH 7.0) into the required number of test tubes.
2. Incubate the tubes at 37°C for 5 min.
3. Add 1 ml of enzyme preparation to start the reaction.
4. At timed intervals remove 50 μl aliquots for reducing sugar determination.

Protocol 18. Methylumbelliferyl-β-D-cellobioside (MUC) fluorescence assay (2)[a]

1. Inoculate agar plates with dilutions of the organism and incubate to obtain single colonies.
2. Overlay the plate with soft agar containing 0.2% MUC in potassium citrate buffer (50 mM, pH 6.5).
3. Incubate for 5–6 h at 37°C.
4. Examine plates with a UV transluminator. Hydrolysis of MUC results in the release of the fluorescent compound methylumbelliferone.

[a] See *Protocol 4* for a microtitre tray method of screening cellulase activity using MUC. Alternatively, 4-methylumbelliferyl-β-D-glucoside can be used in place of MUC to detect β-glucosidase activity.

10. Measurement of microbial activity *in situ*

Extrapolation of *in vitro* studies of microbial metabolism is difficult, even though media have been devised to simulate the microbial habitat (see Chapter 10). Fortunately, it is feasible to measure certain microbial activities *in situ* through isotope-tracing techniques. This approach has been used to quantify volatile fatty acid production and aspects of nitrogen metabolism by the intestinal flora of cattle, sheep, goats, and rabbits (22, 23).

The animal body may be represented as a collection of pools, each composed of identical molecules, which tend to be enclosed by anatomic boundaries. Body pools usually remain constant in size while undergoing replacement by input, which equals output. This dynamic equilibrium is known as steady state. Compartmental analysis is based on the assumption that specific pools can be identified and that the discharge from the pools can be described by exponential equations. The tracer is assumed to behave chemically and physiologically exactly like the tracee. Since the tracer dose can be very small in terms of the number of existing natural atoms the dose should not perturb the system under observation or create a non-steady state (24).

When radioactive ^{14}C-acetate is injected into the rumen or hind gut of a surgically modified animal the labelled acetate mixes with the endogenous acetate and is subsequently metabolized to CO_2, entering the bicarbonate pool. Therefore, the rate of ^{14}C–CO_2 appearance is a measure of microbial degradation of acetate. Further information can be obtained from the kinetics of isotope appearance including metabolite pool size and turnover rate. Nitrogenous compounds labelled with the heavy isotope ^{15}N can be used to follow nitrogen metabolism by the intestinal flora, including ammonia assimilation and the proportion of microbial nitrogen derived from urea entering the gut.

The label is quantified using mass spectrometry or emission spectroscopy, which unfortunately restricts the number of samples that can be analysed due to labour-intensive sample preparation and availability of equipment. The rate of urea entry into the gut can be measured by infusing ^{15}N-urea into the bloodstream and recording the rate of ^{15}N-ammonia appearance in the intestinal contents. An alternative, more rapid, technique is to infuse ^{14}C-urea into the blood system and acetohydroxamate (a urease inhibitor) into the gut and measure the rate of ^{14}C-urea appearance in the intestines (25).

10.1 Methods of sampling intestinal contents

The majority of methods for analysing intestinal material require the use of surgically modified animals and therefore must only be used in accordance with the legal regulations concerning animal experimentation.

Various surgical modifications of the animal may be necessary in order to administer and sample a tracer in whole-animal experiments. Regular blood samples can be obtained by catheterizing a major vein or artery. Rumen samples can be obtained by attaching a fistula to the rumen so that the inner plug can be removed to give access to the rumen contents. Membrane filtration can be used to sample low-molecular-weight metabolites from intestinal contents. This is achieved by implanting into the intestines a dialysis probe (23). The principle of the dialysis is that a semipermeable membrane sheath is attached to a plastic rod with a double helix such that saline can be pumped through the probe (see *Figure 5*). Low-molecular-weight metabolites, such as the volatile acids, acetate, propionate, and butyrate, and nitrogenous compounds such as ammonia, urea,

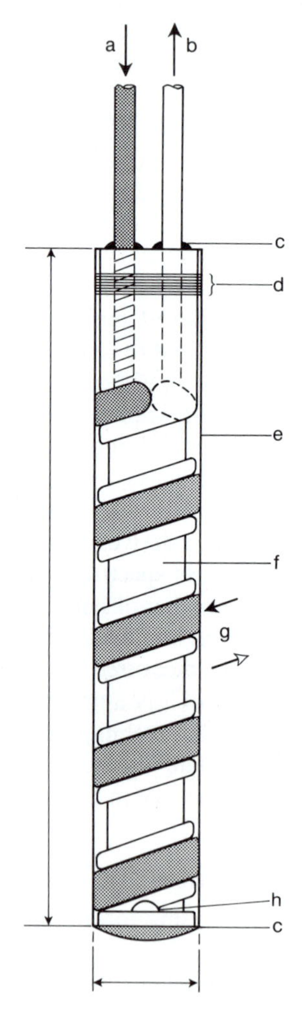

Figure 5. Dialysis probe: (a) inlet tube from infusion pump; (b) outlet tube for collection of dialysate; (c) Araldite cement; (d) suture thread; (e) Visking tubing; (f) plastic shaft with a two-start thread (length 55 mm, width 6 mm); (g) direction of flow; (h) hole.

and amino acids, diffuse into the saline and are collected for analysis. For a fuller treatment of the isotope dilution technique refer to Nolan and Leng (24).

These techniques have been of particular benefit in demonstrating the role of the epithelial microflora on the rate of urea entry into the gut and in quantifying carbohydrate metabolism by the lumen bacteria.

References

1. Gunsalus, R. P., Tandon, S. M., and Wolfe, R. S. (1980). *Anal. Biochem.* **101**, 327.
2. Faure, E., Bagnara, C., Belaich, A., and Belaich, J-P. (1988). *Gene* **65**, 51.
3. Bradford, M. (1976). *Anal. Biochem.* **72**, 248.
4. Lowry, O., Rosebrough, N., Farr, A., and Randall, R. (1951). *J. Biol. Chem.* **193**, 265.
5. Shah, H. and Gharbia, S. (1988). *J. Gen. Microbiol.* **134**, 327.
6. Dickie, P. and Weiner, J. (1979). *Can. J. Biochem.* **57**, 813.
7. Smith, N., Bryant, C., and Boreham, P. (1988). *Int. J. Parasitol.* **18**, 991.
8. Britz, M. and Wilkinson, R. (1979). *Antimicrob. Agents Chemother.* **16**, 19.
9. Forsythe, S., Dolby, J., Webster, A., and Cole, J. (1988). *J. Med. Microbiol.* **25**, 253.
10. Yordy, D. and Delwiche, E. (1979). *J. Bacteriol.* **137**, 905.
11. Kaspar, H. (1982). *Arch. Microbiol.* **133**, 126.
12. Kinouchi, T. and Ohnishi, Y. (1983). *Appl. Environ. Microbiol.* **46**, 596.
13. Braton, A. and Marshall, E. (1939). *J. Biol. Chem.* **128**, 537.
14. Zeikus, J., Fuchs, G., Kenealy, W., and Thauer, R. (1977). *J. Bacteriol.* **132**, 604.
15. van der Meijden, P., Heythuysen, H., Sliepenbeek, H., Houwen, F., van der Drift, C., and Vogels, G. (1983). *J. Bacteriol.* **153**, 6.
16. Butsch, B. and Bachofen, R. (1984). *Arch. Microbiol.* **138**, 293.
17. Hilpert, W. and Dimroth, P. (1983). *Eur. J. Biochem.* **132**, 579.
18. Brock, F., Forsberg, C., and Buchanan-Smith, J. (1982). *Appl. Environ. Microbiol.* **44**, 561.
19. Kopecny, J. and Wallace, R. (1982). *Appl. Environ. Microbiol.* **43**, 1026.
20. Sahm, B., Lamed, R., and Zeikus, J. (1989). In *Clostridia* (ed. N. Minton and D. Clarke), pp. 227–63. Plenum, New York.
21. Wood, T. and Bhat, M. (1988). In *Methods in Enzymology* (ed. W. A. Wood and S. T. Kellogg), Vol. 160, pp. 87–113. Academic Press, London.
22. Allen, S. A. and Miller, E. L. (1976). *Br. J. Nutr.* **36**, 353.
23. Parker, D. S. (1976). *Br. J. Nutr.* **36**, 61.
24. Nolan, J. V. and Leng, R. A. (1974). *Proc. Nutr. Soc.* **33**, 1.
25. Forsythe, S. J. and Parker, D. S. (1985). *Br. J. Nutr.* **54**, 285.

8

Genetics and molecular biology

B. W. WREN, P. MULLANY, and F. I. LAMB

1. Introduction

Genetic and molecular techniques used for the study of anaerobes are essentially the same as those used for aerobes. The major obstacles to the molecular characterization of anaerobes are their stringent growth requirements and the lack of appropriate host–vector systems for the genetic manipulation of cloned determinants.

Research on the genetics of anaerobes has centred on two genera, *Bacteroides* (1) and *Clostridium* (2), and on the methanogens (3). The impetus for this research stems from the importance of anaerobes as human and animal pathogens and the potential biotechnological exploitation of their gene products. Work on the molecular basis of transferable antibiotic resistance in anaerobes has identified a variety of extrachromosomal elements. In the past few years these studies have led to the development of host–vector systems for the manipulation of clostridial and *Bacteroides* genes.

This chapter will focus on genetic and molecular techniques which have been successfully applied to *Bacteroides*, *Clostridium*, and the methanogens. These include DNA extraction procedures, construction of shuttle vectors, and methods for the introduction of cloned DNA into appropriate host strains. Other gene-cloning and analysis procedures, developed for aerobes but equally applicable to anaerobes, are adequately covered elsewhere (4, 5) and will not be considered in this chapter. Genetic techniques are also valuable tools in taxonomic studies. The isolation of DNA for determination of base composition, DNA–DNA hybridization, and ribosomal RNA gene restriction patterns is described in Chapter 5.

Most recently, the polymerase chain reaction (PCR) has allowed the convenient amplification of DNA fragments when nucleotide sequence data are available. The final part of the chapter will consider the applications of PCR to the sensitive, specific, and rapid detection of anaerobes, and the exploitation of the technique for research purposes.

2. Extraction and purification of genomic bacterial DNA

The first step in the cloning and subsequent molecular analysis of a gene is usually the extraction and purification of genomic DNA from the organism of interest. However, direct cloning of PCR products can, in some cases, avoid this requirement by using boiled bacterial cells as the source of template DNA. The following procedures for the extraction of genomic DNA work well for most clostridial, *Bacteroides*, and methanogen species, although additional steps may be required to ensure complete lysis of cells and purification of DNA extracts. Further details of the preparation and treatment of solutions can be found in laboratory manuals for DNA cloning (4, 5).

2.1 Extraction of bacterial genomic DNA by SDS lysis

This simple and straightforward procedure is based on: lysozyme digestion of the cell wall; lysis with SDS; disruption of protein–nucleic acid complexes with pronase; followed by phenol/chloroform extraction to remove proteins.

Protocol 1. Extraction of bacterial genomic DNA by SDS lysis

1. Grow 10–25 ml bacteria to late log phase in an appropriate medium.
2. Harvest cells by centrifugation at 5000 *g* for 5 min at 4°C.
3. Resuspend pellet in 2 ml of 50 mM Tris–HCl pH 8.0, containing 25% sucrose.
4. Add 1 ml of freshly prepared lysozyme (10 mg/ml) and incubate on ice for 30 min.
5. Add 250 μl 5% SDS in 50 mM Tris–HCl pH 8.0, followed by 0.75 ml TE (10 mM Tris–HCl pH 8.0 and 1 mM EDTA) and 0.5 ml pronase (20 mg/ml). Incubate at 56°C for 1 h.
6. Add an equal volume of phenol/chloroform, mix gently and centrifuge at 5000 *g* for 5 min at room temperature. Remove aqueous (top) layer and repeat extraction and centrifugation until there is no interface material.
7. Repeat extraction with chloroform.
8. Add 0.1 vol 3 M sodium acetate pH 6.0.
9. Precipitate with 2 vol cold 100% ethanol.
10. Lift the clot of DNA out with a loop or pipette tip and wash three times with 70% ethanol. If a clot does not form, centrifuge at 8000 *g* for 10 min prior to washing in 70% ethanol.
11. Dry *in vacuo*; do not overdry, to avoid difficulty in redissolving.
12. Dissolve in 50–500 μl of TE buffer, depending on yield.

2.2 Extraction of bacterial DNA using guanidine hydrochloride

In this method enzyme-digested bacteria are lysed by addition of the chaotropic agent guanidine hydrochloride (GCl) and deproteinized with chloroform. Like other chaotropes, GCl disrupts structures linked with hydrogen bonds, resulting in denaturation and inactivation of proteolytic and nucleolytic enzymes. After removal of GCl by ethanol precipitation, further deproteinization is achieved by digestion with proteinase K and successive phenol and chloroform extractions. This method is useful when difficulty is experienced lysing cells or if the resultant DNA is degraded or of insufficient purity for enzymatic digestion.

Protocol 2. Extraction of bacterial DNA using GCl

1. Grow 100 ml bacteria to late log phase in an appropriate medium.
2. Harvest cells by centrifugation at 5000 *g* for 15 min at 4°C.
3. Wash pellet once with SET buffer (10% sucrose, 20 mM EDTA, 50 mM Tris–HCl pH 8.0). This is only necessary for bacteria which possess extracellular nucleases, as is often the case for clostridia.
4. Resuspend in 1–2 ml SET containing 2 mg/ml lysozyme.
5. Incubate at 37°C with gentle shaking for from 30 min to 16 h depending on difficulty of lysing cells; the time necessary has to be determined empirically.
6. Add 0.5–1.0 g GCl per ml suspension and invert a few times until GCl dissolves.[a]
7. Shake gently at 37°C until cells lyse.
8. Extract once with an equal volume of chloroform and centrifuge at 8000 *g* for 10 min at room temperature.
9. Layer 2 vol of ethanol on the surface and mix carefully.
10. Lift the clot of DNA out with a loop or pipette tip. If a clot does not form, centrifuge at 8000 *g* for 10 min at room temperature.
11. Wash the clot or pellet three times in 5–10 ml 70% ethanol; leave for 5–10 min each time to allow the GCl to diffuse out.
12. Remove as much liquid as possible and add 0.5–2.0 ml, depending on size of clot, of TENS-PK buffer (20 mM Tris–HCl pH 8.0, 10 mM EDTA, 100 mM NaCl, 0.5% SDS, 0.5 mg/ml proteinase K) to the clot.
13. Shake gently at 37°C until completely dissolved; this may have to be left overnight.
14. Extract with an equal volume of phenol/chloroform until no material remains at the interface. If the solution is very viscous, dilute with TE buffer (see step 5 in *Protocol 1*).

Protocol 2. *Continued*

15. Follow steps 6 to 12 in *Protocol 1* and dissolve in 0.1–1 ml TE buffer, depending on yield.

[a] The addition of the non-ionic detergent Triton X-100 to 1% often aids lysis of refractory cells at this stage of the procedure.

As an alternative to using liquid culture, these methods can be applied to cells grown on solid media by scraping them off and resuspending them in the first buffer used in the DNA preparation. This may be particularly useful for anaerobes which often grow very slowly or not at all in liquid culture. Since different bacterial species vary in the ease of cell wall degradation, alternative measures may be necessary to ensure lysis of cells. Although digestion with lysozyme at 4°C works with many species, different times or temperatures may be required. If lysozyme fails, a variety of other lytic enzymes, including lysostaphin and mutanolysin, may be tried. Digestion with lipases may also help with species containing large amounts of lipid, as may digestion with amylases for species producing large amounts of extracellular polysaccharides.

The physiological state of the cells can sometimes affect the ease with which DNA can be extracted. For example, it is possible to induce autolysis in some methanogens by hydrogen starvation of cells resuspended in minimal medium (6). Alteration of growth conditions can make it easier to lyse cells; the addition of glycine to a concentration of 0.1 M towards the end of the log phase often weakens cell walls. Another problem may be the production of polysaccharides (e.g. capsule material) which may coprecipitate with DNA. As DNA itself is a modified polysaccharide, it is often difficult to separate from other polysaccharides. Since extracellular materials are produced mostly in physiologically limiting conditions, a useful approach is to work with very young cultures, even though large volumes may be required for harvesting enough cells.

2.3 Extraction of plasmid DNA from anaerobic bacteria

Plasmids appear to be widespread in many *Bacteroides*, clostridial, and methanogen species. The extraction of plasmid DNA from anaerobic bacteria may be desirable for a number of reasons.

(a) Plasmids may contain genes of interest.

(b) Plasmids may contain replicons and selectable markers useful for construction of shuttle vectors.

(c) It may provide information on mechanisms by which plasmids replicate.

(d) It is useful in epidemiological studies.

(e) It can be used to study the evolution and dissemination of plasmid-borne genes.

Perhaps the most extensively characterized plasmid from an anaerobe, which exemplifies the usefulness of the molecular characterization of plasmids, is the 10.2-kb plasmid pIP404 isolated from *Clostridium perfringens* (7). The complete nucleotide sequence has been determined. Ten potential open reading frames have been identified, of which some have been assigned functions. These include: a bacteriocin, BCN5, and the associated immunity and secretion gene products (8), *rep* and *cop*, which encode proteins involved in replication and copy number control (9), and a recombinase, *res* (10).

Many different methods have been used to isolate plasmid DNA from bacteria which depend on the ease of lysis of bacterial cells, plasmid size, and plasmid copy number. The alkaline lysis enrichment method (4), with suitable adaptations, has been generally applied for the extraction of plasmid DNA from a variety of *Bacteroides* and *Clostridium* species (11–13).

Protocol 3. Alkaline lysis enrichment method for extraction of plasmid DNA

1. Grow 150 ml bacteria to late log phase in appropriate medium.[a]
2. Harvest cells by centrifugation at 5000 *g* for 15 min at 4°C.[b]
3. Wash cells by resuspending in 20 ml of TE buffer (see step 5 in *Protocol 1*).
4. Re-pellet cells by centrifugation at 5000 *g* for 5 min.
5. Resuspend cells in 20 ml TE buffer, add 0.4 ml diethyl pyrocarbonate, and heat at 50°C for 1 h to inhibit extracellular nuclease activity.
6. Re-centrifuge cells at 5000 *g* for 5 min and suspend cells in 3 ml of lysozyme (20 mg/ml in 40 mM Tris–HCl pH 8.0 and 40 mM EDTA) and 3 ml of 1 M sucrose and incubate at 37°C for 1 h.
7. Re-centrifuge cells and resuspend pellet in 1.5 ml of 50 mM Tris-HCl pH 8.0, 0.5 ml EDTA, and 0.125 ml pronase (20 mg/ml water) and incubate at 37°C for 30 min.
8. Add 0.5 ml of 20% SDS and invert gently until lysis occurs (the cell suspension clears and becomes viscous). If lysis is not apparent leave at 37°C for 30 min and add further SDS (0.25 ml) if necessary.
9. Denature the chromosomal DNA with 0.25 ml freshly prepared 3 M NaOH and gently invert 20 times per min for 3 min.[c]
10. Neutralize the lysate by rapidly mixing with 0.5 ml 2 M Tris–HCl (pH 7.0) and gently invert 10 times over a period of 30 sec.
11. Add 0.5 ml 2 M Tris–HCl (pH 7.0) as described in step 10.
12. Add 0.65 ml of 20% SDS in TE buffer and 1.25 ml of 5 M NaCl and invert tube 20 times to precipitate chromosomal DNA.
13. Store on ice for 1 to 2 h (or overnight).
14. Centrifuge at 15 000 *g* at 4°C for 30 min.
15. Remove supernatant liquid and add an equal volume of sterile water.

Protocol 3. *Continued*

16. Add RNase A to a final concentration of 100 μg/ml and incubate for 1 h at 37°C.
17. The plasmid DNA is purified and precipitated with ethanol as described in steps 6 to 11 of *Protocol 1*. A clot of DNA is unlikely to form and it will be necessary to recover the DNA by centrifugation.
18. Dissolve in 20–100 μl TE buffer, depending on yield.

[a] To recover enough plasmid DNA for CsCl density gradient centrifugation scale up initial culture volume to 1–2 l.

[b] Steps 3–5 may be omitted for most *Bacteroides* species which do not generally possess extracellular nucleases.

[c] The pH of the lysate (between 12.2 and 12.5) is crucial for the selective denaturation of chromosomal DNA. It may be necessary to measure the pH of a lysate and adjust the molarity of NaOH accordingly for other lysate samples.

2.4 Purification of bacterial DNA

Often, despite careful efforts, genomic DNA is of insufficient purity for enzymatic analysis. Digestion with RNase (final concentration 100 μg/ml) for 1 h at 37°C followed by further phenol/chloroform extractions, ethanol precipitation, and washing with 70% ethanol may help. The best method for purification of chromosomal or plasmid DNA (> 50 μg) is CsCl gradient ultracentrifugation (4). However, this method is time-consuming, inconvenient, and often produces sheared DNA. The following protocol has been found useful for the purification of DNA from some clostridial species, without resorting to CsCl gradient ultracentrifugation procedures. It is based on high salt precipitation of contaminating macromolecules.

Protocol 4. Purification of DNA using CsCl without ultracentrifugation

1. Add 1 g CsCl per ml DNA solution and dissolve.
2. Keep at room temperature for 30 min: a precipitate will form.
3. Centrifuge at 8000 *g* for 10 min at room temperature and decant supernatant liquid.
4. Dilute with 4 vol TE buffer.
5. Ethanol precipitate at room temperature and dissolve in appropriate volume of TE buffer as in steps 8 to 12 of *Protocol 1*.

3. Development of vectors for anaerobic bacteria

An understanding of the role of any given gene (gene X), requires that defined mutations are introduced into the gene and that the effect of these mutations on phenotype can be measured and compared to that of the wild-type strain. Using this approach, the gene function and regulation of gene X can be studied in the species of origin. The use of recombinant DNA technology makes such an approach feasible by the availability of suitable vector systems which can reintroduce wild-type or mutant allele(s) of gene X into the host strain.

3.1 Shuttle plasmids

A shuttle plasmid is a plasmid vector that can function in more than one host. Usually, one of the hosts is *Escherichia coli* or *Bacillus subtilis*, or both, because genetic manipulations can be performed easily in these organisms. Shuttle plasmids should possess origins of replication and selectable markers functional in all host strains, as well as unique restriction endonuclease recognition sites for convenient cloning manipulations. The origins of replication are usually derived from a characterized host plasmid, or by screening collections of plasmids for their ability to replicate in the host strain of interest. The selectable markers, which are usually antibiotic resistance determinants, also often originate from the host strains. The origins of replication and selectable markers can be derived from heterologous species, which generally have the advantage of broad host-range specificity. In fact, the origins of replication and selectable markers of all *Clostridium acetobutylicum–E. coli* shuttle vectors constructed thus far, are of heterologous origin (2). A number of *C. acetobutylicum–E. coli* shuttle vectors have been based on pAMβ1 (a streptococcal plasmid), whose minimum replicon has been cloned into the *E. coli* vector pMTL to form a shuttle vector which is selected in the clostridial host by an erythromycin resistance gene, also derived from pAMβ1 (2). The important shuttle vectors used for *Bacteroides–E. coli* and *Clostridium–E. coli–B. subtilis* are summarized in *Table 1*. Shuttle vectors for methanogens have not been reported. However, derivatives of pME2001, a 4.5-kb high copy number plasmid from *Methanobacterium thermoautotrophicum*, have the ability to replicate in *E. coli*, *B. subtilis*, *Staphylococcus aureus*, and *Saccharomyces cerevisiae*, but have not been reintroduced into methanogens (3).

3.2 Potential transposon-based vector systems

A number of conjugative transposons have been introduced into *Bacteroides* (1) and *Clostridium* (2) species. However, their size and lack of characterization has hampered their use as genetic delivery systems.

The conjugative transposon Tn*925* has been used as a vector to introduce the non-conjugative transposon Tn*917* into *C. acetobutylicum* (25). Another conjugative transposon Tn*916* has been used to generate mutations in fermentation pathway genes (26); inter-generic transfer of this transposon has been demonstrated between *Streptococcus faecalis* and *C. tetani* (27). Indigenous conjugative transposons are often present in bacteroides (1) and clostridia (2). One such element has been shown to undergo bi-directional transfer between

Table 1. Shuttle vectors and genetic exchange systems in *Clostridium* and *Bacteroides*

Plasmid	Size (kb)	Replicon	Selection[a]	Alternative host(s)	Transfer method	Reference no.
C. perfringens						
pJU12	11.6	pJU121	Tc^r	*E. coli*	Autoplast/L-form transformation	14
pHR106	7.9	pJU122	Cm^r	*E. coli*	L-form transformation	15
pHR106	7.9	pJU122	Cm^r	*E. coli*	Electroporation	16
pAK201	8.0	pHB101	Cm^r	*E. coli*	Electroporation	17
C. acetobutyliticum						
pMTL500E	6.43	pAMβ1	Em^r	*E. coli, B. subtilis*	Electroporation	18
pCB3	7.03	pCB101	Em^r	*E. coli, B. subtilis*	Electroporation	As described in ref. (2)
pAT187[b]	10.5	pAMβ1	Km^r	*E. coli, B. subtilis*[c]	*oriT* mobilization	As described in ref. (2)
pKNT11	6.5	pIM13	Em^r	*E. coli, B. subtilis*	Protoplast transformation	19
***Bacteroides* sp.**						
pDP1[b]	19	pCPI	Em^r	*E. coli, B. fragilis, B. uniformis*	*oriT* mobilization/conjugation	20
pES-2	17	pB8-51	Em^r	*E. coli, B. fragilis, B. uniformis*	Conjugation	21
pDP1	19	pCPI	Em^r	*E. coli, B. fragilis, B. uniformis*	Electroporation	22
pFD176	7.3	pBI143	Em^r	*E. coli, B. fragilis*	PEG transformation	23
pVALRX	17	pT1-1	Tc^r	*E. coli, B. fragilis, B. uniformis*	pP751 mobilization/conjugation	24

[a] Tc^r, Cm^r, Em^r, and Km^r: tetracycline, chloramphenicol, erythromycin, and kanamycin resistance.
[b] Contain the *oriT* region of RK2.
[c] Has been transferred to a wide range of Gram-positive bacteria.

C. difficile and *B. subtilis* (28). These elements are generally introduced into host strains by filter-mating procedures (see *Protocol 5*).

4. Methods for the introduction of genetic material into anaerobic bacteria

The introduction of genetic material into anaerobic bacteria has been achieved by conjugation and transformation. Alternative methods, such as transduction, have rarely been explored.

4.1 Conjugation

Exploitation of mobilizable plasmids, or their origins of transfer, can allow transfer of shuttle-vectors directly from *E. coli* to anaerobic bacteria. For example, vectors containing the origin of transfer (*oriT*) from the broad-host-range Gram-negative plasmid RK2 can be transferred to *Clostridium* (2) and *Bacteroides* species (20). If the transfer functions of RK2 are present in *trans*, any vector that contains an *oriT* region should be transferable. The provision of an origin of replication from pAMβ1, coupled with the RK2/*oriT* transfer system, can allow efficient conjugal transfer directly from *E. coli* to a variety of anaerobic bacteria by the filter-mating procedure (2). The filter-mating method is simple to perform and does not require specialized equipment, unlike electroporation (see Section 4.2).

There are a number of variations on the basic filter-mating procedure depending on the choice of donor/recipient strains and the element(s) to be transferred. A general procedure is presented for the conjugal transfer of DNA between *E. coli*/*B. subtilis* and *C. difficile* (28).

Protocol 5. Filter-mating method for conjugative plasmids and transposons

1. Grow 10 ml each of donor and recipient bacterial cells overnight to late log phase in appropriate media.
2. Use the cells prepared in step 1 to inoculate 500 ml of medium with donor and recipient cells and grow to mid-log phase.
3. Measure the optical density of the cells and mix donor and recipient cells in a 1:1 ratio.[a]
4. Add 20 ml of this mix on to sterile Millipore filter holder unit (Nalgene) containing 0.45-μm cellulose nitrate filters (Whatman) under reduced pressure. Repeat for further filters.
5. Place filters cell side up on to blood agar plates and incubate anaerobically for 18 h at 37°C.
6. Harvest the cells by washing and vortexing filters in Wilkens–Chalgren broth.

Protocol 5. *Continued*

7. Plate out serial dilutions of the mating mix on to antibiotic-containing Wilkens–Chalgren agar and incubate anaerobically for 1–2 days.[b]
8. Select transconjugants and confirm the presence of cloned determinants and transferable elements by hybridization analysis (4).

[a] The ratio of donor and recipient cells should be optimized for each set of bacterial strains and transferable element(s) studied.
[b] Perform viable counts on donor and recipient strains to calculate transfer efficiency, and plate out donor and recipient strains individually to check for spontaneous resistance to antibiotics.

4.2 Transformation of anaerobic bacteria

Natural transformation, the process whereby bacterial cells readily take up exogenous DNA from the surroundings, has not been reported for anaerobes. Physical methods are necessary to make the cells competent for the uptake of foreign DNA. Although protoplast transformation has been successfully achieved with some *Clostridium* (14) and *Bacteroides* spp. (23), the most simple and versatile method is transformation of whole cells by electroporation. Another advantage of electroporation is that large plasmids or transposons are readily transferable. The technique has been used successfully to transform cells from *C. perfringens* (16, 17, 29), *C. acetobutylicum* (18), *B. ruminicola* (22), and *B. uniformis* (22).

Electroporation involves the application of a brief (2.5–10 msec) high-voltage pulse (2–12 kV/cm) to a suspension of cells, which results in a transient increase in membrane permeability. Under appropriate conditions the transient pores allow DNA present in the surrounding medium to enter the bacteria. Although the process appears to be universally applicable (DNA has been introduced into animal, plant, and fungal cells by electroporation), optimum conditions have to be determined empirically for each species, and even for different strains within the same species. For example, all *C. perfringens* type A strains examined can be transformed by electroporation, in contrast to type C strains, which cannot (29).

Protocol 6. Transformation of *Clostridium perfringens* cells by electroporation with pAMβ1 using a Bio-Rad Gene Pulser® (29)

1. Grow 100 ml of *C. perfringens* type A cells (e.g. strain 3624A) to late-log phase in an appropriate medium.
2. Harvest cells by centrifugation (5000 *g*, 4°C for 15 min) and wash the pellet once in 10 ml cold electroporation buffer (0.27 M sucrose, 1 mM $MgCl_2$, 5 mM Na_2HPO_4; pH 6.4).[a]
3. Recentrifuge cells and resuspend in 5 ml cold electroporation buffer.

Protocol 6. *Continued*

4. Mix 1 ml of cells with 0.5–1.0 μg of DNA (1 mg/ml TE buffer) in a cold 1.5-ml microcentrifuge tube.
5. Mix well and keep on ice for 5 min.
6. Set the electroporator at 25 μF and 2.5 kV, and the pulse controller at 200 Ω.
7. Put the cells/DNA mixture in a cold cuvette (0.2 cm electrode separation) and tap the suspension to the bottom of the well.
8. Place the chilled cuvette between the contacts in the base of the chamber.
9. Pulse once by pressing the two discharge buttons. This should produce a pulse with a time constant of 4.5–5.0 msec (field strength 12.5 kV/cm).
10. Remove cuvette and immediately add 1 ml of electroporation buffer and leave under anaerobic conditions for 3 h.
11. Plate out serial dilutions of the cells/electroporation buffer mixture on TGY-agar plates (pH 6.4) containing 25 μg of erythromycin per ml and incubate anaerobically for 1–2 days.[b]

[a] For organisms particularly sensitive to oxygen the electroporation buffer should be prepared anaerobically and, with care, the electroporator can be set up in an anaerobic environment with subsequent steps performed under anaerobic conditions (22).

[b] Transformation efficiencies of 5.9×10^3 transformants per μg pAMβ1 DNA have been reported for *C. perfringens* using a similar protocol (29). An alternative procedure, where *C. perfringens* cells (strain 13) are pre-treated with lysostaphin prior to electroporation, has resulted in 3.0×10^5 transformants per μg pHR106 DNA (16).

5. Application of the polymerase chain reaction (PCR) to anaerobes

The PCR promises to make an enormous impact both on molecular studies of anaerobes and on the diagnosis of anaerobic infections. The PCR is a technique which allows the very sensitive detection of a given DNA fragment even in a complex mixture of molecules. The technique is based on the *in vitro* enzymatic amplification of a DNA sequence residing between two oligonucleotide primers, which define the ends of the amplified fragment (30).

Many factors affect the efficiency of a PCR reaction, the most important being:

- primer design (sequence, and degree of homology)
- primer concentration
- template DNA concentration and purity
- annealing and denaturing temperatures
- duration of each step and rate of temperature change
- Mg^{2+} concentration

A good starting reaction mixture for PCR experiments is shown in *Protocol* 7. Reaction conditions can be optimized by individually varying different parameters.

Protocol 7. General procedure for PCR reactions

1. PCR reaction mixture

• Sample (DNA from any source, 0.001–100 ng)[a]	1 μl
• Primers (1 μl each of both primers, each at 1 μM)[b]	2 μl
• 10 × buffer[c]	5 μl
• 4 × dNTP (1 μl each at 10 mM)	4 μl
• Taq polymerase (5 U/μl)[d]	0.5 μl
• Water	to 50 μl

2. Overlay reaction mixture with 50–100 μl mineral oil and place in the microprocessor-controlled heating block (Hybaid Ltd).

3. Programme microprocessor:

(a) One cycle of 5 min at 94°C to provide initial complete denaturation.

(b) 30 cycles of:

(i) 2 min at 30–35°C for annealing;
(ii) 2 min at 72°C for synthesis;
(iii) 2 min at 94°C for denaturation;

(c) One cycle of 5 min at 72°C for completion of products.

4. Start heating cycles, which should take approximately 5 h, depending on thermal cycler used.

5. Load samples onto agarose gel: 0.5–2.0% normal agarose for fragments of 150–7000 bp and 4% Nusieve agarose (ICN Biomedicals) for fragments of <150 bp.

6. Run samples at 100 V for 60 min and visualize amplified DNA products under UV light.

[a] DNA should be free of chelating agents, e.g. EDTA, because of the low Mg^{2+} concentration used in the reaction.

[b] Primers taken directly from an oligonucleotide synthesizer and diluted 1 : 10 generally work, even without ethanol precipitation. Precipitated primers are dissolved in water to 0.5–1.0 mg/ml.

[c] 500 mM KCl, 100 mM Tris HCl pH 8.3, 15 mM $MgCl_2$, and 0.1% gelatin.

[d] Amersham International.

PCR can be a very useful adjunct to conventional cloning procedures. A plethora of PCR techniques are now available which will aid in genetic and molecular characterization of anaerobic micro-organisms (30). More details of PCR protocols and the applications of the technique can be found elsewhere (30, 31).

In this chapter, the diagnostic applications of PCR for anaerobes and PCR with mixed oligonucleotide primers (PCRMOP) will be considered to illustrate the versatility of the technique.

5.1 Diagnostic applications of PCR

The molecular characterization of anaerobes has provided a wealth of nucleotide sequence data which can be used for the design of oligonucleotide primers to allow the specific amplification of target DNA by PCR. An important property of PCR is the capacity to amplify target sequences from crude DNA preparations (e.g. boiled cells). As the technique is sensitive, specific, and rapid, it is ideally suited for the detection and identification of anaerobic bacteria, which have fastidious growth requirements and tend to be slow-growing micro-organisms. organisms.

The amplification of the toxin A gene from toxigenic *C. difficile* is a useful example of the application of PCR for the detection and identification of an anaerobe in mixed sample preparations. The primers are based on a 63-bp tandem repeat sequence found in the toxin A gene (32). The results of a typical PCR with toxigenic strains, non-toxigenic strains, and clinical specimens are shown in *Figure 1*, with positive samples displaying a characteristic profile of bands of multiples of the 63-bp unit length. The PCR amplification was successful

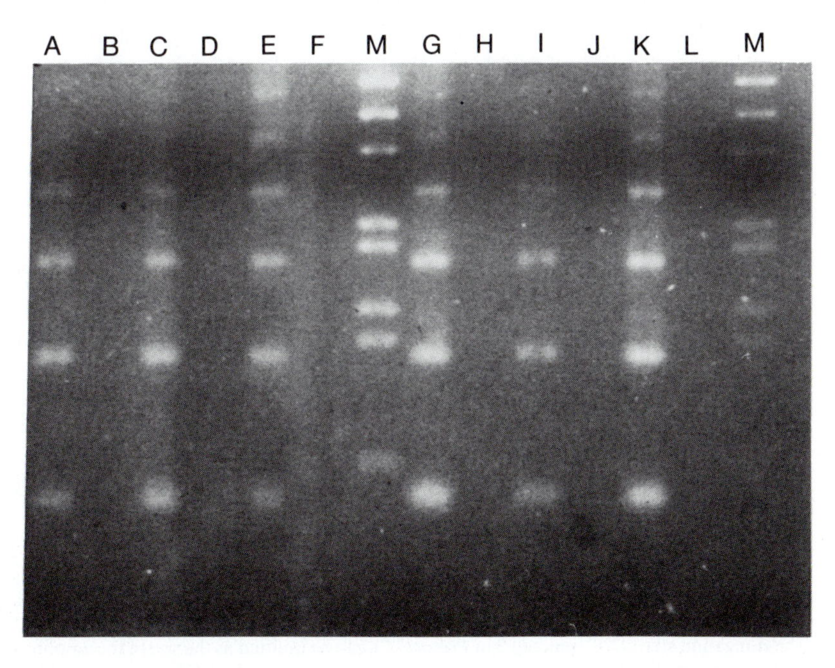

Figure 1. Lanes: A, C, and E, toxigenic strains; B, D, and F, non-toxigenic strains; G, I, and K, positive faecal samples; H and J, negative faecal samples; L, negative water control; and M, DNA size marker (1-kb ladder, visible bands 394, 344, 298, 220, 201, 154, 134, and 75 bp).

when as few as 20 toxigenic bacteria were used and the method can detect this level of toxigenic *C. difficile* in mixed culture and from clinical specimens.

The toxin A gene contains a perfect tandem repeat of 63 bp with which the primers, 5′-GAAGCAGCTACTGGATGGCA-3′ and 5′-AGCAGTGTTAG-TATTAAAGT-3′, anneal under the conditions of the PCR to amplify products of 63 and 126 bp. The extra bands are due to the staggered annealing of tandemly repeated units, followed by primer-independent extension, generating a series of steadily longer DNA fragments as the number of PCR cycles increases (33). This series of products results in an instantly recognizable ladder of eight or more bands which is advantageous over conventional single-product PCR reactions. As many important prokaryotic pathogens have virulence determinants containing at least a single tandem repeat, the use of PCR technology with primers based on repeat nucleotide sequences could have wide applications.

Protocol 8. Identification of toxigenic *Clostridium difficile* by PCR

1. Prepare samples for the PCR as follows.
   (a) *Bacterial samples*. Pick a single colony or take a sweep of colonies from a blood agar plate and add to 0.5 ml water, boil for 5 min, and centrifuge at 14 000 *g* for 5 min. Use the supernatant as the source of DNA.
   (b) *Samples from faecal specimens*. Add approximately 0.1 g of stool to 2.5 ml water and vortex for 30 sec. Boil sample for 10 min, cool, and add proteinase K and pronase (Sigma), both at a concentration of 0.5 mg/ml. Incubate the sample for at least 1 h at 56°C. Then boil for 5 min and centrifuge at 14 000 *g* for 5 min. Use the supernatant as the source of DNA.
2. Add 1 μl of sample to PCR reaction mixture (see step 1 of *Protocol 7*)[a]
3. Include appropriate positive (toxigenic strain) and negative (non-toxigenic strain) control samples.
4. Overlay with mineral oil and place in thermal cycler. Run at 94°C for 5 min (92°C for 30 sec, 45°C for 30 sec)[b] × 35, and 72°C for 5 min.
5. Load samples (25 μl) on to a 2.0% normal agarose gel and run samples at 100 V for 60 min.
6. Visualize under UV light. A recognizable profile of bands of multiples of 63 bp should be observed for positive samples.

[a] Note that it is common practice to batch and aliquot the components of the PCR once ideal conditions have been established, as this reduces the chances of cross-examination. These can be stored at −70°C for several months.

[b] The elongation step, 72°C (see step 3 in *Protocol 7*), can be omitted as the size of the amplification products is very small. This allows for the rapid amplification of desired DNA products. The total time for the whole procedure from picking of cells to visualization of amplification products is 3–4 h.

5.2 PCR with mixed oligonucleotide primers (PCRMOP)

The application of PCRMOP allows the detection of genes in one species using sequence data from an homologous gene or protein from another species. This approach for the amplification of gene sequences will prove increasingly popular as the availability of nucleotide and amino-acid sequence data from micro-organisms continues to increase. The necessity for mixed primers arises from the fact that genes in different species will vary, due to evolutionary change and codon preference in a particular species, so that, although the protein sequence may be conserved, the DNA sequence is likely to differ. One approach is to use fully degenerate mixtures of primers, representing every sequence capable of encoding the specified amino acids; another is to use limited degeneracy based on the codon usage and G + C content of the species of interest, if these are known (e.g. clostridia invariably have a low G + C content). A concept fundamental to PCRMOP is that of patch homology. When homologous genes from several species are compared, patches of 4–10 amino acids are often found to be highly conserved, the intervening regions differing to a greater extent. Primers for PCRMOP are designed using these homologous patches.

The reaction conditions for PCRMOP are essentially the same as those described in *Protocol* 7. However, special attention should be paid to the annealing temperature and the number of amplification cycles. Since a fully degenerate primer could contain more than a million different sequences, the molar concentration of the fully matching sequence is very low; closely related sequences will anneal, but with lower efficiency. The consequence is that the annealing temperature should be low; 30–37°C is usually appropriate. The variability in primer sequence allows priming from sequences other than that desired, resulting in the amplification of incorrect products. This problem cannot be avoided, but can be limited to the early rounds of synthesis by increasing the annealing temperature to 40–55°C after the first 5–10 heat cycles.

An example where PCRMOP could be applied is in the detection of genes encoding vacuolar (V-type) H^+-ATPase regulatory subunits which appear to be universally present in archaebacteria and eukaryotes. *Protocol 9* shows a typical approach that may be taken in the design of fully degenerate oligonucleotide primers to amplify the corresponding gene fragments from other archaebacteria using the PCR.

Protocol 9. Amplification of a V-type H^+-ATPase gene fragment from an archaebacterium by PCRMOP

1. Primer design, based on published sequence data (34) using fully degenerate primers (see *Table* 2).[a,b]

Protocol 9. *Continued*

```
                      V
                      L
      Pst1^c     F    I    N    L    A    N   D
5'  ACGTCTGCAG  TTT  ATA  AAT  CTA  GCA  AAT  GA  3'
                  C  C C    C  T C    C    C
                     G G         G    G
                     T T         T    T

                              S
     M    Y    T    D    L    A   T    HindII^c
3'  TAC  ATG  TGA  CTG  GAA  ACA  TG^d TTCGAATGCA  5'
           A    C    A  A C  CGC
                G         G  T G
                T         T    T
```

2. Prepare reaction mixtures as described in *Protocol* 7 using 0.1–1.0 μg DNA[e]
3. As steps 2–4 of *Protocol* 7, starting with an annealing temperature of 30°C and raising the temperature gradually to 50°C.
4. Run PCRMOP products on 2% agarose gel with suitable DNA size markers and visualize DNA products. A 235-bp amplification product should be observed, among other potential bands, depending on the degeneracy of the primers and the PCR annealing conditions.

[a] The annealing reaction for PCR can tolerate mismatches more readily at the 5′ end of the primer sequence (35). Therefore, it is more desirable to use codon preference and/or less degenerate amino acids at the 3′ end of the primer sequence.

[b] Alternatively, less degenerate primers (particularly at the 3′ end), based on actual sequence data from the anaerobic archaebacterium *Methanococcus thermolithotrophicus* (34), could be used if the required gene fragment is a V-type H^+-ATPase from a closely related species.

[c] Restriction endonuclease recognition sites can be tagged on to the 5′ end of the primer sequence, which will be incorporated into the amplified product. This will make it possible to directly clone the amplified product into a cloning vector with the same restriction endonuclease recognition sites.

[d] Complementary sequence.

[e] It is preferable to use concentrated DNA (0.1–1.0 mg/ml) from lysed cells for PCRMOP. A crude preparation using the SDS lysis method (see *Protocol 1*) is usually sufficient.

The advantage of PCRMOP is that it is quicker and easier than the use of degenerate oligonucleotide probes and filter hybridization methods, and can tolerate far greater degeneracies than these conventional techniques. Once the correct size band has been amplified this could be used as a probe in Southern and colony blot analysis to clone the whole of the gene. Alternatively, the region flanking the DNA insert can be cloned and identified by inverse PCR (30). If restriction endonuclease recognition sites are tagged on to the primers, the amplified fragment can be cloned directly into a suitable vector. Also, direct sequencing of PCR products is possible (30), which results in the rapid acquisition of sequence data. In fact, a gene can be detected, isolated, and

Table 2. Patch homology between V-type H^+-ATPase regulatory subunits of *Methanococcus thermolithotrophicus* (mt), *Sulfolobus acidocaldarius* (sa), *Neurospora crassa* (na), *Arabidopsis thaliana* (at), *Saccharomyces cerevisiae* (sc) (34)

NH2		**147 154**		**213 219**	**COOH**(mt)
mt	———	FINLAND	———	MYTDLAT	———
sa	———	FVNLAND	———	MYTDLAT	———
at	———	FLNLAND	———	MYTDLAT	———
na	———	FLNLAND	———	MYTDLST	———
sc	———	FLNLAND	———	MYTDLST	———

sequenced without the necessity for nucleic acid extraction procedures and cloning manipulations.

The use of the PCR allied with conventional cloning technology and the development of appropriate genetic exchange systems, represents the dawn of an exciting new era for the genetic and molecular characterization of anaerobes.

Acknowledgements

We gratefully acknowledge the contributions of our colleagues in this Department to the development of various aspects of the experimental methodology. We wish to thank Mark Wilks and Mark Pallen for their critical review of manuscript preparations.

References

1. Salyers, A. A., Shoemaker, N. B., and Guthrie, E. P. (1987). *CRC Crit. Rev. Microbiol.* **14**, 19.
2. Young, M., Minton, N. P., and Staudenbauer, W. L. (1989). *FEMS Microbiol. Rev.* **63**, 301.
3. Brown, J. N., Daniels, C. J., and Reeve, J. N. (1989). *CRC Crit. Rev. Microbiol.* **16**, 287.
4. Sambrook, J., Fritsch, E. F., and Maniatis, T. (ed.) (1990). *Molecular Cloning, A Laboratory Manual.* Cold Spring Harbor Press, Cold Spring Harbor, NY.
5. Glover, D. M. (ed.) (1985). *DNA Cloning, A Practical Approach*, Vols I, II, and III. IRL Press, Oxford.
6. Weil, C. F., Cram, D. S., Sherf, B. A., and Reeve, J. N. (1988). *J. Bacteriol.* **170**, 4718.
7. Garnier, T. and Cole, S. T. (1988). *Plasmid* **19**, 134.
8. Garnier, T. and Cole, S. T. (1988). *Mol. Microbiol.* **2**, 607.
9. Garnier, T. and Cole, S. T. (1988). *Plasmid* **19**, 151.
10. Garnier, T., Saurin, W., and Cole, S. T. (1987). *Mol. Microbiol.* **1**, 371.
11. Smith, C. J. and Macrina, F. L. (1984). *J. Bacteriol.* **158**, 739.
12. Strom, M. S., Eklund, M. W., and Poysky, F. T. (1984). *Appl. Environ. Microbiol.* **48**, 956.
13. Roberts, I., Holmes, W. M., and Hylemon, P. B. (1986). *Appl. Environ. Microbiol.* **52**, 197.

14. Hefner, D. L., Squires, C. H., Evans, R. J., Kopp, B. J., and Yarus, M. J. (1984). *J. Bacteriol.* **159**, 460.
15. Roberts, I., Holmes, W. M., and Hylemon, P. B. (1988). *Appl. Environ. Microbiol.* **54**, 268.
16. Scott, P. T. and Rood, J. I. (1989). *Gene* **82**, 327.
17. Allen, S. P. and Blaschek, H. P. (1988). *Appl. Environ. Microbiol.* **54**, 2322.
18. Oultram, J. D., Loughlin, M., Swinfield, M. T., Brehn, J. K., Thompson, D. E., and Minton, N. P. (1988). *FEMS Microbiol. Lett.* **56**, 83.
19. Truffaut, N., Hubert, J., and Reysset, G. (1989). *FEMS Microbiol. Lett.* **58**, 15.
20. Guiney, D. G., Hasegawa, P., and Davis, D. E. (1984). *Proc. Natl Acad. Sci. USA*, **81**, 7203.
21. Shoemaker, N. B., Guthrie, E. P., Salyers, A. A., and Gardner, J. F. (1985). *J. Bacteriol.* **162**, 626.
22. Thomson, A. M. and Flint, H. J. (1989). *FEMS Microbiol. Lett.* **61**, 101.
23. Smith, C. J. (1985). *J. Bacteriol.* **164**, 294.
24. Whitehead, T. R. and Hespell, R. B. (1990). *FEMS Microbiol. Lett.* **66**, 1.
25. Bertram, J. and Durre, P. (1989). *Arch. Microbiol.* **151**, 551.
26. Bertram, J., Kuhn, A., and Durre, P. (1990). *Arch. Microbiol.* **153**, 373.
27. Volk, W. A., Bizzini, B., Jones, K. F., and Macrina, F. L. (1988). *Plasmid* **19**, 225.
28. Mullany, P., Wilks, M., Lamb, I., Clayton, C. L., Wren, B. W., and Tabaqchali, S. (1990). *J. Gen. Microbiol.* **136**, 1343.
29. Allen, S. P. and Blaschek, H. P. (1990). *FEMS Microbiol. Lett.* **70**, 217.
30. Erlich, H. A. (1989). *PCR Technology: Principles and Applications for DNA Amplification.* Stockton Press, New York.
31. Innis, M. A., Gelfand, D. H., Sninsky, J. J., and White, T. J. (1989). *PCR Protocols: A guide to methods and applications.* Academic Press, London.
32. Wren, B. W., Clayton, C. L., and Tabaqchali, S. (1990). *FEMS Microbiol. Lett.* **70**, 1.
33. Wren, B. W., Pallen, M. J., and Tabaqchali, S. (1990). *Lancet*, **1540**, 335.
34. Bernasconi, P., Rausch, T., Gogarten, J. P., and Taiz, L. (1989). *FEBS Lett.* **251**, 132.
35. Sommer, R. and Tautz, D. (1989). *Nucl. Acids Res.* **17**, 6749.

9

Toxins of anaerobes

DAVID M. LYERLY and TRACY D. WILKINS

1. Introduction

A large number of the anaerobic bacteria, including members of the genera *Clostridium*, *Bacteroides*, *Fusobacterium*, and *Actinomyces*, are pathogenic for man and animals. Members of the *Bacteroides*, *Fusobacterium*, and *Actinomyces* produce capsules, haemagglutinins, and a variety of proteolytic enzymes but the exact role of these substances as virulence factors is not clear. Diseases caused by the clostridia, on the other hand, have been shown to result from the production of tissue-damaging toxins (see references 1–5 for general reviews of clostridial toxins). In this chapter, properties and assay procedures for the better known clostridial toxins are described. Some of these procedures have been described in detail in other literature and in those instances, the reader is referred to those references.

2. Clostridial toxins

Bergey's Manual of Systematic Bacteriology (6) describes 83 species of clostridia and, of these, at least 14 species of clostridia are known to produce biologically active products that are believed to be involved in pathogenesis (see *Tables 1* and *2*). Numerous schemes have been developed for the purification of these diverse molecules and it is beyond the scope of this chapter to discuss these purification schemes. The reader is referred to several selected references which describe purification methods as well as properties for some of the major clostridial toxins, including the neurotoxins from *C. botulinum* and *C. tetani*, *C. perfringens* alpha-toxin and enterotoxin, and toxins A and B from *C. difficile* (7–12).

2.1 *C. botulinum* toxins

Strains of *C. botulinum* have been classified into seven different groups (A through G) depending on the type of neurotoxin the organism produces (see *Table 1*). All the neurotoxins from the seven groups are approximately the same size (relative molecular mass, M_r, of 150 000) but these toxins tend to complex with other proteins produced by the organism, thus explaining the various sizes

Table 1. Major toxins produced by toxigenic clostridia

Species	Toxin
C. argentinense	Type G neurotoxin
C. botulinum	Neurotoxin (types A through G)
	C2 toxin (lethal and enterotoxic with ADP-ribosylating activity)
C. baratii	Type F neurotoxin
C. butyticum	Type E neurotoxin
C. chauvoeii/	Alpha-toxin (oxygen-stable haemolysin that is necrotizing and lethal)
C. septicum	Beta-toxin (deoxyribonuclease)
	Gamma-toxin (hyaluronidase)
	Delta-toxin (oxygen-stable haemolysin)
C. difficile	Toxin A (tissue-damaging enterotoxin which is lethal and cytotoxic)
	Toxin B (lethal cytotoxin)
C. histolyticum	Alpha-toxin (lethal and necrotizing toxin related to *C. septicum* alpha-toxin)
	Beta-antigen (collagenase)
	Gamma-antigen (thiol-activated protease)
	Delta-antigen (protease)
C. novyi	Alpha-toxin (lethal and necrotizing)
	Beta-toxin (phospholipase C)
	Gamma-toxin (phospholipase C)
	Delta-antigen (oxygen-labile haemolysin)
	Epsilon-antigen (lipase)
	Type A: alpha-toxin, gamma-toxin, delta-antigen, epsilon-antigen
	Type B: alpha-toxin, beta-toxin
	Type C: no toxin
	Type D: beta-toxin (Type D also is referred to as *C. haemolyticum*)
C. perfringens	Alpha-toxin (phospholipase C that is lethal and necrotizing)
	Beta-toxin (lethal and necrotizing)
	Epsilon-toxin (lethal and necrotizing)
	Iota-toxin (lethal and necrotizing toxin with ADP-ribosylating activity)
	Enterotoxin (lethal and emetic protein found in sporulating cells)
	Type A: alpha-toxin
	Type B: alpha-toxin, beta-toxin, epsilon-toxin
	Type C: alpha-toxin, beta-toxin
	Type D: alpha-toxin, epsilon-toxin
	Type E: alpha-toxin, iota-toxin
C. sordellii	Alpha-toxin (phospholipase C related to the α-toxin of *C. perfringens)*
	Toxin HT (haemorrhagic toxin related to toxin A of *C. difficile*)
	Toxin LT (lethal toxin related to toxin B of *C. difficile*)
C. spiroforme	Iota-toxin (binary toxin with ADP-ribosylating activity)
C. tetani	Tetanospasmin (the neurotoxin responsible for the clinical features of tetanus)
	Tetanolysin (heat- and oxygen-labile haemolysin)

reported for the toxins. The neurotoxins are believed to act by interfering with the release of the neurotransmitter acetylcholine, resulting in the clinical signs associated with botulism. Type A, B, E, and occasionally type F strains cause botulism in humans with types A and B being the more common. Types A, B, C, D, and E strains can produce botulism in animals and fowl, with types C and D being the more common cause. Type C strains are divided further into C_α and C_β strains. Both types of strains produce the type C neurotoxin, designated C1, and

Table 2. Biological assays for toxins produced by anaerobic bacteria

Mouse lethality	Vascular permeability	Haemolytic assay
C. botulinum neurotoxins *C. tetani* neurotoxin *C. perfringens* alpha, beta, epsilon toxins *C. difficile* toxins A and B	*C. botulinum* C2 toxin *C. perfringens* enterotoxin, iota toxin *C. spiroforme* iota toxin *C. difficile* toxins A and B	*C. chauvoeii* alpha, delta toxins *C. tetani* tetanolysin *C. perfringens* alpha, theta toxins *C. novyi* delta toxin
Enterotoxic activity	**Cytotoxic activity**	**Haemagglutination assay**
C. botulinum C2 toxin *C. difficile* toxin A *C. perfringens* enterotoxin, iota toxin *C. spiroforme* iota toxin	*C. botulinum* C2 toxin *C. difficile* toxins A and B *C. perfringens* enterotoxin	*C. difficile* toxin A

an ADP-ribosylating binary toxin designated C2. Type C_α strains produce more C1 neurotoxin than C2, whereas type C_β strains produce more C2 than C1. Type D strains (i.e. those producing type D neurotoxin) also produce some C2 toxin.

In addition to being classified according to the type of toxin produced, strains of *C. botulinum* also are divided into four distinct groups based on their different biochemical and serological properties and DNA homologies. Group I strains (types A, B, and F) and group IV strains are proteolytic but not saccharolytic. Group II strains (types B, E, and F) and group III strains (types C and D) are non-proteolytic but are saccharolytic. Despite the fact that the botulinum toxins share the same activity, they are immunologically distinct molecules. Antibodies against one type of toxin do not neutralize the other toxins. For example, antitoxin against type A neurotoxin is not effective against the other types of neurotoxin. Type A botulinum neurotoxin is the most toxic naturally occurring substance known to man. It has been estimated that as little as 100 ng of toxin may be lethal to humans. Fortunately, the toxin is effectively inactivated by boiling. The toxin is also inactivated by alkaline solutions; therefore, 1 M NaOH should be kept nearby to decontaminate spills. Persons working directly with botulinum neurotoxin should be vaccinated with the appropriate toxoid. The botulinum toxoid currently used is distributed by the Centers for Disease Control (Atlanta, GA) and contains formalin-inactivated toxins of type A, B, C, D, and E adsorbed to aluminium phosphate.

The neurotoxins are produced in relatively high amounts when the organism is grown in brain–heart infusion dialysis flasks (see *Protocol 9*). When this procedure is used, it is important that the flask be properly vented (usually through a cotton-plugged tube through the stopper) to prevent pressure buildup within the flask. A number of purification schemes have been developed for preparing botulinum toxin, but most are complicated and tedious. Rather than growing the organism and purifying the toxin, researchers may prefer purchasing the toxin. Purified botulinum toxin (types A, B, D, and E) is available from Sigma Chemical Co (St. Louis, MO) and less pure preparations of types A, B, C, D, E, and F are available from Calbiochem (San Diego, CA). Types A through F also are available from Wako Chemicals USA Inc (Dallas, TX).

Protocol 1. Detection of botulinum toxin[a]

1. Incubate the isolate overnight at 37°C in chopped meat–carbohydrate broth.
2. Remove the cells by centrifugation at 10 000 *g* for 10 min.
3. Mix the supernatant with an equal volume of 10% $CaCl_2$.
4. Inject 0.5 ml of the mixture into the hind leg of a guinea-pig. Observe the animal for hind-leg paralysis and death.
5. Alternatively, inject either 0.5 ml of the supernatant intraperitoneally, or 0.05–0.1 ml retro-orbitally, into adult mice and then observe over several days for toxicity.
6. If no toxicity is observed with culture filtrates prepared from overnight cultures, filtrates from 3- and 5-day cultures should be tested.

[a] With non-proteolytic strains of *C. botulinum* (some type B strains and all type F strains) the culture filtrate should first be treated with trypsin to activate the toxin (see *Protocol 2*).

Protocol 2. Trypsin activation of botulinum toxins

1. Mix nine parts of culture filtrate with one part of 1% trypsin (Difco 1/250 trypsin) diluted in sterile distilled water.
2. Incubate the mixture at 37°C for 1 h.
3. After incubation, inject animals as described in *Protocol 1*.

The animal assay is the most sensitive assay described to date for botulinum toxin. As little as 6 pg of toxin injected intravenously into mice is lethal. The toxin type is determined by neutralization tests with specific antitoxins (see *Protocol 5*). Monoclonal antibodies against the botulinum neurotoxins have been described but enzyme immunoassays employing these antibodies are not as sensitive as the animal assay.

Detection of botulinum toxins in food and in specimens of serum and faeces from affected patients is fundamental in confirming a diagnosis of botulism (see Chapter 2 for details of isolation techniques). In urgent cases neutralization of toxin with specific antitoxin (see *Protocol 5*) should be attempted at the same time as the toxin detection assay.

Protocol 3. Detection of botulinum toxin in serum

1. Inject 0.4 ml of patient's serum intraperitoneally into each of two adult mice.
2. Observe mice for signs of toxicity, frequently over the first 24 h, and continuing for up to 4 days.
3. If mice show signs of toxicity, repeat the test using specific antitoxins to determine toxin type (see *Protocol 5*).

Protocol 4. Detection of botulinum toxin in food and faeces

1. Grind food or faeces in a sterile pestle and mortar, together with an equal volume of sterile gelatin phosphate buffer (0.2% (w/v) in 0.05 M phosphate buffer).
2. Incubate at 4°C for 18–24 h for toxin extraction.
3. Centrifuge at 12 000 *g* for 30 min.
4. Decant supernatant fluid and adjust pH to 6.2–6.5. At least 12 ml supernatant is required for toxin assay and neutralization tests.
5. Add 0.9 ml supernatant to 0.1 ml streptomycin sulphate solution (10 mg/ml). Inject 0.5 ml intraperitoneally into each of two mice as described in *Protocol 1*.
6. Treat 0.9 ml supernatant with 0.1 ml trypsin as described in *Protocol 2*.
7. Add 0.9 ml of this trypsinized supernatant to 0.1 ml streptomycin sulphate solution and inject mice as described in step 5.
8. Boil 1 ml neat supernatant for 10 min; then inject mice as described in step 5.
9. Boil 1 ml neat supernatant for 10 min; then treat with trypsin as described in steps 6 and 7.
10. Repeat steps 5–9 using a 1/5 dilution of supernatant in sterile physiological saline.

Protocol 5. Determination of botulinum toxin type

1. Mix 0.6 ml supernatant or serum with 0.6 ml specific botulinum antitoxin[a] (TechLab Inc).
2. Incubate the mixtures at room temperature for 30 min.
3. Inject 0.5 ml of each mixture intraperitoneally into each of two mice and observe the animals for toxicity. Neutralization of toxicity with specific antitoxin identifies the toxin type.

[a] Antitoxins are also available from Diagnostics Pasteur, Paris (only types A, B, and E); and Centers for Disease Control, Atlanta.

The C2 toxin produced by type C_β strains of *C. botulinum* consists of two different components (iota$_a$ and iota$_b$, with molecular weights of 100 000 and 50 000, respectively) and is active only when both components are present. C2 is not a neurotoxin; it has ADP-ribosylating activity and transfers the ADP-ribose moiety of cofactor NAD to actin in tissue culture cells. The toxic activity of C2 is enhanced following trypsinization, suggesting that it is produced as a pro-toxin. Caution must be exercised, however, since C2 may be inactivated if exposed to too much trypsin. The toxin can be assayed by its ability to increase vascular permeability (see *Protocol 6*) as well as its enterotoxic activity in the ligated intestinal loop assay (see *Protocol 10*).

Protocol 6. Measurement of increased vascular permeability

1. Prepare the samples to be tested in gelatin phosphate buffer (see step 1 of *Protocol 4*).
2. Shave the back of a New Zealand white rabbit, or a guinea-pig, and remove stubble with a depilatory cream. Follow the manufacturer's instructions; take care not to leave the cream on the skin for too long a time as this will irritate the skin.
3. Inject 0.1-ml volumes intradermally, causing the skin at the injection site to bleb.
4. Measure the diameters of any skin lesions at regular intervals (e.g. 6 and 18 h). Lesions may be erythematous, necrotic, or both.
5. To determine an effect on vascular permeability, inject intravenously a solution of Evans blue dye in phosphate-buffered saline, at a dose of 40 mg/kg body weight. This should be injected 18 h after injection of samples and bluing diameters should be measured 3 h later.
6. A 0.02% solution of cholera toxin (Sigma) can be used as a positive control.

2.2 *C. tetani* toxins

C. tetani neurotoxin, also referred to as tetanospasmin, is the toxin responsible for the signs of clinical tetanus. This toxin has an M_r of about 150 000, making it about the same size as the botulinum neurotoxins. Like the botulinum neurotoxins, tetanus toxin binds to neurons and interferes with the release of the neurotransmitter acetylcholine. Tetanus toxin causes a rigid paralysis, resulting in the severe spasms and death. Thus, the clinical signs are distinct from those caused by botulinum poisoning. The incidence of tetanus in the United States has decreased dramatically because of the extensive vaccination program. In Third World countries, tetanus is still a major problem, causing over one million deaths per year. The disease results from the production of the toxin in wounds infected with the actively growing organism. Like botulinum toxin, tetanus toxin is extremely dangerous. Fortunately, vaccination against the toxin is effective. Persons working with the toxin should be appropriately vaccinated. The most common medium used for the production of tetanus neurotoxin is the protein-free medium described by Latham *et al.* (13). Purified tetanus neurotoxin can be purchased from Calbiochem Brand Biochemicals or Massachusetts Public Health Biologic Laboratories (Boston, MA).

Only one serotype of tetanus toxin has been described. This may be due to the limited number of *C. tetani* isolates that have been examined. The toxin is assayed using a procedure similar to that used for the detection of the botulinum toxins.

Protocol 7. Detection of tetanus toxin

1. Incubate the isolate overnight at 37°C in chopped meat broth.
2. Remove the cells by centrifugation at 10 000 *g* for 10 min.
3. Inject 0.1–0.2 ml of the supernatant into the hind leg of a mouse. In the presence of tetanus toxin, the animal develops severe muscle spasms in its hind leg and dies within a day or two.
4. To confirm the presence of tetanus toxin, protect a second mouse by intraperitoneal injection of antitoxin[a] (500–1500 units) 1 h before the test.

[a] Tetanus antitoxin can be obtained from Sclavo Inc.

The C-fragment of tetanus toxin, which is located at the COOH-terminal portion of the molecule, elicits neutralizing antibody against the toxin and it is believed that this portion represents the binding section. For this reason, the C-fragment is being considered as a possible vaccine and has been expressed in recombinant form. Rabbit antiserum against the C-fragment of tetanus toxin is available from Calbiochem.

In addition to the neurotoxin, *C. tetani* produces a haemolysin designated tetanolysin. Tetanolysin, which is produced as a single polypeptide with a M_r of 45 000, is oxygen-labile and is related to a group of oxygen-labile haemolysins including *C. perfringens* theta toxin (perfringolysin), *C. septicum* delta toxin (septicolysin O), *C. histolyticum* epsilon toxin (histolyticolysin O), *C. novyi* delta toxin, and haemolysins from several other genera of bacteria, including *Streptococcus*, *Bacillus*, and *Listeria*. All of these act by affecting the integrity of the cell membrane, resulting in cell lysis. All are inhibited by cholesterol. Tetanolysin is assayed using sheep erythrocytes in a format similar to the haemolytic assay used for the alpha toxin of *C. perfringens*.

2.3 *C. perfringens* toxins

C. perfringens is the most commonly isolated clostridial pathogen. It causes gas gangrene and intestinal disease. The organism produces five major toxins (alpha, beta, epsilon, and iota toxins and an enterotoxin) and at least nine minor toxins (delta, theta, kappa, lambda, mu, nu, gamma, eta, and neuraminidase). Isolates are divided into types A through E, depending on the major toxin produced (*Table 1*). All the major toxins and many of the minor toxins are lethal when injected into animals.

Protocol 8. Typing of *Clostridium perfringens*[a]

1. Incubate the isolate to be tested in chopped meat–carbohydrate broth at 37°C overnight. Cultures of suspected type A *C. perfringens* should be tested in the logarithmic phase since alpha toxin (phospholipase C) decreases in stationary cultures.
2. Remove the cells by centrifuging at 10 000 *g* for 10 min.
3. Sterilize the supernatant by passing through a 0.45 μm filter.
4. Mix 1.2-ml aliquots of the filtrate with 0.3 ml of each typing antitoxin and incubate the mixtures at 37°C for 30 min.
5. To activate toxins produced by types B, D, and E, treat the filtrate with trypsin (see *Protocol 2*) before adding the antitoxins.
6. Inject 0.5 ml of each mixture intraperitoneally into each of two mice.
7. Death of the mice injected with the culture filtrate and survival of the mice protected with specific antitoxin demonstrates neutralization of the toxin and identifies the toxin type (see *Table 1*).

[a]See also the VPI Manual (14) for additional information.

Alpha-toxin, which is available from Sigma Chemical Co and Worthington Biochemical Corp, is a phospholipase C (also called lecithinase) that catalyses the hydrolysis of lecithin into 1,2-diglyceride and phosphorylcholine and it was the first toxin to have its enzymatic activity identified. Detailed methods for the assay of phospholipase C activity against lecithin are described in the Worthington Enzyme Manual (15). Alpha-toxin is necrotizing and haemolytic because of its activity on cell membranes. Of the toxins of *C. perfringens*, alpha-toxin has been the most intensively studied. The production of alpha-toxin by isolates of *C. perfringens* can be determined by growing the organism on egg-yolk agar. Isolates that produce alpha-toxin will develop zones of opacity around the colony due to the hydrolysis of the lecithin in the egg yolk (see Chapter 3). Alternatively, turbidimetric and titrimetric methods measuring the action of alpha-toxin on egg lecithin may be used (16).

Different types of *C. perfringens* give different types of haemolysis on blood agar. Blood agar is prepared using blood agar base supplemented with 7% sterile blood. Different species of erythrocytes should be tested since the toxins of *C. perfringens* differ in their activity against various red blood cells. Alpha-toxin is active against most erythrocytes except those from horse and goat and usually causes a hazy zone on blood agar. The beta and iota toxins are non-haemolytic. Theta toxin (perfringolysin), which is an oxygen-labile haemolysin related to other oxygen-labile haemolysins such as streptolysin O and tetanolysin, is active against sheep and horse erythrocytes and causes complete clearing zones but is not highly active against mouse erythrocytes. Haemolytic activity may be more accurately quantitated using an *in vitro* test tube assay in which the amount of haemoglobin released from toxin-treated erythrocytes is determined spectrophotometrically. Haemolytic assays have been used routinely to study a number of bacterial toxins and a variety of haemolytic assays employing different concentrations of erythrocytes, various buffer solutions, and divalent cations have been described for alpha-toxin. The reader is referred to reference (16) for a detailed description of a haemolytic assay used for alpha-toxin.

Beta-toxin has been implicated as an important virulence factor in enterotoxaemia of farm animals. It is lethal and necrotizing. The epsilon and iota toxins, both of which are lethal and necrotizing, are produced as protoxins and are activated with trypsin. The iota toxin is a binary toxin indicating that it requires two distinct proteins for expression of toxicity. *C. perfringens* iota toxin is antigenically very similar to iota toxin of *C. spiroforme*, which causes enterotoxaemia in rabbits. The iota toxin from *C. spiroforme* also consists of two distinct components, both of which are necessary for toxic activity. Both toxins possess ADP-ribosylating activity and transfer the ADP moiety of NAD to actin. Thus, they share the same activity as the C2 toxin of *C. botulinum*, although the C2 toxin is antigenically distinct from iota toxin.

C. perfringens is a major cause of food poisoning due to the production of enterotoxin (M_r of 34 000). This toxin is not detected under normal growth conditions because it is a spore-associated protein. Cultures containing high levels of spores must be used as the starting material in order to obtain sufficient levels of the toxin. The Duncan–Strong medium (17) is routinely used for the sporulation of *C. perfringens* and a detailed description for the preparation of this medium is presented by McDonel and McClane (10).

The enterotoxin of *C. perfringens* can be detected by its cytotoxic activity against tissue culture cells such as Vero cells (see *Protocol 11*). These can be purchased from the American Type Culture Collection. Alternatively, the toxin can be assayed immunologically. Oxoid markets a reversed passive latex agglutination kit (RPLA) for the detection of the enterotoxin. The kit, which is manufactured by Denka Seiken Ltd, consists of latex particles coated with rabbit antiserum against the enterotoxin. Latex particles coated with neutral rabbit serum serve as a control.The results are read after 24 h and the sensitivity is reportedly in the 2 ng/ml range. Detailed methods are included with the kit. Rabbit antiserum against the enterotoxin is available from Calbiochem Corporation.

2.4 *C. difficile* toxins

C. difficile causes gastrointestinal disease ranging from mild diarrhoea to life-threatening pseudomembranous colitis. The organism produces two toxins, designated toxin A and toxin B. Toxin A is referred to as the enterotoxin because it is highly active in the gastrointestinal tract and causes the accumulation of fluid. It also has cytotoxic activity and is lethal when injected into animals. Toxin B is referred to as the cytotoxin because it is much more cytotoxic than toxin A against most mammalian cells. Like toxin A, toxin B is lethal when injected into animals. However, toxin B does not have enterotoxic activity. One of the most unusual characteristics of these toxins is their large size. *Clostridium sordellii*, which is pathogenic for animals, produces two toxins that are very similar to toxins A and B. The *C. sordellii* toxins are designated haemorrhagic toxin (toxin HT) and lethal toxin (toxin LT). HT and LT are neutralized by toxin A and toxin B antibodies, respectively.

The dialysis flask system is the most efficient means for the production of large amounts of the *C. difficile* and *C. sordellii* toxins. This method, which was actually developed in the 1940s for the large-scale production of botulinum toxin, offers several advantages. In the dialysis flask, the organism usually grows more slowly, resulting in the accumulation of intracellular toxin prior to autolysis. As a result, higher yields of toxin are achieved. In addition, the dialysis tubing allows only the low-molecular-weight constituents of the media to diffuse into the organism. Thus, the culture and toxin are not contaminated with the large-molecular-weight components present in the media. The following procedure is an adaptation of these earlier methods and has been utilized by us for the production of toxins by *C. difficile* and by other clostridia, including *C. botulinum*, *C. spiroforme*, *C. perfringens*, and *C. sordellii*.

Protocol 9. Brain–heart infusion dialysis flasks

You will need

- Three 2 l conical flasks
- Dialysis membrane tubing (28.6 mm dry cylinder diameter)
- A no. 10 rubber stopper with a 2-cm hole bored in the centre, a 3-mm hole bored off centre, and a 7.5-cm piece of 2 cm diameter glass tubing inserted into the centre hole
- An aluminium foil, steel, or plastic cap to cover the 2-cm glass tubing
- A 5 cm piece of glass tubing to fit into the 3-mm hole, one end of which is plugged with cotton wool
- String

1. Add 74 g dehydrated brain–heart infusion and 1 l of distilled water to a 2-l flask.

Protocol 9. *Continued*

2. Pour approximately 1 l distilled water into a second flask.
3. Soak a piece of dialysis tubing, about 30 cm in length, in distilled water. Carefully tie a knot in the end of the tubing and push the knot toward the centre of the tubing in order to tighten it (this reduces the likelihood of tearing the tubing).
4. Place the open end of the tubing over the 2-cm glass tubing and secure it with string.
5. Fill the assembly with distilled water to check for any holes in the membrane and then place the assembly in a conical flask to be autoclaved; make sure to cover the assembly with distilled water.
6. Autoclave the following items separately
   - Flask containing brain–heart infusion
   - Flask containing water
   - Dialysis tubing assembly
   - Aluminium, plastic, or steel cap
   - A 5-cm piece of 3-mm glass tubing
7. Assemble the apparatus while it is still hot; this helps to create an anaerobic atmosphere within the flask.
8. Pour about 800 ml of sterile distilled water into the flask containing brain–heart infusion (the smaller volume of medium does not boil over during autoclaving).
9. Place the dialysis tubing assembly inside the flask of medium. There should be 50–100 ml distilled water inside the dialysis sac. Make sure that the rubber stopper is securely inserted into the top of the flask.
10. Insert the 3-mm tubing, plugged with cotton wool, into the 3-mm hole in the rubber stopper. This tubing serves as a vent for gas-producing clostridia.
11. Place the metal or plastic cap over the glass tubing and allow the flask to cool overnight. The assembled flask is shown in *Figure 1*.
12. Inoculate the flask with an overnight broth culture, placing several drops through the glass tubing into the 'dialysis medium' within the dialysis sac.
13. Incubate for an appropriate time (3 days at 37°C). This apparatus maintains an environment that is sufficiently anaerobic for many clostridia, without requiring incubation in an anaerobic chamber.
14. To harvest the culture, remove the dialysis tubing assembly and pour the culture through the centre tube into an appropriate sterile container.
15. Remove the cells by centrifugation at 10 000 g for 30 min.
16. Sterilize the supernatant by passing through a 0.45 μm filter.

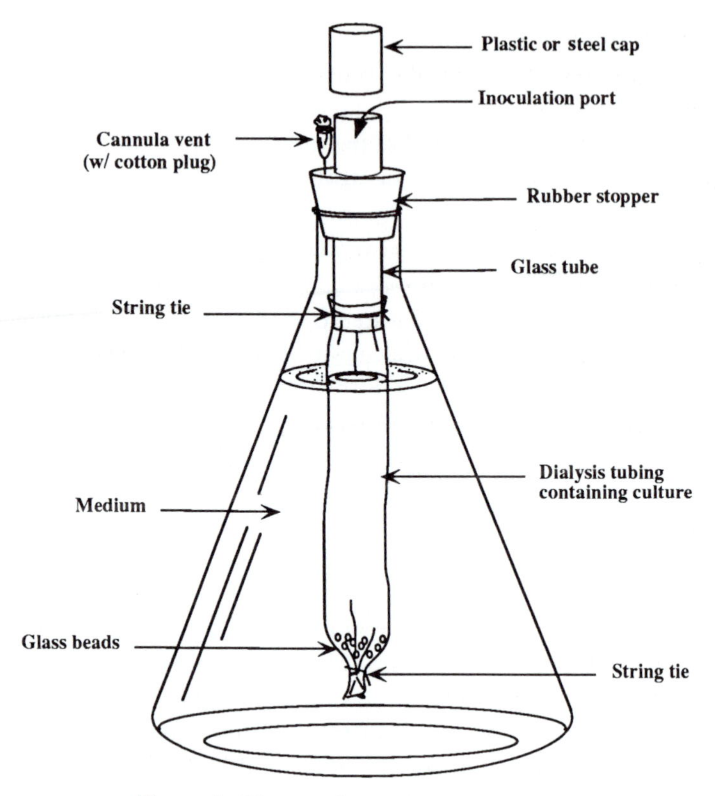

Figure 1. Diagram of assembled dialysis flask.

Toxin A and enterotoxins produced by other bacteria can be assayed for biological activity in the intestine using the ligated loop procedure. The toxic activity is detected by measuring the volume of fluid accumulated within the ligated loop. Results are recorded as a ratio of the volume of fluid (ml) per length of loop (cm). Unlike the ricewater fluid accumulation caused by other types of enterotoxins, such as cholera toxin, toxin A will cause the accumulation of a viscous, haemorrhagic field, since it causes extensive tissue damage to the intestinal wall. Thus, if the ligated loop assay is used, the researcher also should note the type of fluid in addition to fluid volume. The assay is tedious and requires extensive surgery. Only persons with the appropriate training should perform the assay. Toxin A may also be detected by ELISA, using specific antitoxin (see *Protocol 12*).

Protocol 10. Intestinal loop assay for enterotoxic activity

1. The rabbit should be fasted overnight and properly anaesthetized. In our assay, rabbits are anaesthetized with sodium pentothal and halothane

Protocol 10. *Continued*

according to the recommended dosages. The animal should be kept warm during the entire operation.

2. Shave the midsection of the abdomen and swab with antiseptic solution.
3. Make a 10-cm incision, exposing the intestine.
4. Locate the ileocaecal junction and withdraw the ileum through the incision. The portion of the intestine to be used for the assay may be rinsed internally with neomycin sulphate solution to help obtain consistent results.
5. Ligate the exposed intestinal tract every 5 to 10 cm. Leave a blank loop between each loop to be injected with test samples. These blank loops aid as controls.
6. When injecting the samples, inject at one end of the loop and then tie off the injection site to prevent leakage of the sample. It helps to use different colours of thread for each loop (except for red, because of the blood staining). For example, yellow thread for each end of the first loop, blue for the ends of the second loop, etc. This assists identification of the loops.
7. Once all the samples have been injected, gently place the exposed intestine back into the abdominal cavity. Close the muscle wall and skin with sutures and revive the animal.
8. Place the rabbit in a warm area and inject the animal intramuscularly with an appropriate analgesic (e.g. Numorphan) at the appropriate time intervals. Make sure that the animal has water to drink.
9. To read the results, kill the animal with the appropriate euthanasia solution. Most loop reactions occur within 12 h and can be measured 10–12 h after injection of the samples.
10. Carefully cut out the portion of the intestinal tract that contains the series of ligated loops. Measure the volume of fluid in each loop and the length of the loop after the fluid has been removed.
11. Determine the volume/length ratio (ml/cm). A ratio of 0.5 or greater is usually considered positive. Note the type of fluid response. Cholera toxin, which is used as a positive control, elicits a ricewater fluid. Toxin A from *C. difficile* elicits a viscous, haemorrhagic fluid response, with as little as 1 μg injected.

Toxin B usually is assayed by tissue culture since it is highly active against all mammalian cells tested to date. WI-38 and MRC-V lung fibroblasts, Chinese hamster ovary (CHO) K-1 cells, and human foreskin cells are the most commonly used cell lines. As little as 1 pg of toxin B causes a characteristic rounding effect. Toxin A, which is also cytotoxic, can be assayed by tissue culture and causes a rounding effect like that seen with toxin B. Against most cell lines, toxin A is considerably less active than toxin B. However, with P19, F9, and OTF-9 cell lines, toxin A is as active as toxin B. If CHO K-1 cells are used, they can be grown in F12 medium supplemented with 2% fetal calf serum.

2.4.1 Detection of *C. difficile* toxins in faeces

The tissue culture assay is used routinely for detection of *C. difficile* toxin in faecal specimens from patients with gastrointestinal disease (see Chapter 2 for culture methods). The assay involves the detection of cytotoxic activity and its neutralization by specific antitoxin. A commercial kit is available from Baxter Scientific. Alternatively, toxin and antitoxin can be obtained from TechLab Inc.

Protocol 11. Detection of *Clostridium difficile* toxin in faecal specimens by tissue culture assay

1. Mix 1 ml (1 g wet weight) of faecal specimen with 1 ml phosphate-buffered saline (PBS) by gently vortexing the suspension for 10 sec.
2. Remove insoluble material by centrifuging at 10 000 *g* for 5 min.
3. Pass the supernatant fluid through a 0.45-μm membrane filter and store the filtrate at 4°C until used.
4. Prepare a series of decimal dilutions (10^{-1} to 10^{-5}) of the toxin filtrate in PBS.
5. Prepare a series of decimal dilutions (10^{-1} to 10^{-5}) of the faecal filtrate in PBS.
6. Dilute 0.1 ml of *C. difficile* antitoxin in 2.4 ml PBS.
7. In one set of microtitre wells, add 0.1 ml of each toxin dilution to 0.1 ml PBS.
8. In a second set of wells, add 0.1 ml of each toxin dilution to 0.1 ml of diluted *C. difficile* antitoxin.
9. In a third set of wells, add 0.1 ml of each faecal filtrate dilution to 0.1 ml PBS.
10. In a fourth series of wells, add 0.1 ml of each faecal filtrate dilution to 0.1 ml of diluted *C. difficile* antitoxin.
11. Cover, and incubate the microtitre plate at room temperature for 30 min.
12. Transfer 20 μl from each well to a tissue culture well (in a microtitre plate) containing the appropriate tissue culture cells (seeded at a density of about 1000 cells/well) and 180 μl of medium.
13. Incubate the tissue culture plate in a CO_2 incubator overnight and observe the reactions the following day.
14. Positive control wells containing toxin dilutions mixed with PBS should show a typical cytopathic effect (CPE), of rounded cells, to a titre of 10^5. Wells containing toxin dilutions mixed with *C. difficile* antitoxin should show no rounding, due to the neutralization of the toxin.

Protocol 11. *Continued*

15. Test wells containing faecal filtrate dilutions mixed with PBS that show a typical CPE contain cytotoxic activity. Record the titre by noting the highest dilution that causes rounding of >90% of the cells (the titre is expressed as the reciprocal of the dilution, e.g. 10^{-4} = titre of 10^4). If the CPE is neutralized in wells containing dilutions of faecal filtrate mixed with antitoxin, the activity is due to *C. difficile* toxin in the stool specimen.[a]

[a] Non-specific reactions are not uncommon, and may be due to several causes. One of these is the enterotoxin of *C. perfringens* (see Section 2.3). *C. perfringens* enterotoxin may be detected by the tissue culture assay method described here, using Vero cells and specific antitoxin.

Protocol 12. ELISA for detection of *Clostridium difficile* toxin A in faeces

You will need

- Microtitre plates (Immulon type 2; Dynatech Industries)
- Rabbit antiserum against crude *C. difficile* toxin preparation (19)
- Affinity-purified goat toxin A antibodies (18)
- Rabbit anti-goat IgG alkaline phosphatase conjugate (Sigma)
- *p*-nitrophenyl phosphate (Sigma phosphatase 104 substrate)
- 0.05% (v/v) Tween 20 in phosphate-buffered saline (PBS–T)
- 0.01% (v/v) neutral rabbit serum in PBS–T (PBS–T–RS)
- Diethanolamine buffer, pH 9.8
- Carbonate buffer, pH 9.6
- 5 M NaOH

1. Dilute crude rabbit antiserum 1/1000 in carbonate buffer.
2. Add 0.3 ml diluted antiserum to each well in microtitre plate. Do not use the wells on the outer perimeter of the plate.
3. Incubate microtitre plates overnight at 37°C; then empty.
4. Wash each well with 0.3 ml PBS–T.
5. Add 0.2-ml test samples to wells (see *Protocol 13* for sample preparation).
6. Incubate for 1 h at 37°C.
7. Wash wells four times with 0.3 ml PBS–T. Incubate the third wash for 2 min at room temperature.
8. Add to each well 0.2 ml of a 1/500 dilution of purified antitoxin in PBS–T–RS.
9. Incubate for 1 h at 37°C.

Protocol 12. *Continued*

10. Wash wells as described in step 7.
11. Add to each well 0.2 ml of a 1/500 dilution of anti-goat IgG conjugate in PBS–T–RS.
12. Incubate for 1 h at 37°C.
13. Wash wells as described in step 7.
14. Add to each well 0.2 ml of a 1 mg/ml solution of phosphatase substrate in diethanolamine buffer.
15. Incubate for 30 min at room temperature.
16. Stop the reaction by adding to each well 20 μl of 5 M NaOH.
17. Add contents of each well to 0.8-ml volume of deionized water.
18. Read absorbance at 405 nm.
19. Use the following controls.
    (a) Coat wells with neutral rabbit serum in place of rabbit antiserum.
    (b) Add PBS–T in place of test sample.
    (c) Add PBS–T in place of purified antitoxin.

Protocol 13. Preparation of samples for toxin A ELISA

1. If stool is watery, dilute 1 ml stool with an equal volume of PBS–T.
2. If stool is solid, add 0.2 g of stool to 1 ml of PBS–T, and vortex to mix thoroughly.
3. Centrifuge stool suspension (from step 1 or 2) at 15 000 *g* for 30 min.
4. Pass the resulting supernatant through a 0.45-μm membrane filter.
5. Use sterile filtrate as test sample (step 5 in *Protocol 12*).

2.5 Toxins produced by 'gas-gangrene' clostridia

Gas gangrene, also referred to as clostridial myonecrosis, is caused by a number of clostridia. *C. perfringens* type A, *C. septicum*, and *C. novyi* type A are the most common causes but *C. histolyticum*, *C. sporogenes*, *C. sordellii*, *C. bifermentans*, and *C. fallax* can also cause the disease. Usually more than one species is involved. The disease results from the various tissue-damaging toxins produced by these organisms. These toxins, including phospholipases from *C. perfringens* and *C. novyi* and collagenases of *C. perfringens* and *C. histolyticum*, are capable of degrading the surrounding tissue following infection of a wound by these clostridia.

C. histolyticum produces a variety of proteases and collagenases, and partially purified preparations containing mixtures of these proteases are used routinely to digest tissue. *C. histolyticum* collagenase ranging from crude material to highly purified enzyme is available from Boehringer Mannheim Biochemicals, Calbiochem Brand Biochemicals, Sigma Chemical Co, and Worthington Biochemical Corporation. Various assays employing different substrates (e.g. native collagen, *N*-[3-(2-Furyl)acryloyl]-Leu-Gly-Pro-Ala, Pz-Pro-Leu-Gly-Pro-Arg) are used to detect collagenase activity. The Worthington Enzyme Manual (15) provides detailed methods for the assay of *C. histolyticum* collagenase. Methods for detection of *C. septicum* toxins using tissue culture neutralization were described recently (20).

3. *Bacteroides fragilis* enterotoxin

B. fragilis, which is an obligate anaerobe found in the intestine of animals and man, causes bacteraemia and abscess formation. Recently, it has been demonstrated that some isolates of *B. fragilis* from lambs and calves with diarrhoeal disease are enterotoxigenic. In addition, enterotoxigenic *B. fragilis* have been isolated from humans with diarrhoea. These isolates have been shown to cause the accumulation of fluid when injected into ligated intestinal loops in calves and lambs. The activity appears to be due to a heat-labile molecule (estimated M_r of 19 500) in culture filtrate prepared from *B. fragilis* brain–heart infusion cultures. The filtrate is concentrated by ultrafiltration in order to detect the activity. The assay for the toxin is done in lambs, calves, or rabbits (21) using the general methodology described for the rabbit ileal loop assay (see *Protocol 10*).

4. Cloning of toxin genes from clostridia

A number of toxin genes from clostridia have been cloned and expressed in *E. coli* and, in most instances, the proteins expressed by the cloned genes retained their toxic activity. For the most part, the general recombinant DNA methodology presented by Maniatis *et al.* (22) was utilized and clones were identified using DNA probes or immunoassay procedures with toxin-specific antibodies. Clostridial toxin genes cloned to date include the tetanus neurotoxin gene, genes for the alpha-toxin, theta-toxin, and enterotoxin from *C. perfringens*, and genes for toxins A and B of *C. difficile* (23–29).

Many clostridial toxins are extremely toxic and guidelines in toxin research involving recombinant DNA must be followed. In the United States, a booklet containing guidelines for research involving recombinant DNA molecules is available from the Department of Health and Human Services of the National Institutes of Health. It is important that toxicity tests be done to identify the expression of any toxic fragments. With highly toxic molecules, alternative strategies for cloning may be necessary. Tetanus neurotoxin, for example, was cloned under L3B1 containment with DNA fragments of 3 kb or smaller to avoid cloning the entire fragment.

5. Conclusions

The toxins produced by the clostridia represent a diverse range of biologically active molecules that have been studied for many years by investigators around the world. Because of the vast amount of literature available on these toxins, the reader is encouraged to use general references, including references 1–5, for additional information. Studies on the mechanism of action of these toxins and how they participate in the disease process continue to be an integral part of research in microbial pathogenesis. These toxins are being used as tools in many types of basic and applied research and in some instances, they are being used as therapeutic agents. Botulinum toxin, for example, is being used to treat certain types of eye disorders. The methods used to study these toxins will help scientists to identify biologically active molecules from other anaerobic bacteria and these types of studies should give us valuable information of how other anaerobes cause disease.

Acknowledgements

We wish to thank Roger Van Tassell for supplying the diagram shown in *Figure 1*. We also thank Roger Van Tassell and Kenneth Tucker for their helpful suggestions during the preparation of this chapter.

References

1. Hatheway, C. L. (1990). *Clin. Microbiol. Rev.* **3**, 66.
2. Lyerly, D. M., Krivan, H. C., and Wilkins, T. D. (1988). *Clin. Microbiol. Rev.* **1**, 1.
3. Rolfe, R. D. and Finegold, S. M. (ed.) (1988). *Clostridium difficile: Its Role in Intestinal Disease.* Academic Press, New York.
4. Simpson, L. L. (1989). *Botulinum Neurotoxin and Tetanus Toxin.* Academic Press, New York.
5. Willis, A. T. (1969). *Clostridia of Wound Infection.* Butterworths, London.
6. Cato, E. P., George, W. L., and Finegold, S. M. (1986). In *Bergey's Manual of Systematic Bacteriology*, Vol. II (ed. P. H. A. Sneath), pp. 1141–200. Williams & Wilkins, Baltimore.
7. Simpson, L. L., Schmidt, J. J., and Middlebrook, J. L. (1988). In *Methods in Enzymology* (ed. S. Harshman), Vol. 165, pp. 76–85. Academic Press, New York.
8. Robinson, J. P. (1988). In *Methods in Enzymology* (ed. S. Harshman), Vol. 165, pp. 85–90. Academic Press, New York.
9. Jolivet-Reynaud, C., Moreau, H., and Alouf, J. E. (1988). In *Methods in Enzymology* (ed. S. Harshman), Vol 165, pp. 91–94. Academic Press, New York.
10. McDonel, J. L. and McClane, B. A. (1988). In *Methods in Enzymology* (ed. S. Harshman), Vol. 165, pp. 94–103. Academic Press, New York.
11. Krivan, H. C. and Wilkins, T. D. (1987). *Infect. Immun.* **55**, 1873.
12. Sullivan, N. M., Pellett, S., and Wilkins, T. D. (1982). *Infect. Immun.* **35**, 1032.

13. Latham, W. C., Bent, D. F., and Levine, L. (1962). *Appl. Microbiol.* **10**, 146.
14. Holdeman, L. V., Cato, E. P., and Moore, W. E. C. (1977). *Anaerobe Laboratory Manual* (4th edn). Virginia Polytechnic Institute and State University, Blacksburg, VA.
15. Worthington, C. C. (1988). *Worthington Enzyme Manual.* Worthington Biochemical Corp, Freehold, NJ.
16. Jolivet-Reynaud, C., Moreau, H., and Alouf, J. E. (1988). In *Methods in Enzymology* (ed. S. Harshman), Vol. 165, pp. 293–7. Academic Press, New York.
17. Duncan, C. L. and Strong, D. H. (1968). *Appl. Microbiol.* **16**, 82.
18. Lyerly, D. M., Sullivan, N. M., and Wilkins, T. D. (1983). *J. Clin. Microbiol.* **17**, 72.
19. Ehrich, M., Van Tassell, R. L., Libby, J. M., and Wilkins, T. D. (1980). *Infect. Immun.* **28**, 1041.
20. Knight, P. A., Tilleray, J. H., and Queminet, J. (1990). *Biologicals* **18**, 181.
21. Myers, L. L., Shoop, D. S., and Collins, D. E. (1990). *J. Clin. Microbiol.* **28**, 1658.
22. Maniatis, T., Fritsch, E. F., and Sambrook, J. (ed.) (1982). *Molecular Cloning: A Laboratory Manual.* Cold Spring Harbor Press, Cold Spring Harbor, NY.
23. Dove, D. H., Wang, S.-Z., Price, S. B., Phelps, C. J., Lyerly, D. M., Wilkins, T. D., and Johnson, J. L. (1990). *Infect. Immun.* **58**, 480.
24. Eisel, U., Jarausch, W., Goretzki, K., Henscehn, A., Engels, J., Weller, U., Hudel, M., Habermann, E., and Niemann, H. (1986). *EMBO J.* **5**, 2495.
25. Iwanejko, L. A., Routledge, M. N., and Stewart, G. (1989). *J. Gen. Microbiol.* **135**, 903.
26. Johnson, J. L., Phelps, C., Barroso, L., Roberts, M. D., Lyerly, D. M., and Wilkins, T. D. (1990). *Curr. Microbiol.* **20**, 397.
27. Makoff, A. J., Ballantine, S. P., Smallwood, A. E., and Fairweather, N. F. (1989). *Biotechnology* **7**, 1043.
28. Tso, J. Y. and Siebel, C. (1989). *Infect. Immun.* **57**, 468.
29. Tweten, R. K. (1988). *Infect. Immun.* **56**, 3228.

10

Methods for the study of anaerobic microflora

B. S. DRASAR and APRIL K. ROBERTS

1. Introduction

This chapter is about the study of the microbial ecology of the bacteria associated with the human body. This is among the most intensively studied of microbial habitats and includes a number of distinct ecosystems of particular interest. Attention has focused on the microbial flora of the mouth (1), the skin (2), the genital tract (3), and the intestine (4–6). These themes have on occasion been drawn together in more systematic treatments of the whole system (7–9). Systematic studies of animals other than man have concentrated on the bovine rumen (10, 11). Many studies have been concerned to describe these habitats in terms of their bacterial inhabitants while others have concentrated on examining the metabolic potential of the bacterial microflora and its impact on the host (12, 13).

In ecological studies it is of particular importance to be aware of the nature of the question being asked, or the hypothesis studied, in the context of the ecosystem being investigated and the methods being used. Thus, for example, the intestinal flora normally includes some 300–400 distinct bacterial species and the effort needed to isolate and identify them is large and probably not justified unless the various bacterial species are themselves the primary object of study. The microflora of the body can be considered as occupying three distinct zones or ecologies. There is probably considerable overlap between the membership of these communities but distinctions between them when considered in terms of the primary ecological determinants are conceptually useful.

(a) *Cell-surface ecologies*, for example, flora of mucosal surfaces, such as the mouth, large intestine, skin, and vagina. It is thought that specific attachments to cell-surface receptors are of primary importance in such situations. On the skin, products of bacterial breakdown of various substrates, and naturally produced antimicrobials are also important.

(b) '*Biofilm ecologies*' is probably a better description of the flora of most mucosal surfaces in that bacteria attached directly to cell surfaces are only part of such ecologies. Bacterial communities, mutually adherent one to the other, are of major significance. The best studied example of such an ecology is plaque on teeth, but studies in model systems indicate that adhesion is probably an important determinant of the composition of the gut flora.

(c) *Luminal ecologies* may be thought of as those ecologies filling cavities associated with the body. In practice these ecologies have been widely studied; they include faeces, saliva, and vaginal secretions. It has been hypothesized that in such situations microbial competition for available nutrients and sensitivity to metabolic products, colicines, phage, and naturally occurring antibiotics, are ecologically important. Some investigators have doubted the reality of such interactions and have regarded bacterial mixtures isolated from faeces and saliva as the result of shedding from mucosal biofilms.

Each of these types of ecology requires different and specific methods for their study. In this chapter many of these methods will only be touched upon. This conceptual background is presented so that the limitations of the methods described here and of the results obtained with them can be understood. The methods described are for the isolation, enumeration, and identification of bacteria together with some investigations of bacterial metabolism. The immediate focus is on the microbial ecology of the gut but many of the methods described are applicable to other systems.

2. Sampling

Obtaining adequate samples is crucial to the study of the ecology of the human body. It must be remembered that sampling may be an invasive procedure and it is important to ensure that proper ethical guidelines are approved, that informed consent is obtained, and that sampling procedures are informed by good clinical practice. Microbiologists will find active collaboration with the appropriate clinical specialists to have major advantages.

The methods used for sampling depend on the site to be sampled. In the mouth samples other than saliva should be obtained by a dentist, and intestinal samples, other than faeces, by a gastroenterologist.

Dental plaque may be collected with dental instruments (from accessible surfaces) and with dental floss (from between the teeth). Special devices have been produced for sampling material from peridontal pockets. Biofilms are best investigated on extracted teeth or removable test pieces that simulate tooth surfaces and can be held in the mouth (14).

Faeces are relatively simple to obtain but other samples are usually obtained using various intubation procedures (4, 15). Stomach and small intestinal samples are usually obtained by per-oral intubation and large bowel samples during colonoscopy. Tissue samples examined are most often biopsy materials obtained during diagnostic procedures. Some investigators have examined material obtained during surgery or at post-mortem examination. All these procedures have disadvantages and all are likely to disturb the system under study; however, it is thought that intubation procedures produce the least disturbance.

Sampling the genital tract is an invasive procedure and sampling is usually performed during gynaecological examination. Studies have concentrated on the bacteria present in the vagina. Samples have been collected on paired pre-weighed swabs, using calibrated wire loops and by direct collection of secretions (3). The adequacy of the results may be directly affected by the amount of sample obtained.

Though apparently the most accessible of sites, the skin presents major problems. Swabs are the most usual method of sampling but it should be recognized that only about 10% of the bacteria at a site are taken up and only about 10% of these are deposited on to plates. Greater yields of bacteria can be achieved by scrubbing the skin with a suitable fluid including a non-ionic detergent. Skin biopsies of various sorts have also been used to study distribution of bacteria (16). It should be remembered that anaerobic bacteria tend to be found in follicles.

2.1 Preservation of samples

When growing in or on the human body, anaerobic bacteria are part of complex ecological systems and are probably protected from the toxic effects of oxygen. Almost all sampling procedures increase the exposure of anaerobic bacteria to oxygen. Attempts have been made to overcome this problem, e.g. in the mouth by gassing the sampling site with oxygen-free gas, and in the gut by passing faeces directly into containers filled with oxygen-free gas, but such procedures are not widely applicable.

Whenever possible, samples should be cultured immediately after collection. Samples should be placed in an anaerobic atmosphere and transported to the laboratory as rapidly as possible.

However, there is more usually a considerable time before samples can be cultivated. Studies on the mouth and genital tract have often used transport media to preserve the viability of the bacteria in the samples. Studies on the intestinal flora have relied on direct culture of samples or preservation in the frozen state. If samples are frozen, it is imporant to use a suspending medium with a cryoprotective agent, such as DMSO or glycerol (17).

Protocol 1. Preservation and transport of faeces

1. Prepare brain–heart infusion (BHI) containing 10% (v/v) glycerol.
2. Dispense 4.5-ml aliquots into autoclavable screw-capped tubes suitable for freezing. Autoclave at 121°C for 15 min. Polycarbonate or polypropylene centrifuge tubes are suitable.
3. Add about 0.5 g faeces (about the size of a pea) to 4.5 ml sterile BHI–glycerol.

Protocol 1. *Continued*

4. Emulsify the mixture using a sterile swab until no further lumps remain.
5. Immediately freeze the mixture by immersion in liquid nitrogen. If liquid nitrogen is unavailable the tubes must be packed in dry-ice inside a vacuum flask. Once the specimens have been diluted they must be frozen immediately and kept frozen until cultured. The bacteria will survive for several months when stored in this way. Weigh and record weight of bottle + broth + sample before cultivation.

3. Microscopy

Strictly speaking, microscopy is not a method for the detection of anaerobic bacteria, but it should be remembered that the development of methods for the isolation of anaerobes from the rumen and gut was stimulated by the observation that many more bacteria could be seen under the microscope than could be isolated by normal methods. Even now there remain morphologically distinct bacterial cells, in many habitats, that can be visualized but not cultivated.

Electron microscopy has been used for the study of adherent biofilms, particularly in the mouth. The structure of plaque has been examined using both transmission and scanning electron microscopy. Though, at present, identification of bacterial types is made on the basis of shape, it is likely that the increased availability of antisera suitable for use with immune electron microscopic techniques will extend the range of possible studies.

Whether or not the availability of suitable antisera makes the use of immune electron microscopy or fluorescent antibody techniques possible it is important to undertake microscopic examinations of the bacteria present in the sample studied.

Protocol 2. Direct microscopic (total) count of bacteria in faeces

1. Dispense 4.5-ml aliquots of sterile saline into sterile test tubes. Add 3 drops of Tween 80 to each tube and mix with the saline.
2. These tubes are used to prepare 10^{-3}–10^{-5} dilutions of the faecal sample under study. It is convenient to use 0.5 ml of the 10^{-2} dilution prepared for plating (see *Protocol 6*) to start this dilution series.
3. Spread 10 μl of the 10^{-3}, 10^{-4}, and 10^{-5} dilutions on to three separate 0.5-cm^2 circles, on special pre-printed slides (Dynatech Laboratories Ltd). The addition of Tween 80 to the diluent prevents the liquid from 'shrinking back' on the slides and allows it to dry more evenly.
4. Dry the slides, fix with acetone (three times), Gram-stain, and allow to air-dry.
5. Under oil immersion, count the cells seen in 10 fields, five centre fields and five

Protocol 2. *Continued*

perimeter fields. Count at least 100 cells; then calculate the average number of cells per field.

6. Calculate the direct microscopic count, using the formula[a]

Number of bacteria/g (ml) sample =

$$\frac{\text{average number of bacteria}}{\text{field}} \times \frac{\text{number of microscope fields}}{0.5\ \text{cm}^2} \times \text{dilution counted} \times 10^2.$$

7. If the sample is not exactly 0.5 g, a correction factor must be introduced, i.e.

$$\text{Correction factor} = \frac{\text{weight of sample} + \text{weight of broth}}{\text{weight of sample}} \times \frac{1}{10}.$$

This factor is used to correct the figure obtained using the equation in step 6. The total counts are expressed as $\log_{10}$ organisms/g of wet faeces.

[a] Number of microscope fields/0.5 cm^2 = 0.5 cm^2/field area. Determine the field area for the particular objective used. Measure the field diameter with a micrometer slide and calculate field area.

4. Bacterial cultivation

There is no ideal method for the culture of anaerobic bacteria from natural habitats. The method used should be selected to answer specific questions related to the particular ecosystem under study. It is essential that the investigative strategy be formulated to take account of the methods used. Anaerobic culture methods are the subject of Chapter 1 and it is assumed that the reader is familiar with these basic procedures.

Identification procedures are described in Chapter 3. However, it should be noted that, in contrast to clinical isolates, many of the anaerobes isolated in studies of the microflora will be difficult to identify. Reference should be made to the *Colour Atlas of Anaerobic Bacteria* (18) and to *Bergey's Manual of Systematic Bacteriology* (19) although, even using these resources, not all isolates will be assigned to recognized species. If a particular bacterial group is sought reference to *The Prokaryotes* (20) will usually yield any specific techniques that have been tried. The *Anaerobe Laboratory Manual* (21) and *Wadsworth Anaerobic Bacteriology Manual* (22) give details of methods for normal flora studies; they also illustrate the two approaches to the problem that have been adopted.

(a) In clinical practice bacteriologists usually use a range of selective culture media to improve the chance of isolating particular species. In an analogous fashion a battery of media have been employed to study the microflora. This approach has the advantages that a wide range of bacteria can be detected and preliminary identification made on the basis of differential growth on the media used. However, the apparent quantitative relationships between the bacterial groups may reflect the differential toxicity of the media rather than the ratio of organisms in the sample.

(b) Other investigators have advocated the use of non-selective media. Thus the proportions of the bacterial groups present are elucidated by the identification of a large number of randomly picked strains. It may be argued that this method will give a more accurate picture of the quantitative relationships between the bacteria present in the samples.

4.1 Culture media

Media used in studies of the anaerobic microflora are often designed to simulate the habitat in broad terms. Selective agents may be added to help in the isolation of particular types of organism (see *Table 2* in Chapter 2). Other media may be designed to isolate bacteria of a particular metabolic type or from a particular biochemical niche. For this type of study a minimal medium to which the substrates of interest are added may be used.

4.1.1 Habitat-simulating media

Habitat-simulating media are most easily recognized on the basis of the natural fluids or extracts that are major constituents. Rumen fluid has been widely used not only for studies on the rumen but also on the gut microflora of man and other animals.

Protocol 3. Preparation of rumen fluid

1. Take the contents of the rumen of a cow, sheep, or bullock and filter through several layers of cheesecloth.
2. Bottle the filtrate under oxygen-free CO_2; then autoclave at 121°C for 15 min. Store at 4°C until required.
3. Before use, clarify by centrifuging at 10000 *g* for 15 min.

A mixture of volatile fatty acids, as used in minimal media, are sometimes used in place of rumen fluid.

Faecal extracts made from faeces from a variety of species (including man) have been used as the basis of media for gut bacteria. Faecal extracts are made by autoclaving mixtures of equal quantities of faeces and water. The mixture is clarified by centrifugation and supernatant collected. The pH of the supernatant is adjusted to 7.2. Skin lipids extracted by washing the skin with diethylether have been used in some studies of skin bacteria but Tween 80 (1 g/l) can be substituted. Addition of mucus to media used for isolation of bacteria from mucous membrane has also been advocated. Because of the problems in obtaining other material hog gastric mucin (1 g/l) has been used.

A habit-simulating medium for gut flora studies would include rumen fluid or faecal extract, an energy source, a nitrogen source, a buffering system, mineral salts, and reducing agents. A redox potential (Eh) indicator (usually resazurin) is normally included. The amount and type of energy source will differ, depending on the use of the medium; cellulose, cellobiose, maltose, lactose, glucose, glycerol, lactate, and fumarate can be used. Ammonium salts can be used as sources of nitrogen as many gut bacteria can utilize them, but peptones can be used as appropriate. When oxygen-free carbon dioxide is used, as with pre-reduced, anaerobically sterilized (PRAS) media, a CO_2/HCO_3 buffer is best: 0.5% (w/v) $NaHCO_3$ is added to the medium. Cysteine hydrochloride (0.05%) is a useful reducing agent; addition of a second agent such as sodium formaldehyde sulphoxylate (0.03%) may act as an Eh buffer.

Habitat-simulating media may be used as conventional plating media but are more usually used as PRAS roll tubes. Though such media can be made selective by the addition of various agents it is more usual to determine the identity of a number of randomly selected isolates.

4.1.2 Minimal media

Although complex media are usual in microecological studies, many of the bacteria isolated have simple growth requirements and thus the use of minimal media may be of value in metabolic studies. The composition of the medium used will depend on the bacteria sought; thus, among the anaerobic cocci, *Gemmiger*, *Coprococcus*, and *Ruminococcus* require fermentable carbohydrates for growth, while others require amino acids, usually arginine or threonine, for growth.

The media can be prepared in the normal way and used as plates in an anaerobic chamber. Alternatively, they may be prepared by the Hungate technique as PRAS agar or broth (see Chapter 1). In either case the substrate to be studied is added to the medium. Heat-stable substrates may be added before the medium is autoclaved, but more usually a sterile solution is added to the sterile molten agar.

4.2 The culture of a sample using an anaerobic cabinet

In this section the culture of a sample from an ecologically important site is described. It must be remembered that, although anaerobes are a major concern, such samples almost invariably will also contain aerobic and facultatively anaerobic bacteria. The protocol presented is based upon that used for the examination of faeces, but can with a little modification be used for other samples.

Protocol 4. Cultivation of a sample of faeces or intestinal contents using an anaerobic chamber

You will need

- An anaerobic chamber (see Chapter 1)

Protocol 4. *Continued*

- Sterile spreaders (made from bent glass rods)
- Automatic micropipettes (100 and 500 μl) with sterile tips
- Sterile 500-μl tips with their ends removed (wide bore)
- Pasteur pipettes (wide bore) and a 0.5-ml syringe pipette
- Automatic dispensing pipette (4.5 ml) with tips
- Marker pens or grease pencils
- Autoclave or masking tape
- Empty discard jar
- Vortex mixer
- Heating block (for thawing frozen samples)
- Sterile capped tubes
- Sample diluent (brain–heart infusion)
- Fresh palladium catalyst
- Fresh silica gel and atmospheric scrubber
- Anaerobic jars
- Plates of culture media (see *Tables 1* and *2*).

1. The apparatus and reagents just listed should be placed in the chamber at least 3 days before the cultivation of specimens. Anaerobic jars should be evacuated and filled with gas mixture before transferring into the chamber.
2. Plates of culture media may become dessicated under these conditions. An alternative is to pre-reduce plates in anaerobic jars at room temperature and to introduce them into the chamber 18–24 h before culturing.
3. Dispense the diluent in 4.5-ml aliquots.
4. Prepare serial decimal dilutions of the specimen. Use a wide-bore Pasteur pipette for the first dilution; then wide-bore 500 μl tips or an automatic micropipette for the remainder. Mix each dilution well on the vortex mixer.
5. Spread 100 μl aliquots of the appropriate dilutions on to the pre-reduced media (see *Table 2* for dilutions and media). The inoculated plates are stacked together, bound with adhesive tape, and labelled.
6. If a commercial cabinet is used, it may have space to incubate all plates inside the chamber. If, however, a flexible isolator-type chamber is used, it is necessary to incubate the plates in anaerobic jars, outside the chamber. Place the stacks of plates in anaerobic jars, seal, and remove from the chamber via the airlock.
7. After removing from the chamber, evacuate the anaerobic jars, and refill with gas mixture (see Chapter 1). Incubate for 3–4 days at 37°C.

Table 1. Culture media for isolation of anaerobic bacteria from faeces

Brain–heart infusion agar+volatile fatty acids (BHF)

Brain–heart infusion (Oxoid CM225)	37 g
Yeast extract (Oxoid L21)	5 g
Liver digest (Oxoid L27)	5 g
Sodium formaldehyde sulphoxylate	0.3 g
L-cysteine hydrochloride	0.5 g
Purified agar (Oxoid L28)	15 g
Distilled water	883 ml
Haemin solution (0.5 mg/ml)	10 ml
Vitamin K_1 solution (1 mg/ml)	1 ml
Volatile fatty acid (VFA) mixture	12.5 ml
Defibrinated horse blood	100 ml

Dissolve the solid reagents in distilled water by heating, then autoclave at 115°C for 20 min. Cool to 50°C then add the horse blood, haemin, vitamin K_1, and VFA mixture before pouring plates.

Volatile fatty acid mixture

Acetic acid	67 ml
Propionic acid	24 ml
n-Butyric acid	16 ml
n-Valeric acid	4 ml
isobutyric acid	4 ml
isovaleric acid	4 ml
2-methyl butyric acid	4 ml

Kanamycin vancomycin agar (KV)

Brain–heart infusion agar (Oxoid CM375)	23.5 g
Distilled water	450 ml
Kanamycin (10 mg/ml)	5 ml
Vancomycin (1 mg/ml)	3.75 ml
Defibrinated horse blood	50 ml

Aseptically add the kanamycin, vanomycin, and horse blood to the molten agar before pouring plates.

***Bacteroides* agar (BAC)**

Columbia agar (Oxoid CM331)	39 g
Ox bile (Oxoid L50)	10 g
L-cysteine hydrochloride	0.5 g
Sodium formaldehyde sulphoxylate	0.3 g
Distilled water	872 ml
Haemin solution (0.5 mg/ml)	10 ml
Vitamin K_1 solution (1 mg/ml)	1 ml
Kanamycin (10 mg/ml)	5 ml
Vancomycin (1 mg/ml)	3.75 ml
Defibrinated horse blood	100 ml

Dissolve the solid reagents in distilled water by heating, adjust pH to 7.2, and then autoclave at 115°C for 20 min. Cool to 50°C then add the horse blood, haemin, vitamin K_1, kanamycin, and vancomycin before pouring plates.

Table 1. *(Continued)*

***Bifidobacterium* agar (BIF)**

Columbia agar (Oxoid CM331)	39 g
Yeast extract (Oxoid L21)	5 g
Maltose	5 g
Fructose	5 g
L-cysteine hydrochloride	0.5 g
Sodium formaldehyde sulphoxylate	0.3 g
Distilled water	880 ml
Haemin solution (0.5 mg/ml)	10 ml
Vitamin K_1 solution (1 mg/ml)	1 ml
Tomato juice (Oxoid)	100 ml
Kanamycin (10 mg/ml)	5 ml
Nalidixic acid (10 mg/ml)	5 ml

Dissolve the solid reagents in distilled water by heating; then autoclave at 115°C for 20 min. Cool to 50°C then add the tomato juice, haemin, vitamin K_1, kanamycin, and nalidixic acid before pouring plates.

Rifampicin agar (RIF)

Brain–heart infusion agar (Oxoid CM375)	23.5 g
Distilled water	450 ml
Rifampicin (5 mg/ml)	5 ml
Defibrinated horse blood	50 ml

Aseptically add the rifampicin and horse blood to the molten agar before pouring plates.

***Veillonella* agar (V)**

Tryptone (Oxoid L42)	5 g
Yeast extract (Oxoid L21)	3 g
Sodium thioglycollate	0.75 g
Tween 80	1 ml
Sodium lactate (70% solution)	18 ml
Basic fuchsin (1% solution)	0.2 ml
Purified agar (Oxoid L28)	12 g
Distilled water	973 ml
Vancomycin (1 mg/ml)	7.5 ml

Asceptically add the vancomycin to the molten agar before pouring plates.

Willis and Hobbs medium (WH)

Brain–heart infusion (Oxoid CM225)	37 g
Yeast extract (Oxoid L21)	5 g
Liver digest (Oxoid L27)	5 g
Lactose	9.6 g
Neutral red (1% solution)	4 ml
L-cysteine hydrochloride	0.5 g
Sodium formaldehyde sulphoxylate	0.3 g
Purified agar (Oxoid L28)	12 g
Distilled water	886 ml
Haemin solution (0.5 mg/ml)	10 ml
50% fresh egg yolk suspension	100 ml

Asceptically add the haemin and the egg yolk suspension to the molten agar before pouring plates.

Table 1. *(Continued)*

Crystal violet erythromycin agar for fusobacteria (CVE)

Trypticase	10 g
Yeast extract (Oxoid L21)	5 g
Sodium chloride	5 g
Glucose	2 g
Tryptophane	0.2 g
Agar	15 g
Crystal violet	5 mg
Distilled water	950 ml
Defibrinated sheep blood	50 ml
Erythromycin	4 mg

Asceptically add the erythromycin and sheep blood to the molten agar before pouring plates.

Habitat simulatory medium[a]

Natural product	Rumen fluid (30 ml), faecal extract (10 ml), skin lipid (1 g), or mucin (1 g)
Energy sources	0.5–5 g of appropriate sources
Nitrogen source	Peptone (10 g) or ammonium sulphate (0.4 g)
Buffer	Sodium bicarbonate (5 g)
Reducing agents	L-cysteine hydrochloride (0.5 g) or sodium formaldehyde sulphoxylate (0.3 g)
Mineral salts	See minimal media (5 ml)
Distilled water	Make up total volume to 1 litre
Agar	20 g (if required)

Minimal media[a]

Basal agar

$(NH_4)_2SO_2$	0.4 g
$NaHCO_3$	5 g
L-methionine	0.08 g
L-cysteine hydrochloride	0.5 g
Sodium formaldehyde sulphoxylate	0.3 g
Agar	20 g
Distilled water	1 l
Haemin (0.5 mg/ml)	10 ml
Vitamin K_1 (1 mg/ml)	1 ml
Mineral salts solution (list follows)	5 ml
Volatile acid mixture (list follows)	4.5 ml

Mineral salt solution

KH_2PO_4	18 g
NaCl	18 g
$CaCl_3 \cdot 2H_2O$	0.53 g
$MnCl_2 \cdot 4H_2O$	0.2 g
$CoCl_2 \cdot 6H_2O$	0.2 g
Distilled water	1 l

Volatile acid mixture

Glacial acetic acid	36 ml
Isobutyric acid	1.8 ml
N-valeric acid	2 ml
D-L-2 methyl butyric acid	2 ml
Isovaleric acid	2 ml
Distilled water to total volume of	100 ml

Test substrate

[a] These media can be made up as conventional media but are more usually used as PRAS media.

Table 2. Culture media and incubation conditions for isolation of anaerobic bacteria from faeces

Medium[a]	**Dilutions cultured**	**Bacteria counted**	**Incubation**[b]
BHF	10^{-2}, 10^{-4}–10^{-8}	Total obligate and facultative anaerobes	3 days at 37°C
KV	10^{-2}, 10^{-4}–10^{-7}	*Bacteroides* spp.	
BAC	10^{-2}, 10^{-4}–10^{-7}	*B. fragilis* group	
BIF	10^{-2}, 10^{-4}–10^{-7}	*Bifidobacterium* spp.	
RIF	10^{-1}–10^{-7}	*Fusobacterium* spp., *Clostridium ramosum, Eubacterium* spp.	4 days at 37°C
V	10^{-1}–10^{-5}	*Veillonella* spp.	
WH[c]	10^{-1}–10^{-5}	Sporing clostridia	
CD[d]	10^{-1}–10^{-5}	*C. difficile*	2 days at 37°C

[a] See *Table 1* for key to media.
[b] Incubate anaerobically plus 10% CO_2.
[c] Alcohol shock for 1 h before plating (see *Protocol 5*).
[d] *C. difficile* agar; for preparation see *Protocol 5* in Chapter 2.

Protocol 5. Isolation of sporing clostridia

1. Prepare sterile absolute alcohol by Tyndallization.
2. Prepare a 50% (v/v) suspension of the 10^{-1} dilution (from step 4 in *Protocol 4*) in absolute alcohol. Incubate at room temperature in the chamber for 60 min.
3. Prepare dilutions of this mixture to 10^{-5} and plate 100 μl of each dilution on to Willis and Hobbs plates. Allow to dry; then incubate as described in *Protocol 4* for 3 days.

Protocol 6. Identification of non-sporing anaerobes[a]

1. Count the number of colonies of each colony type growing on each of the media used.
2. Pick colonies from the final dilution(s) showing growth[b] on to blood agar plates for incubation aerobically and anaerobically. After incubation for 3 days, strains which grow aerobically are discarded.
3. Pure cultures, as judged by colonial morphology and Gram-staining, are inoculated into broth medium for gas–liquid chromatography (GLC; see Chapter 3), and incubated anaerobically for 3 days.
4. After incubation the broth is subcultured to a blood agar plate for aerobic incubation; aerobes are again discarded. The broth culture is analysed by GLC (see Chapter 3).

Protocol 6. *Continued*

5. The anaerobic bacteria are assigned to genera on the basis of the colonial appearance on the isolation media and the results of this examination (see *Table 3* and *Table 6* in Chapter 3). The counts of anaerobic bacteria are then based upon the identity and proportion of the colonial types. Isolated strains are kept for further study.

[a] The procedure outlined is a minimalist approach to identification.
[b] Colony types not represented at the highest dilutions should be sampled from lower dilutions. Thus all colony types are described, subcultured, and purified. The number of colonies of each type to pick is determined by experience, personal preference, and resources.

Protocol 7. Calculation of viable counts

1. All the isolated organisms are placed into an appropriate group or genus (see *Protocol 6*).
2. The viable counts are determined using the formula

   Viable count (c.f.u./g wet faeces)

   $$= (\text{number of colonies counted on plate}) \times (\text{dilution}^{a}\ \text{counted}) \times 10 \times (\text{correction factor})$$

   where c.f.u. are colony-forming units.
3. The correction factor is calculated by simple proportion on the basis of the weight of the sample (see step 7 in *Protocol 2*). Thus, if the first dilution is an exact 1/10 dilution (0.5 g faeces to 4.5 ml diluent), the correction factor is 1.
4. All counts are then converted to $\log_{10}$ c.f.u./g.

[a] The dilution occurring when 100 μl is spread on the plate.

5. Bacterial metabolism

Many studies on the bacterial microflora have focused on the metabolic potential of the bacteria and their role in the metabolism of drugs, food additives, and other xenobiotics (23, 24). Similar considerations have led to the introduction of the glycocholate breath test, for measurement of bile acid breakdown, and the study of the production of hydrogen and methane. An important stimulus to normal flora studies was the suggestion that the metabolic impacts of the microflora might include the causation of cancer. These and other matters have been extensively reviewed (12, 13, 25, 26).

In practical terms attention has focused on two aspects of the problem, the investigation of the bacterial metabolism of xenobiotics and the investigation of metabolism as an indicator of changes in the microflora.

Table 3. Initial identification of some obligate anaerobes isolated from faeces

Organism	Initial isolation medium[a] and colonial appearance	Gram stain
Anaerobic cocci	BHF: 1 mm, cream/clear	Gram-positive cocci in chains and clumps
Bacteroides spp.	BHF: 2–3 mm, grey, semimucoid KV: 2–5 mm, grey/green/fawn, semimucoid	Gram-negative cocco-bacilli; long rods; often pleiomorphic Gram-negative, straight/curved rods
B. fragilis group	BAC: 3 mm, flat, brown, dull. Some haemolytic strains	
Bifidobacterium spp.	BHF: 2–3 mm, grey, shiny BIF: 2 mm, white, shiny	Gram-positive rods, 2 μm long × 0.25 μm wide Usually straight rather than curved; or branched with bulbous ends
	BIF: pinpoint, cream, transparent	
Clostridium spp.	BHF: variety of colonial forms, 2–10 mm, yellow/cream/grey, some strains haemolytic WH: variety of colonial forms, 2 mm–>15 mm, cream/pink, lecithinase positive or negative CD: 2–4 mm, flat, grey-green, dull; phenolic odour[b] RIF: 2–4 mm, grey/pink, shiny[c]	Gram-positive rods, rapidly becoming Gram-negative; variable morphologies; with or without detectable spores
Eubacterium spp.	RIF: 0.5–1.5 mm, cream, shiny, usually non-haemolytic	Gram-positive but irregularly staining; short rods, sometimes cocco-bacilli; straight to curved
Fusobacterium spp.	BHF: 1–2 mm, water-clear RIF: 2–3 mm, 'fried-egg', and tan, translucent colonies	Gram-negative, long thin rods with pointed ends and spherical forms
Veillonella spp.	BHF: 2 mm, dull, grey, brittle V: <3 mm, pale/olive green, shiny	Gram-negative cocci, 0.25 μm in diameter; irregular masses

[a] CD, *C. difficile* agar; other media abbreviations are found in *Table 1*.
[b] *C. difficile*.
[c] *C. ramosum*.

5.1 The metabolism of xenobiotics

Direct evidence that any particular biotransformation is mediated by a single species of bacteria present among the gut flora is difficult to obtain. This problem is well illustrated by consideration of digoxin inactivation. Digoxin is reduced by the gut flora of some people. *Eubacterium lentum* is the organism known to perform these reactions but neither the presence of these bacteria nor their concentration in the gut relates simply to digoxin reduction. The extent of possible difficulties can be appreciated when it is realized that the number of *E. lentum* in faeces ranges from $\log_{10}$ 3.6 to $\log_{10}$ 11.7 c.f.u./g. Thus, for some individuals, examination of 10 colonies of non-sporing anaerobes would ensure detection, while in other as many as 10^8 colonies would have to be studied. For this reason studies of the metabolism of xenobiotics often rely on the short-term incubation of the compound under study with a suspension of faecal or caecal bacteria, or the use of animal models (see *Table 4*). In such systems anaerobic metabolism can be selected for by treatment with aminoglycoside antibiotics (such as kanamycin). Metronidazole can be used to select for aerobic metabolism.

However, such mixed incubations do not enable the isolation of the particular bacteria able to metabolize specific substrates. In some instances it may be possible to use the substrate in a minimal medium to select for the bacteria able to undertake its metabolism. PRAS media for this purpose are prepared by following *Protocol 3* in Chapter 1, using the recipe shown in *Table 1*. PRAS media are most often prepared as habitat-simulating media; minimal media are more difficult to prepare in this way and tend to become oxygenated.

Table 4. Methods for the study of xenobiotic metabolism by the bacterial flora

Method of study	Example: reaction studied
Metabolism by isolated strains	Digoxin reduction by *Eubacterium lentum*
Metabolism in faecal or caecal suspensions	Tartrazine breakdown
Whole-animal studies and antibiotic treatment	Steroid hormones metabolism
Germ-free animal and conventional animal comparison	Nitrazepam breakdown
Ex-germ free animal with ex-human bacteria	Parachlor metabolism

Protocol 8. Culture of a sample in PRAS minimal medium

1. Melt the tube of PRAS minimal agar, taking care that the stopper does not come off.
2. Remove the stopper and insert a sterile gas cannula emitting a gentle flow of oxygen-free CO_2 into the tube.
3. Add the required amount of test substrate[a] and a dilution of the test sample.[b]

Protocol 8. *Continued*

4. Replace the stopper while removing the cannula taking care to prevent the entry of air.
5. Mix well, then prepare a roll tube (see *Protocol* 7 in Chapter 1).
6. Incubate the tube at 37°C for 3–5 days.
7. After incubation count the colonies, pick for pure cultures, and confirm their ability to metabolize the substrate.

[a] Test substrate should be deoxygenated by preparing as a sterile solution in deoxygenated minimal broth.
[b] Dilutions of sample should be prepared in deoxygenated diluents.

5.2 Metabolic indicators of the bacterial flora

The number of species of bacteria isolated is related to the number of bacterial colonies that are isolated and screened. It is seldom possible to screen more than 100 colonies from any sample; even if time and money are not limited, 10 colonies is a more reasonable estimate for most investigators. When it is realized that 300–400 species are normally present, the problem of detecting changes in the gut flora can be appreciated. It is likely that similar considerations apply to the microfloras of other habitats. To overcome this difficulty investigators have concentrated on the study of the metabolism of administered xenobiotics (e.g. cyclamate), or studied levels of bacterial enzymes (see Chapter 7). The API ZYM system can be used to develop enzyme profiles of the bacterial flora present in a sample (27). The enzyme glucuronidase has been used to monitor changes in the flora induced by diet (28).

Protocol 9. Measurement of faecal glucuronidase

1. Prepare a 1/10 dilution of a well-mixed sample of stool in PRAS diluent.
2. Centrifuge at 500 *g* for 30 min at 4°C to remove coarse particles. Discard the deposit.
3. Centrifuge the supernatant fraction (containing bacterial cells) at 15 000 *g* for 30 min at 4°C.
4. Wash the pellet three times with 10 ml sterile normal saline, by resuspension and centrifugation. The washings are pooled and retained.
5. Resuspend the final sediment in 10 ml cold phosphate buffer (0.025 M) and ultrasonicate for 15 min at 4°C.
6. Assay β-glucuronidase in both the washings and the ultrasonicated bacterial pellet. The incubation mixture contains
   - 0.01 M phenolphthalein glucuronide, pH 5.5 0.1 ml

Protocol 9. *Continued*

- 0.01 M acetate buffer, pH 5.8 — 0.8 ml
- Sample (washings or pellet) — 0.1 ml

Set up a control without the glucuronide. Incubate both test and control at 37°C for 5 h.

7. Terminate the reaction by adding

- 0.1 M glycine solution — 2.5 ml
- 5% trichloracetic acid — 1 ml

Make up the volume to 6 ml by adding 1.5 ml distilled water. Check that the reaction mixture is alkaline.

8. Measure free phenolphthalein at 540 nm.

References

1. Marsh, P. and Martin, M. (1984). *Oral Microbiology* (2nd edn). Van Nostrand Reinhold, Wokingham.
2. Noble, W. C. (1981). *Microbiology of Human Skin* (2nd edn). Edward Arnold, London.
3. Ison, C. A. (1989). In *Human Microbial Ecology* (ed. M. J. Hill and P. D. Marsh), pp. 111–13. CRC Press, Boca Raton, FL.
4. Drasar, B. S. and Hill, M. J. (1974). *Human Intestinal Flora*. Academic Press, London.
5. Clarke, R. T. J. and Bauchoup, T. (ed.) (1977). *Microbial Ecology of the Gut*. Academic Press, London.
6. Hentges, D. J. (ed.) (1983). *Human Intestinal Microflora in Health and Disease*. Academic Press, NY.
7. Rosebury, T. (1962). *Microorganisms Indigenous to Man*. McGraw-Hill Book Co., NY.
8. Skinner, F. A. and Carr, J. G. (ed.) (1974). *The Normal Microbial Flora of Man*. Academic Press, London.
9. Hill, M. J. and Marsh, P. D. (ed.) (1989). *Human Microbial Ecology*. CRC Press, Boca Raton, FL.
10. Hungate, R. E. (1966). *The Rumen and its Microbes*. Academic Press, NY.
11. Barnes, E. M. and Mead, G. C. (ed.) (1986). *Anaerobic Bacteria in Habitats Other than Man*. Blackwell Scientific Publications, Oxford.
12. Hill. M. J. (ed.) (1986). *Microbial Metabolism in the Digestive Tract*. CRC Press, Boca Raton, FL.
13. Rowland, I. R. (ed.) (1988). *Role of the Gut Flora in Toxicity and Cancer*. Academic Press, London.
14. Theilade, E. (1989). In *Human Microbial Ecology* (ed. M. J. Hill and P. D. Marsh), pp. 2–56. CRC Press, Boca Raton, FL.
15. Hill, M. J. (1989). In *Human Microbial Ecology* (ed. M. J. Hill and P. D. Marsh), pp. 57–85. CRC Press, Boca Raton, FL.

16. Noble, W. C. (1989). In *Human Microbial Ecology* (ed. M. J. Hill and P. D. Marsh), pp. 131–53. CRC Press, Boca Raton, FL.
17. Crowther, J. S. (1971). *J. Appl. Bacteriol.* **34**, 477.
18. Mitsuoka, T. (1980). *A Colour Atlas of Anaerobic Bacteria*. Sobunsha, Tokyo.
19. Kreig, N. R. and Holt, J. G. (ed.) (1984–90). *Bergey's Manual of Systematic Bacteriology*, 4 volumes. Williams and Wilkins, Baltimore, MD.
20. Starr, M. P., Stolp, H., Trüper, H. G., Balows, A., and Schlegel, H. G. (ed.) (1981). *The Prokaryotes, a Handbook on Habitats, Isolation and Identification of Bacteria*, 2 volumes. Springer-Verlag, Berlin.
21. Holdeman, L. V., Cato, E. P., and Moore, W. E. C. (1977). *Anaerobe Laboratory Manual* (4th edn). Virginia Polytechnic Institute and State University, Blacksburg, VA.
22. Sutter, V. L., Citron, D. M., Edelstein, M. A. C., and Finegold, S. M. (1985). *Wadsworth Anaerobic Bacteriology Manual* (4th edn). Star Publishing Co., Belmont, CA.
23. Scheline, R. R. (1968). *J. Pharm. Sci.* **57**, 2021.
24. Goldman, P. (1973). *New Engl. J. Med.* **289**, 623.
25. Banbury Report (1981). *Gastrointestinal Cancer: Endogenous Factors*. Cold Spring Harbor Laboratory, Cold Spring Harbor, NY.
26. Hill, M. J. (1986). *Microbes and Human Carcinogenesis*. Edward Arnold, London.
27. Drasar, B. S., Montgomery, F., and Tomkins, A. M. (1986). *J. Hyg.* **96**, 59.
28. Reddy, B. S., Weisburger, J. S., and Wynder, E. L. (1974). *Science* **183**, 416.

11

Sulphate-reducing bacteria

GEORGE T. MACFARLANE and GLENN R. GIBSON

1. Introduction

Dissimilatory sulphate reduction is carried out by a diverse group of strictly anaerobic bacteria that share the ability to use sulphate as a terminal electron acceptor in the oxidation of organic matter. In these reactions, sulphate is stoichiometrically reduced to sulphide (1) according to the equation

$$2CH_2O + SO_4^{2-} \rightarrow 2HCO_3^- + H_2S.$$

Sulphate-reducing bacteria (SRB) are particularly active in ecosystems that are high in sulphate, but they can be isolated from most aquatic and terrestrial environments that are depleted in oxygen, such as sediments, sewage sludge digestors, water-logged soils, and the gastrointestinal tracts of man and animals. In the biosphere, they play a significant role in the mineralization reactions that enable cycling of nitrogen, carbon, and phosphorus to occur. In some anaerobic environments, they are the principal oxidizers of organic carbon. SRB are particularly important in the sulphur cycle where the production of H_2S facilitates the growth of other groups of chemotrophic and phototrophic bacteria that participate in sulphur-cycling processes.

Largely through their destructive potential, for example, the corrosion of pipes and pumps and the spoiling of coal, oil, and gas, the activities of SRB are recognized as being of considerable economic as well as scientific importance and this has provided a major stimulus to research on these bacteria in recent years. The purpose of this chapter is to outline the methods used for the cultivation of SRB and the study of their activities in laboratory and natural systems.

2. Growth media

2.1 General description

SRB grow best under slightly alkaline conditions over a relatively restricted pH range (pH 7.2–7.8). The media used contain an energy source or electron donor, which can be either a reduced carbon compound or molecular H_2. The electron donor and carbon source are usually provided by the same substrate, but some species can grow mixotrophically, in which case the electron donor is provided by

H_2, or a carbon source other than that used for incorporation into cellular material. Examples are the growth of *Desulfovibrio vulgaris* on H_2/CO_2 and acetate (2) and *Desulfotomaculum acetoxidans* on formate and acetate (3) where acetate is used for the synthesis of cell carbon. Sulphate is included as the terminal electron acceptor, but thiosulphate, sulphite, tetrathionate, and elemental sulphur also serve this function in some species (4, 5). NaCl is obligately required by some marine isolates (6) and iron is included in many growth media as a nutrient and as an indicator of sulphide formation, during which a black precipitate is formed with ferrous ions. SRB need anaerobic and reducing conditions for growth and the initial redox potential (Eh) of growth media should be in the region of −100 mV. Thus, reducing agents such as ascorbate, thioglycollate, dithionite, or sodium sulphide are added to media to exclude O_2 and maintain conditions of low Eh. Cysteine is avoided, since precipitation of ferrous sulphide may occur due to H_2S formation from the amino acid, rather than dissimilatory sulphate reduction. Although many SRB are nutritionally undemanding and grow on simple mineral-salt-based media, yeast extract is stimulatory to some species and a requirement for others (7). Traditionally, SRB growth media contained lactate and, to a lesser degree, pyruvate or malate as the electron donor (8). As a result, the limited number of types of SRB isolated led to the belief that dissimilatory sulphate reduction was restricted to a few species (7). However, over the last 10–15 years, many new SRB have been discovered that utilize a wide range of reduced carbon sources as electron donors. A variety of basically similar media are available for their enrichment, enumeration, and isolation, but the most widely used are those of Postgate (7) and Pfennig *et al.* (9) which are described here.

2.2 Media for the growth of desulfovibrios and desulfotomacula

A range of media has been developed by Postgate (7) for the culture of these SRB (see *Table 1*). For growth of marine strains, 20 g/l NaCl and 3 g/l $MgCl_2$ are added or, alternatively, sea water is used instead of tap water. The media are always acidic and the pH must be adjusted with 2 M NaOH before autoclaving. If the media are to be used at once, all ingredients can be added and autoclaved together, but if the constituents are retained as stock solutions, the ascorbate and thioglycollate should be added immediately before autoclaving. Some workers separately sterilize the thioglycollate by filtration (10). Medium B is autoclaved (121°C, 15 min) in screw-capped bottles. The bottles are completely filled to exclude air and aseptically topped up with medium as required, after autoclaving. Medium C can be made in 10- or 20-l batches. The medium contains no reducing agents and large volumes are autoclaved (121°C, 60 min), then cooled, and maintained under a suitable anaerobic gas (e.g. 80% N_2, 20% CO_2). See *Protocol 2* for a description of enumeration methods using Medium E. Postgate's media are generally suitable for the culture of desulfovibrios and desulfotomacula

Table 1. Media for the growth, enrichment, enumeration, and isolation of sulphate-reducing bacteria

	Postgate's medium[a]		
	B	**C**	**E**
KH_2PO_4	0.5	0.5	0.5
$CaSO_4$	1.0	—	—
Na_2SO_4	—	4.5	1.0
NH_4Cl	1.0	1.0	1.0
$MgSO_4 \cdot 7H_2O$	2.0	0.06	2.0
$CaCl_2 \cdot 6H_2O$	—	0.06	1.0
Sodium lactate	3.5	6.0	3.5
Trisodium citrate	—	0.3	—
Yeast extract	1.0	1.0	1.0
$FeSO_4 \cdot 7H_2O$	0.5	0.004	0.5
Ascorbate	0.1	—	0.1
Thioglycollate	0.1	—	0.1
Purified agar	—	—	15.0
Tap water	1000 ml	—	1000 ml
Distilled water	—	1000 ml	—
pH	7.0–7.5	7.5	7.6
Notes	Contains a precipitate	Clear medium	—
Use	Batch culture, enrichment, and growth of pure cultures	Chemostat culture	Enumeration

[a] Constituents are given in g/l.

but, partly due to the presence of thioglycollate, they do not support good growth of other types of SRB, for which the medium of Pfennig *et al.* (9) is preferred.

2.3 Media for the growth of other SRB

The medium of Pfennig *et al.* (9) is more complex and time-consuming to prepare than those shown in *Table 1*. However, depending on the type of electron donor used (see *Table 3*), the complete medium facilitates growth of all SRB currently recognized. The medium is made by sterilizing five separate stock solutions (solutions 1 to 5) and aseptically combining them under anaerobic conditions to produce the basal medium. Various additions, e.g. electron donors or growth factors, are added later.

Protocol 1. Preparation of the medium of Pfennig *et al.* (9)

You will need

- A suitable medium reservoir (e.g. 1–5 l) fitted with a tap and facilities for aseptic addition of medium components and for flushing the vessel with the anaerobic gas mixture
- A Teflon-coated magnetic stirring bar

Protocol 1. *Continued*

- A shrouded glass connector
- 0.2-μm bacteriological filters (e.g. from Gallenkamp)
- Screw-capped bottles with lids fitted with butyl rubber seals
- 80% N_2, 20% CO_2 anaerobic gas mixture
- 100% oxygen-free nitrogen gas (OFN)
- 100% carbon dioxide gas
- A magnetic stirrer
- Filtration apparatus
- 1 M HCl in distilled water
- 2 M Na_2CO_3 in distilled water

1. The individual components of the medium are shown in *Table 2* and the assembled apparatus is displayed in *Figure 1*.
2. It is convenient to prepare solution 1 (the mineral salts base) in batches of 1–5 l. Autoclave at 121°C for 30–60 min, depending on the quantity made. After autoclaving, stir while cooling under an atmosphere of 80% N_2/20% CO_2.
3. Sterilize solutions 2 and 3 separately by autoclaving for 20 min at 121°C.
4. Sterilize solution 4 by filtration after saturating with CO_2.
5. Autoclave solution 5 (121°C, 20 min) in screw-capped bottles with an OFN atmosphere.
6. Mix the solutions in the following proportions
   - Solution 1 970 ml
   - Solution 2 1 ml
   - Solution 3 1 ml
   - Solution 4 30 ml
   - Solution 5 3 ml
7. Adjust the pH to 7.2, using either 1 M HCl or 2 M Na_2CO_3, then dispense the basal medium aseptically to completely fill OFN-containing bottles. The medium can be stored in this way for many weeks.
8. Solution 6 contains electron donors recommended by Pfennig *et al.* (9). The composition of this solution will vary, depending upon the species studied (see *Table 3*). Add the components (see *Table 2*) one at a time to 100 ml of distilled water. Adjust to pH 9 with NaOH and autoclave in screw-capped bottles (20 min, 121°C). Add 10 ml of the appropriate electron donor solution to 1 l basal medium.
9. The vitamin solution (solution 7) may contain any of the vitamins shown in *Table 2*, depending upon the requirements of the strain under study (see

Protocol 1. *Continued*

Table 3). Sterilize solution 7 by filtration and add 1 ml to 1 l of the basal medium.

10. Prepare solution 8 by dissolving the components in 100 ml distilled water. Autoclave (121°C, 20 min) and add 1 ml to 1 l of the basal medium.
11. Prepare solution 9 just before use. Deoxygenate the distilled water before adding the sodium dithionite. Filter-sterilize the solution into screw-capped bottles and maintain under an OFN headspace. Add 1 ml to 1 l of the basal medium.

Table 2. Constituents of the medium of Pfennig *et al.* (9)

Solution 1		Solution 2		Solution 3	
Na_2SO_4	3.0 g	$FeCl_2 \cdot 4H_2O$	1.5 g	Na_2SeO_3	3 mg
NaCl	1.2 g[a]	H_3BO_3	60 mg	NaOH	0.5 g
KCl	0.3 g	HCl (25%)	6.5 ml	Distilled water	1000 ml
$MgCl_2 \cdot 6H_2O$	0.4 g[b]	$CoCl_2 \cdot 6H_2O$	120 mg		
NH_4Cl	0.3 g	$MnCl_2 \cdot 4H_2O$	100 mg		
KH_2PO_4	0.2 g	$NaMoO_4 \cdot 2H_2O$	25 mg		
$CaCl_2 \cdot 2H_2O$	0.15 g	$NiCl_2 \cdot 6H_2O$	25 mg		
Distilled water	970 ml	$ZnCl_2$	70 mg		
		$CuCl_2 \cdot 2H_2O$	15 mg		
		Distilled water	993 ml		

Solution 4		Solution 5		Solution 6	
$NaHCO_3$	8.5 g	$Na_2S \cdot 9H_2O$	12 g	Sodium acetate	20 g
Distilled water	100 ml	Distilled water	100 ml	Propionic acid	7 g
				n-Butyric acid	8 g
				n-Palmitic acid	5 g
				Benzoic acid	5 g
				Distilled water	100 ml

Solution 7		Solution 8		Solution 9	
Biotin	1 mg	Isobutyric acid	0.5 g	Sodium dithionite	3 g
p-Aminobenzoic acid	5 mg	Valeric acid	0.5 g	Distilled water	100 ml
Vitamin B_{12}	5 mg	2-Methylbutyric acid	0.5 g		
Thiamine	10 mg	3-Methylbutyric acid	0.5 g		
Distilled water	100 ml	Succinic acid	0.6 g		
		Caproic acid	0.2 g		
		Distilled water	100 ml		

[a] 20 g for marine strains.
[b] 3 g for marine strains.

Table 3. Distinguishing characteristics of sulphate-reducing bacteria

Bacterium	Habitat	Morphology and size (μm)	G:C ratio (mol%)	Spores	Motility	Preferred growth medium	Electron donors[a]	Other additions (g/l)
Desulfovibrio								
desulfuricans	Soils, aquatic muds	Vibrio 0.5–1 × 3–5	55 ± 1	–	+	Postgate B, C, E	L, Py, F, C, M, PA, Py–S, C–S	None
vulgaris	Soils, aquatic muds	Vibrio 0.5–1 × 3–5	61 ± 1	–	+	Postgate B, C, E	L, Py, F, PA, Ha	None
gigas	Estuarine muds	Spiral 1.2–1.5 × 5–10	60.2	–	+	Postgate B, C, E	L, Py, M–S, PA	None
africanus	Salt and fresh water	Vibrio 0.5 × 5–10	61 ± 1	–	+	Postgate B, C, E	L, Py, M, PA	None
salexigens	Seawater, marine muds	Vibrio 0.5–1 × 3–5	46 ± 1	–	+	Postgate B, C, E	L, Py, M	25 g NaCl
baculatus	Manganese ores	Rod 0.6 × 1.3	56.8	–	+	Postgate B, C, E	L, Py, M, F, Ha	None
sapovorans	Freshwater muds	Vibrio 1.5 × 3.5–5	52.7	–	+	Postgate B, C, E	L, Py, B, MB, HFA	0.26 g palmitate + 1.2 g NaCl + 0.4 g $MgCl_2$
baarsii	Freshwater muds	Vibrio 0.5–1 × 2–4	65.9	–	+	Postgate B, C, E	F, A, P, B, MB, HFA	2.8 g stearic acid + 7 g NaCl + 1 g $MgCl_2$
thermophilus	Stratal vents	Rod 0.5 × 2	ND	–	+	Postgate B, C, E	L, Py	None
sulfodismutans	Aquatic muds	Rod 0.8 × 2.5–3.5	64 ± 1	–	+	Pfennig	L, Et, Pr, B	1 g NaCl + 0.4 g $MgCl_2$ + biotin + pantothenate
carbinolicus	Purification plant	Vibrio 0.6–1.1 × 1.5–5	65	–	+	Pfennig	L, Py, S, Fa, O, F, Ha, PA, G, PD	Vitamins (Pfennig)
simplex	Sour whey digester	Vibrio 0.5–1 × 1.5–3	47.5	–	+	Pfennig	L, Py, Et, Fu, PA, Ha	1 g yeast extract
furfuralis	Pulp and paper waste	Vibrio 0.3–1.2 × 0.8–3	61 ± 0.5	–	+	Pfennig	L, Py, Et, Fu, BD, FF, FFA, FuA, HB	Vitamins (Pfennig)
Desulfotomaculum								
orientis	Soils	Vibrio 1.5 × 5	41.7	+	+	Postgate B, C, E	L, Py, PA, H, F, CO, 345–T, TS	None
ruminis	Sheep rumen	Rod 0.5 × 3–6	45.5	+	+	Postgate B, C, E	L, Py, Py–S, Fa, Ha, Al, PA	None
nigrificans	Soils, spoiled food	Rod 0.3–0.5 × 3–6	44.5	+	+	Postgate B, C, E	L, Py, Py–S, Al, Ha, CO, Fa, PA, Fr	None

antarcticum	Antarctica muds	Rod 1–1.2 × 4–6	ND	+	+	Postgate B, C, E	L, Gl, Py	1 g yeast extract
acetoxidans	Animal faeces, muds	Rod 1–1.5 × 3.5–9	37.5	+	+	Pfennig	A, B, Et, B, V, Fa	Biotin
sapomandens	Soils	Rod 1.2–2 × 5–7	48	+	+	Postgate B	A, Py, B, HFA, PS, HA, Et, F	Vitamins (Pfennig)
kuznetsovii	Mineral waters	Rod 1–1.4 × 3.5–5	49	+	+	Pfennig	F, A, P, B, PA, L, Py, H, V, Cap, Pal, M, S, Fu	None
thermoacetoxidans	Thermophilic bioreactor	Rod 0.7 × 2–5	49.7	+	+	Pfennig	L, Py, P, F, M, A, B, S, V, Pr, H, PA, Al	55–60°C incubation
guttoideum	Muds	Rod 1–1.5 × 3–6	52	+	–	Postgate B, C, E	L, H, Py	None
Desulfomonas								
pigra	Human faeces	Rod 0.8–1 × 2.5–10	66 ± 1	–	–	Postgate B, C, E	Py, Et, L	None
Desulfobacter								
postgatei	Brackish and marine muds	Coccobacillus 1–2 × 1.7–3.5	45.9	–	–	Pfennig	A, Et, L	20 g NaCl[b] + 3 g $MgCl_2$[b] + *p*-aminobenzoate + biotin
hydrogenophilus	Marine muds	Coccobacillus 1–1.3 × 2–3	44.6	–	–	Pfennig	A, Et, H, Pty	20 g NaCl[b] + 3 g $MgCl_2$[b] + *p*-aminobenzoate + biotin
curvatus	Aquatic muds	Vibrio 0.5–1 × 1.7–3.5	46.1	–	+	Pfennig	A, Et, H, Pty	7 g NaCl[b] + 1.3 g $MgCl_2$[b] + biotin
latus	Marine muds	Coccobacillus 1.6–2.4 × 4–7	43.8	–	–	Pfennig	A	20 g NaCl + 3 g $MgCl_2$ + thiamine + biotin
Desulfobulbus								
propionicus	Aquatic muds	Lemon-shaped 1–1.3 × 1.5–2	59.9	–	–	Pfennig	P, L, Py, Et, Pr, Ha, Py–S, L–S, PA	20 g NaCl[b] + 3 g $MgCl_2$[b] + *p*-aminobenzoate
elongatus	Anaerobic digester	Curved rod 0.7 × 1.5–2.5	59	–	–	Pfennig	P, L, Py, Et, Pr, Ha, Py–S, L–S	*p*-aminobenzoate
Desulfococcus								
multivorans	Aquatic muds	Cocci 1.5–2.2	57.4	–	–	Pfennig	F, A, P, B, HFA, L, Py, PA, Benz, PS	20 g NaCl + 3 g $MgCl_2$ + vitamins (Pfennig)
niacini	Marine muds	Coccobacillus 1.5 × 3	45.8	–	+	Pfennig	Et, HA, H, F, B, HFA, Py, M, Fu, S, Ni	12 g NaCl + 2 g $MgCl_2$ thiamine + biotin

Table 3. *(Continued)*

Bacterium	Habitat	Morphology and size (μm)	G:C ratio (mol%)	Spores	Motility	Preferred growth medium	Electron donors[a]	Other additions (g/l)
Desulfosarcina								
variabilis	Brackish and marine muds	Random packets 1–1.5 × 1.5–2.5	51.2	–	–	Pfennig	F, A, P, B, HFA, PA, Benz, H, L–S, Py–S, V, MB, PS	13 g NaCl + 2 g $MgCl_2$
Desulfonema								
magnum	Aquatic muds	Filamentous 6–8 (diam.)	41.6	–	+	Pfennig	A, P, B, HFA, PA, Benz, PS, Hi	20 g NaCl + 5 g $MgCl_2$ + 1 g $CaCl_2$ + vit B_{12} + *p*-aminobenzoate
limicola	Aquatic muds	Filamentous 3 (diam.)	34.5	–	+	Pfennig	F, A, P, HFA, H, L, Py, S, Fu	12 g NaCl + 2 g $MgCl_2$ + biotin
Desulfobacterium								
phenolicum	Marine muds	Vibrio 1–1.5 × 2–3	40.6	–	+	Pfennig	AC, F, A, B, PA, Py, M, Fu, S, CY, Benz, GT	20 g NaCl + 3 g $MgCl_2$
indolicum	Marine muds	Coccobacillus 0.7–1.5 × 2–2.5	47.4	–	+	Pfennig	I, F, A, P, Py, M, PA, Fu, MA, S, AN, Q	20 g NaCl + 3 g $MgCl_2$ + Vit B_{12}
autotrophicum	Marine muds	Coccobacillus 1–1.5 × 1.5–2.5	47.6	–	+	Pfennig	H, F, B, HFA, Et, HA, S, M–S, Fu–S, L–S, M, Fu, Py, Py–S, L	20 g NaCl + 3 g $MgCl_2$ + *p*-aminobenzoate + biotin
macestii	Sulphide spring	Coccobacillus 0.7 × 2	58	–	+	Pfennig	L, Py, F, Et, H	13 g NaCl + 2 g $MgCl_2$ + vitamins (Pfennig)
vacuolatum	Marine muds	Coccobacillus 1–1.5 × 2–3	47	–	–	Pfennig	L, H, F, Py	None

catecholicum	Sludge	Coccobacillus 1–1.5 × 2–3	52	–	–	Pfennig	Cat, F, L	None
Desulfomicrobium								
apsheronum	Oil deposits	Rod 0.7–0.9 × 1.4–3	52	–	+	Postgate B	L, Py, M, F, H, Py–S, Fu–S	None
Desulfomonile								
tiedje	Sewage sludge	Rod 0.8–1 × 5–10	49	–	–	Pfennig	Ni, Py, H, Benz, F, Van Isovan, 3-A, 4-A	None
Thermodesulfobacterium								
commune	Volcanic thermal vents	Rod 0.3 × 0.9	34.4	–	–	Postgate B	L, Py, H, Py–S	70°C incubation
Archaeoglobus								
fulgidis	Hydrothermal vents	Cocci 1–1.5	45 ± 1	–	+	Postgate B	L, Py, H, F, Fu	75°C incubation
profundus	Deep sea hydrothemal vents	Cocci 1.3 (diam.)	41	–	–	Pfennig	A, L, Py	80°C incubation

[a] L, lactate; Py, pyruvate; F, formate; Et, ethanol; Pr, propanol; M, malate; H, H_2/CO_2; PA, primary alcohols; O, oxaloacetate; PD, 1,3-propanediol; G, glycerol; Fu, fumarate; BD, 1,4-butanediol; FF, furfural; FFA, furfurylalcohol; FuA, 2 furoic acid; HB, 4-hydroxybutyrate; A, acetate; V, valerate; Gl, glucose; HA, higher alcohols; PS, phenolic acids; Cap, caproate; Pal, palmitate; Fa, formate + acetate; Fr, fructose; Benz, benzoate; Ni, nicotinate; Hi, hippurate; AC, aromatic compounds (phenol, *p*-cresol, 4-hydroxybenzoate, 2-hydroxybenzoate, benzoate, indole, 2-aminobenzoate, phenylalanine); CY, cyclohexane–carboxylate; MA, maleinate; L–S, lactate minus sulphate; Py–S, pyruvate minus sulphate; M–S, malate minus sulphate; C–S, choline minus sulphate; Fu–S, fumarate minus sulphate; B, butyrate; MB, 2-methylbutyrate; HFA, higher fatty acids (up to C18), P, propionate; GT, glutarate; CO, carbon monoxide; I, indole; AN, anthranilate; Q, quinoline; 345-T, 3, 4, 5-trimethoxybenzoate; TS, tetrazolium salts; Ha, H_2/CO_2 + acetate; S, succinate; Al, L-alanine; Van, vanillate; Isovan, isovanillate; 3-A, 3-anisate; 4-A, 4-anisate; Cat, catechol.

[b] Required for strains isolated from a marine environment. Vitamins if needed, are added at concentrations given in *Table 2*.

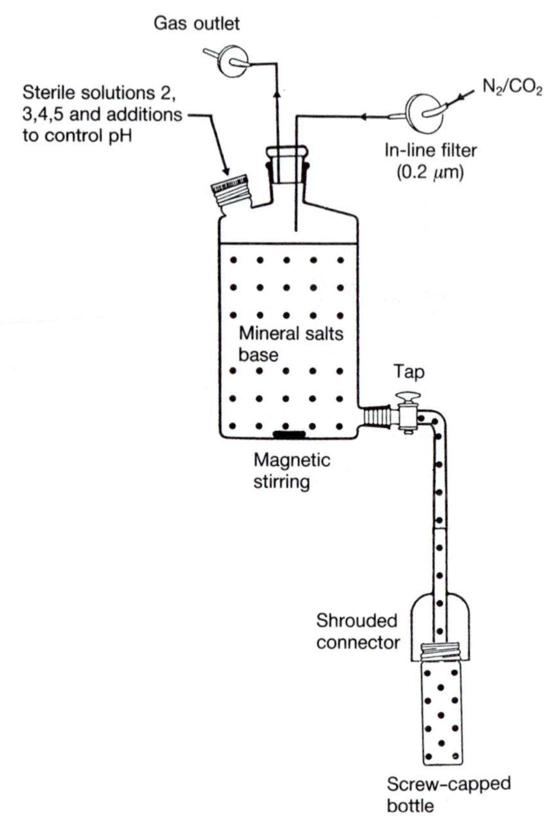

Figure 1. Apparatus used for making the basal medium of Pfennig *et al.* (9). Solution 1, the mineral salts base, is autoclaved in a 5-l medium reservoir fitted with a tap. Immediately after autoclaving, the medium is stored and flushed with the anaerobic gas mixture. When cool, sterile solutions 2–5 and reagents to adjust pH are added aseptically. The basal medium is then dispensed into sterile N_2-filled screw-capped bottles, fitted with rubber septa in the lids. The carbon source and any other desired additions are injected through the rubber seal using sterile syringes.

2.4 Enumeration of SRB

Conventional methods for the counting of viable SRB populations in pure or mixed culture have involved the use of plating techniques (11), most probable number estimates (12), and anaerobic roll tubes (13). However, the agar shake dilution series is most frequently employed (7, 9). Here, the sample is introduced and serially diluted into media containing molten agar in test tubes, mixed, rapidly set, and incubated. A surface-active agent to desorb viable bacteria from particles will enhance recoveries. Practical details are given in *Protocol 2* and *Figure 2*. Some sulphide is included in Pfennig's medium as a reductant and the medium as originally described by these authors (9) contains very low levels of

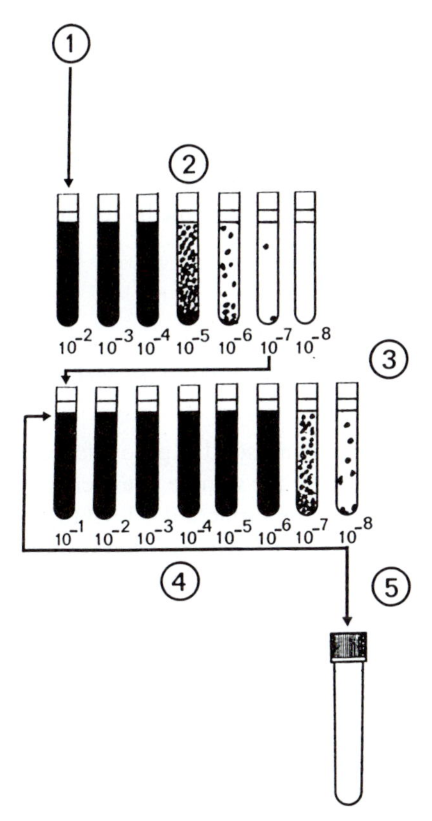

Figure 2. Agar shake dilution procedure. (1) The initial 10-fold dilution of the solid sample is made in anaerobic half-strength Ringer's solution. (2) The slurry is serially diluted in the molten agar (held at 40°C). The tubes are set in a cold-water bath and then incubated at the desired temperature. (3) After growth of the SRB, a single colony is removed using a sterile Pasteur pipette and resuspended in 1 ml of anaerobic half-strength Ringer's solution. The agar shake dilution and incubation are repeated (4) until culture purity is confirmed. (5) When pure, the colony is transferred into a screw-capped bottle filled with a suitable liquid medium.

ferrous ions. This makes it difficult to identify SRB colonies by precipitation of FeS. However, 66 μg $FeSO_4$ added to 9 ml of the complete medium provides enough iron to identify SRB colonies by their sulphide generation, but insufficient to react with the sulphide reductant (14).

Protocol 2. Viable counting of mesophilic SRB using the agar shake dilution method and isolation of pure cultures

You will need

- Half-strength anaerobic Ringer's solution
- Cetyl trimethylammonium bromide (CTAB)

Protocol 2. *Continued*

- Enumeration media (Postgate's medium E or Pfennig's medium with 15 g/l purified agar)
- Sterile paraffin wax (1 part pastillated wax, m.p. 60°C, added to 3 parts liquid paraffin)
- Screw-capped bottles containing liquid media for the storage of pure cultures
- Sterile Pasteur pipettes
- 40°C water bath
- 4°C water bath
- Long (15 cm × 1 cm) thin-walled soda-glass test tubes fitted with rubber stoppers, and racks

1. If the inoculum is solid or semi-solid (e.g. soil, faeces, or sediment), prepare replicate 10-fold dilutions of the sample in anaerobic half-strength Ringer's solution. Add 0.01 mg/l of the surfactant CTAB to the slurry and mix gently for 15 min; this step is not necessary for liquid samples, such as water, or pure cultures.
2. Make a 10-fold serial dilution of the slurry in the sterile test tubes containing 9 ml of autoclaved growth medium, held molten at 40°C.
3. Invert each tube once after inoculation, transfer 1 ml to the next tube, and immediately harden in the cold-water bath.
4. After the agar has set, rapidly seal with a layer (2 cm thick) of liquified paraffin wax and allow to harden.
5. Incubate in the dark at 28°C or 37°C, depending upon the temperature optimum of the bacteria, for 2 weeks.
6. After 24 h incubation, remelt the wax with a Bunsen flame to exclude air and form a better seal.
7. At the end of the incubation time, count black colonies at an appropriate dilution and calculate the original population from the colony count and dilution tube number.
8. To isolate pure cultures of SRB, cut open the test tube with a file and remove well separated colonies from the agar using a sterile Pasteur pipette. Use a single colony, resuspended in 1 ml anaerobic sterile Ringer's solution, for repeated agar shake passage(s).
9. If microscopically homogeneous, transfer the colony to liquid culture for further purity checks (see *Protocol 3*).
10. Maintain pure cultures in screw-capped bottles containing liquid media. SRB grow faster in liquid culture. After 4–6 days incubation they should be stored at 5°C or at room temperature depending upon the strain. Transfer stock cultures every 2–3 months.

Due to inherent limitations in the agar shake dilution method, enrichment of SRB may be necessary if the populations of interest are present in low numbers in the original inoculum. Both batch (9) and chemostat (15) enrichments are used to select for sulphate reducers. However, batch culture enrichments expose bacteria to much higher substrate concentrations than are found in the environment from which they originate. Consequently, slowly growing species in the original sample may be unable to compete with faster growing types that are better able to exploit the growth conditions imposed, and so they will remain undetected. In contrast, chemostat enrichments enable selection of SRB populations at low growth rates and nutrient concentrations, more comparable to those occurring in natural ecosystems. Their principal disadvantages compared to the batch culture method are the requirements for specialized equipment and space.

3. Identification of SRB

The purity of an SRB culture may be assessed initially by microscopic analysis coupled with a qualitative determination of sulphide production (e.g. formation of a distinct black precipitate in media containing sulphate and ferrous ions). Plating methods are used to establish the absence of contaminants.

Protocol 3. Assessment of culture purity (7)

You will need

- Nutrient agar plates containing glucose and peptone
- Peptone–glucose agar + 0.05% (w/v) ferrous ammonium sulphate (pH 7.0–7.6)
- 15 × 1 cm test tubes with rubber stoppers
- Sterile Pasteur pipettes
- 40°C water bath
- 4°C water bath

Aerobic contamination

1. Plate out the culture on nutrient agar plates.
2. Incubate in air at 30°C.
3. No growth should occur.

Qualitative test for anaerobic contamination

1. Aseptically distribute 5 ml of sterile molten peptone–glucose agar containing ferrous ammonium sulphate into each of five test tubes.
2. Cool to 40°C.
3. Make serial Pasteur pipette dilutions from the test culture.

Protocol 3. *Continued*

4. Set in the cold-water bath and incubate at 30°C.

5. The exclusive appearance of black colonies after 4 days indicates the presence of SRB.

The procedure in *Protocol 3* provides no indication as to whether one or more species of SRB are present in an inoculum. Other criteria are needed for this. Many SRB are morphologically distinct. Thus, microscopic examination to determine cell shape, size, and motility provides some information on the identity of the bacteria. The spore-forming genus *Desulfotomaculum* is easily distinguished from other SRB by pasteurization. All SRB are Gram-negative, but desulfonemas, apart from their distinctive large size, are the only known Gram-positive species. Substrate preferences have been extensively used as a taxonomic determinant (6, 7) as have the estimation of DNA base pair ratios (16) and requirements for vitamins and salts (6). *Table 3* outlines how SRB may be recognized using these criteria.

Other diagnostic tests have involved the use of cytochrome profiles, serology, DNA–rRNA homology, temperature requirements, type of flagella, and antibiotic resistance (7). The genera *Desulfovibrio*, *Desulfobacter*, and *Desulfobulbus*, as well as *Desulfotomaculum acetoxidans*, may be distinguished on the basis of specific membrane lipids (17). Immunofluorescent markers have also received attention (18).

The presence of the most commonly occurring species of *Desulfovibrio* (i.e. *Dsv. desulfuricans*, *vulgaris*, *gigas*, *africanus*, *salexigens*) can be assessed rapidly in a crude enrichment culture by the desulfoviridin test (7). The procedure involves centrifuging about 15 ml of an SRB culture and resuspending the pellet in a minimal volume of residual medium. A portion of the suspension is placed on a microscope slide and one drop of 2 M NaOH is added. The desulfoviridin fluoresces a brick red colour under UV light at 365 nm. In the past, this test was useful to distinguish desulfoviridin-positive desulfovibrios from desulfoviridin-negative desulfotomacula but, with the recognition of new types of SRB, results should be interpreted with care. *Desulfomonas pigra*, *Desulfococcus multivorans*, and *Desulfonema limicola* are all desulfoviridin-positive, whilst desulfoviridin-negative desulfovibrios are now known to exist. The isolation of pure cultures and subsequent characterization by other means is by far a more satisfactory alternative.

4. Methods for studying SRB activity

Each of the following analyses can be applied to the study of pure cultures or individually occurring ecosystems.

4.1 Radiotracer analysis of sulphate reduction rates

A solution of ^{35}S–SO_4^{2-} is introduced to the samples, incubated at the ambient temperature, frozen, distilled and the acid-volatile ^{35}S trapped. The portion of radiotracer converted to ^{35}S–S^{2-} is used to determine sulphate reduction rates. Although slurry mixing has been used (19) the core injection method of Jørgensen (20) described here is more widely used.

Protocol 4. Radiotracer estimation of sulphate reduction rates

You will need

- Truncated 5-ml plastic syringes (see *Figure 3*)
- Suba Seal rubber stoppers (BDH)
- One microsyringe (e.g. 10-μl capacity Hamilton syringe)
- Glass distillation apparatus (see *Figure 4*)

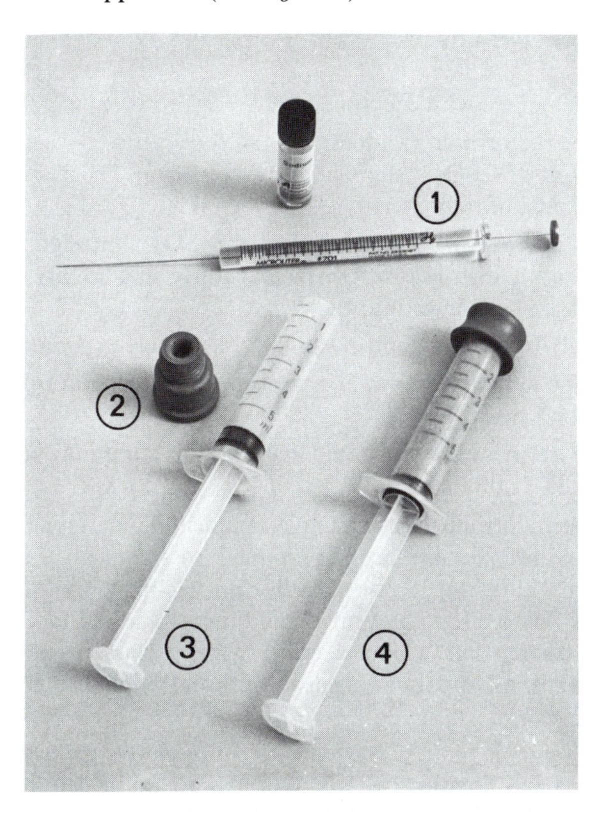

Figure 3. Materials required for radiotracer measurement of sulphate reduction rates. (1) Microsyringe; (2) Suba seal; (3) truncated 5-ml capacity plastic syringe; (4) sealed syringe containing sample.

Protocol 4. *Continued*

- A vortex mixer
- Whatman glass fibre filters (GFF)
- ^{35}S–SO_4^{2-} (Amersham)
- 6 M deoxygenated HCl
- Deoxygenated distilled water
- 10% (w/v) zinc acetate solution
- 10% (w/v) cadmium chloride solution
- Gelling scintillation fluid (e.g. Packard Insta-gel)
- β-emission scintillation counter, preferably with internal quench correction
- Method for measuring sulphate concentrations (see *Protocol 6*)
- A hotplate magnetic stirrer
- Oxygen-free nitrogen gas

1. Distribute samples (5 ml) to be studied into five subcores (truncated syringes).
2. Close each syringe with a Suba Seal.
3. Prepare the radiotracer by adding 1 ml of deoxygenated water to the vial of solid sodium ^{35}S–SO_4^{2-} supplied by Amersham. The resultant solution contains 1 mCi of radioactivity.
4. Using the microsyringe, inject 2 μl of ^{35}S–SO_4^{2-} through each seal. To ensure an even distribution of tracer, slowly and evenly withdraw the microsyringe whilst injecting.
5. Immediately freeze two control samples to provide experimental blanks.
6. Incubate the remaining cores at an appropriate temperature (e.g. 15°C) in the dark for 1 h to 1 day.
7. Terminate the incubation by rapid freezing. The cores may be stored in this manner until distillation.
8. Whilst frozen, introduce each sample to a distillation apparatus similar to that shown in *Figure 4*. Place the sample into a conical 50-ml Pyrex flask fitted with a side-arm, screw thread, and gassing ports. The screw cap contains a Teflon disc to prevent sulphide absorption. Add a Teflon-coated magnetic stirring bar. Connect each flask by glass tubing to a pair of sulphide traps. All apparatus is glass to prevent sulphide binding. All joints are PTFE.
9. Thaw the sample under a stream of oxygen-free nitrogen carrier gas (20 ml/min) and liberate ^{35}S–S^{2-} by adding 5 ml of deoxygenated water followed by 1.6 ml of a solution of 6 M deoxygenated HCl, through the flask side-arm. Heat the flask at 95°C for 40 min with continuous stirring on the magnetic hotplate stirrer.

Protocol 4. *Continued*

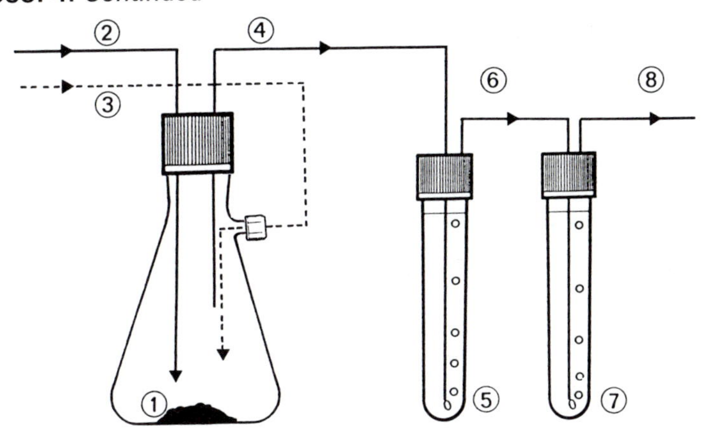

Figure 4. Sulphide distillation apparatus. (1) Sample, heated and stirred; (2) stream of oxygen-free nitrogen gas; (3) addition of deoxygenated water and deoxygenated HCl; (4) transfer of acid-volatile sulphide; (5) zinc acetate trap; (6) gas outlet; (7) cadmium chloride trap to detect sulphide carry-over; (8) gas outlet.

10. Acid-volatile sulphide released is flushed from the reaction flasks and trapped in 20 ml of the zinc acetate solution, forming a ZnS precipitate. The second trap, containing 20 ml of the cadmium chloride solution is used to detect sulphide carry-over. This should not occur, but if it does the solution changes from colourless to yellow.
11. Mix the ZnS precipitate with a vortex mixer, withdraw duplicate 1-ml aliquots, and form an emulsion with 9 ml of scintillation fluid.
12. To determine the efficiency of recovery of the radiolabel, dilute the remaining slurry in the reaction vessel to 100 ml with deoxygenated distilled water, filter through GFF, and remove 1 ml of the filtrate for scintillation counting of residual radioactive sulphate.
13. Determine sulphate reduction rates by the equation

$$\text{Rate} = \frac{[\text{SO}_4^{2-}] \times {}^{35}\text{S–S}^{2-} \times 1.06}{{}^{35}\text{S–SO}_4^{2-} \times t \times d} \text{ nmol SO}_4^{2-} \text{ reduced/ml sample/day}$$

where $[SO_4^{2-}]$ is the initial concentration of sulphate (see *Protocol 6*), $^{35}S–S^{2-}$ the radioactivity of distilled sulphide (d.p.m.), 1.06 the isotopic correction factor between ^{32}S and ^{35}S, $^{35}S–SO_4^{2-}$ the radioactivity of sulphate added (d.p.m.), t the incubation time in days, and d is the dilution factor for conversion to 1 ml of sample.

4.2 Colorimetric determination of H_2S

Methods of measuring hydrogen sulphide, the metabolic end-product of dissimilatory sulphate reduction, may involve the use of sulphide-selective electrodes (21) or gas chromatography (22). However, spectrophotometric techniques such as that of Cline (23) are simple, accurate, and sensitive.

Protocol 5. Determination of H_2S concentrations

You will need

- 10% (w/v) zinc acetate solution
- Mixed diamine reagent: 20 g *N,N*-dimethyl-*p*-phenylenediamine sulphate in 500 ml 50% (v/v) HCl + 30 g anhydrous ferric chloride in 50 ml 50% (v/v) HCl are mixed together immediately before analysis; if the solutions are not made fresh, they should be stored separately in dark bottles
- 10 ml or 250 ml volumetric flasks
- Glass fibre filters (GFF)
- Sulphide standard solutions prepared in deoxygenated distilled water (low range = 0 to 2 mM; high range = $>$2 mM)
- A spectrophotometer

1. Add 4 ml of sample to 1 ml of zinc acetate solution (allows long-term storage as ZnS precipitate).
2. Prepare a suitable range of standard sulphide solutions in zinc acetate.
3. If sample is opaque, syringe through GFF.
4. Place 1 ml of sample or standard into a 10 ml (low range) or 250 ml (high range) volumetric flask.
5. Add 1 ml of mixed diamine reagent, mix, dilute to 10 or 250 ml with deoxygenated distilled water, and leave at room temperature for 20 min.
6. Read at 670 nm in a spectrophotometer against distilled water.
7. Subtract a reagent blank (take 4 ml of distilled water, mix with 1 ml of 10% (w/v) zinc acetate, remove a 1-ml aliquot, add to 1 ml of diamine reagent, and make up to either 10 or 250 ml with distilled water) from all readings and multiply by 1.25 to compensate for the zinc acetate dilution.

4.3 Chemical measurement of sulphate

The disappearance of sulphate from biological material over time may be used as index of SRB activity. Howarth (24) described a rapid and precise method suitable for use in sea water and sediment pore waters, in which barium sulphate is precipitated in low-pH EDTA solution and the precipitate is filtered, dissolved in excess EDTA at high pH, and then titrated with $MgCl_2$.

Protocol 6. Chemical measurement of sulphate concentrations

You will need

- 0.4 M HCl (dilute 33 ml conc. HCl to 1 l with double distilled water)
- 0.01 M EDTA
- 0.05 M HCl (dilute 4 ml conc. HCl to 1 l with double distilled water)
- 10% (w/v) $BaCl_2 \cdot 2H_2O$
- 1.8 cm diameter membrane filters (0.45 μm)
- Ammonium hydroxide solution (29.4%)
- 0.025 M $MgCl_2$
- pH 10 buffer solution (dissolve 7 g NH_4Cl and 57 ml NH_4OH in 25 ml double distilled water and dilute to 100 ml)
- Indicator solution (dissolve 0.4 g Eriochrome black T in 30 ml concentrated triethanolamine solution and 10 ml absolute alcohol)
- Microburette (BDH Ltd)

1. Place 1 ml of clarified sample into an acid-washed 50 ml conical flask.
2. Add 3 ml 0.4 M HCl and 4 ml 0.01 M EDTA.
3. Boil gently for 2 min to chelate any metals in the sample.
4. Add 10 ml 0.05 M HCl and allow to cool.
5. Add 5 ml 10% (w/v) $BaCl_2$ and leave at room temperature for 20 min.
6. Filter through a 0.45-μm membrane filter, rinse with 10 ml double distilled water, rinse with 10 ml 0.05 M HCl, then rinse again with 20 ml double distilled water.
7. Place the filter and precipitate back into the conical flask, add 5 ml 0.01 M EDTA and 4 ml NH_4OH, and leave overnight.
8. Add 0.5 ml of pH 10 buffer and one drop of indicator solution; then titrate with 0.025 M $MgCl_2$.
9. The end-point is indicated by a formation of a red/pink colour from the original purple/blue complex.
10. Compare titration values with those of standard sulphate solutions.

4.4 Ecological studies with sodium molybdate

The addition of molybdate, a specific metabolic inhibitor of SRB (25), can provide information on the types of substrate that they utilize *in situ*. Compounds that accumulate in the presence of the inhibitor are potential electron donors for SRB. If the accumulation is linear, rates of substrate increase can be measured and related to sulphate reduction rates prior to addition of the inhibitor.This enables the relative importance of individual electron donors to be assessed when the relevant reaction stoichiometries are taken into account. The method outlined in *Protocol* 7 describes the procedure for a sediment slurry (26), but similar results with intact sediment cores have been reported (27).

Protocol 7. Determination of electron donors for sulphate reduction

You will need

- 1-mm mesh size sieve
- 500 ml capacity flasks fitted with liquid and gas sampling ports
- 1-l capacity flasks
- A magnetic stirrer and Teflon-coated stirring bar
- Glass fibre filters (GFF)
- Deoxygenated distilled water containing 1 mM Na_2S
- 100% oxygen-free nitrogen gas
- Sodium molybdate
- Methods of measuring organic acids and H_2 (GLC), amino acids (GLC or HPLC), or any other potential electron donor utilized by SRB
- Equipment for the measurement of sulphate reduction rates (see *Protocols 4* and *6*)
- 12-ml capacity serum vials (Jencons)

1. Mix 450 g of sample with 450 ml of deoxygenated sulphide-containing water, whilst maintaining under an atmosphere of oxygen-free nitrogen gas.
2. Pass the resulting slurry through a 1-mm mesh size sieve to remove detritus.
3. Dispense 400-ml amounts in duplicate into 500-ml flasks with thorough mixing and gas out with oxygen-free nitrogen, leaving a slight positive pressure.
4. Incubate at a constant temperature (e.g. 15°C) with continuous mixing. Remove samples regularly for analysis of organic acids.
5. When organic acid concentrations reach a stable level (about 5–6 days for marine sediment samples) remove five aliquots, each of 5 ml, to serum vials pre-gassed with oxygen-free nitrogen. Inject 2 μl ^{35}S–SO_4^{2-} into each vial for the determination of sulphate reduction rates (see *Protocol 4*).
6. Remove samples for gas and chemical analyses from the slurry flask at 30-min intervals during a 4-h incubation period. Subsamples are centrifuged for analysis of organic acids or syringed through GFF and stored frozen until later. Gas samples should be measured immediately.
7. Add the solution of sodium molybdate prepared in deoxygenated water to the slurry to give a final concentration of 20 mM and remove samples periodically over an equivalent 4-h period.

Protocol 7. *Continued*

8. Determine the rate of accumulation of SRB substrates in the presence of molybdate.[a]

[a] **Example.** Butyrate concentration in a marine sediment slurry remained constant before the addition of molybdate. After the inhibitor was introduced, butyrate accumulated at a rate of 0.42 μmol/h. Prior to the addition of molybdate, a sulphate reduction rate of 24.96 μmol/h was recorded. Assuming incomplete oxidation of the electron donor, the stoichiometry of butyrate oxidation by SRB is

$$2\ C_3H_7COO^- + 3\ SO_4^{2-} \rightarrow 3\ HS^- + 4\ HCO_3^- + H^+ + 2\ CH_3COO^-,$$

i.e. there is a 2:3 relationship between butyrate utilized and sulphate reduced. In this example therefore, the equation,

$$\frac{0.42}{24.9} \times \frac{3}{2} \times 100,$$

shows that, of the total sulphate reduced, 2.53% could be attributed to the activity of butyrate oxidizing SRB.

References

1. Berner, R. A. (1974). In *The Sea*, Vol. 5 (ed. E. D. Goldberg), pp. 427–50. Interscience, New York.
2. Badziong, W., Thauer, R. K., and Zeikus, J. G. (1978). *Arch. Microbiol.* **116**, 41.
3. Widdel, F. and Pfennig, N. (1977). *Arch. Microbiol.* **112**, 119.
4. Cypionka, H. (1987). *Arch. Microbiol.* **148**, 144.
5. Biebl, H. and Pfennig, N. (1977). *Arch. Microbiol.* **112**, 115.
6. Pfennig, N. and Widdel, N. (1984). In *Bergey's Manual of Systematic Bacteriology*, Vol. 1 (ed. N. R. Krieg and J. G. Holt), pp. 663–79. Williams and Wilkins, Baltimore.
7. Postgate, J. R. (1984). *The Sulphate-Reducing Bacteria* (2nd edn). Cambridge University Press, Cambridge.
8. Postgate, J. R. (1966). *Lab. Pract.* **15**, 1239.
9. Pfennig, N., Widdel, F., and Trüper, H. G. (1981). In *The Prokaryotes*, Vol. 1 (ed. M. P. Starr, H. Stolp, H. G. Trüper, A. Balows, and H. G. Schlegel), pp. 926–40. Springer-Verlag, New York.
10. Postgate, J. R. (1969). *Lab. Pract.* **18**, 286.
11. Pankhurst, E. S. (1971). In *Isolation of Anaerobes* (ed. D. A. Shapton and R. G. Board), pp. 223–40. Academic Press, London.
12. Battersby, N. S., Stewart, D. J., and Sharma, A. P. (1985). *J. Appl. Bacteriol.* **58**, 425.
13. Hardy, J. A. (1981). *J. Appl. Bacteriol.* **57**, 505.
14. Taylor, J. and Parkes, R. J. (1985). *J. Gen. Microbiol.* **131**, 631.
15. Gibson, G. R. and Macfarlane, G. T. (1988). *Lett. Appl. Microbiol.* **7**, 127.
16. Skyring, G. W., Jones, H. E., and Goodchild, D. (1977). *Can. J. Microbiol.* **23**, 1415.
17. Taylor, J. and Parkes, R. J. (1983). *J. Gen. Microbiol.* **129**, 3303.
18. Smith, A. D. (1982). *Arch. Microbiol.* **133**, 118.
19. Sorokin, Y. I. (1962). *Microbiology* **31**, 329.
20. Jørgensen, B. B. (1978). *Geomicrobiol. J.* **1**, 11.
21. Barcia, J. (1973). *J. Fish. Res.* **30**, 1589.

22. Hawke, D. J., Lloyd, A., Martinson, D. M., and Slater, P. G. (1985). *Analyst* **110**, 269.
23. Cline, J. D. (1969). *Limnol. Oceanogr.* **14**, 454.
24. Howarth, R. W. (1978). *Limnol. Oceanogr.* **23**, 1069.
25. Peck, H. D. (1959). *Proc. Natl Acad. Sci. USA* **45**, 701.
26. Parkes, R. J., Gibson, G. R., Mueller-Harvey, I., Buckingham, W. J., and Herbert, R. A. (1989). *J. Gen. Microbiol.* **135**, 175.
27. Christensen, D. (1984). *Limnol. Oceanogr.* **29**, 189.

12

Methanogens

MAHENDRA K. JAIN, J. GREGORY ZEIKUS, and LAKSHMI BHATNAGAR

1. Introduction

Methanogenic bacteria are a morphologically diverse group of bacteria unified by their ability to produce methane. This property has been used to distinguish these bacteria from the other groups of anaerobic bacteria. Methanogens utilize a very limited number of simple carbon compounds as carbon and energy sources for methanogenesis, i.e. H_2 plus CO_2, formate, methanol, methylamines, acetate, and CO (1). Methanogens also play an important role in interspecies H_2-transfer in anaerobic ecosystems. Under certain environmental conditions, interspecies formate transfer can be quantitatively more important than interspecies H_2 transfer (2), but the general importance of this mechanism has not been established in diverse ecosystems. Methanogens also possess some unique features which distinguish them as a special group of bacteria. These include: the absence of peptidoglycan in the cell wall (3, 4); the presence of mainly ether-linked isoprenoids rather than ester-linked phospholipids in the membranes (5–7); and the presence of unusual coenzymes such as coenzyme M (8, 9), factor F_{420} (10, 11), factor F_{430} (12), methanopterin (13), and methanofuran (14). Based on the unique 16S rRNA structure, methanogens have been grouped under 'archaeobacteria', the third kingdom of life (15–17).

Methanogens are very fastidious anaerobes. Their detailed study requires isolation and cultivation in the absence of oxygen and at low redox potentials. Procedures developed by Hungate have proved most successful for cultivation of methanogens (18–23) and the Hungate technique is described in Chapter 1. The development of improved anaerobic techniques stimulated many laboratories to study methanogens and, as a result, novel biochemical information about these bacteria is accumulating in the literature. The purpose of this chapter is to review the methods commonly used to grow and study methanogens so that the necessary information is available in the form of protocols.

2. Habitat

Methanogenic bacteria are found in a variety of anaerobic habitats including sediments, sludge, and animal waste digestors, the large bowel of man and animals, the guts of insects, wetwood of living trees, rumen, protozoa, and extreme environments. They have so far only been isolated from anoxic environments. They are most abundant below Eh values of −200 mV (24). In general, methanogens get inactivated by the presence of oxygen, although not every species is rapidly killed by oxygen (25). There are no reports which indicate if any attempt has been made to study occurrence and abundance of methanogenic bacteria in oxic environments.

3. Isolation

Protocol 1. Collection and transport of samples

You will need the following materials for sampling

- Anaerobic tubes or vials
- Sodium or cysteine sulphide (2.5%)
- N_2 gas tank with regulator and butyl rubber tubing for gassing the vials
- Syringes (1 ml, 3 ml)
- Needles (23-gauge, 22-gauge)
- Plastic tape
- Marker
- Sampler
- Ice box

1. Gas the tubes and vials with N_2 gas.
2. Collect sediment sample with a sampler and transfer the sample into the gassed tube or vial.
3. Continue to gas the tube or vial containing sample for another 2–3 min. Seal the tube with a butyl rubber stopper and remove the gassing needle at the same moment.
4. Seal the tube with rubber stopper and with aluminium crimp seal, if possible. If not, seal the stopper and tube with plastic tape.
5. Add sodium or cysteine sulphide (2.5%) at 0.1–0.2 ml per 10 ml of sample to further reduce the dissolved oxygen present in the tube.
6. Label the tube containing the sample.
7. Transfer the sample tubes preferably in an ice box to keep them at low temperature for reduced or minimal metabolic activity.
8. Transport the samples to the laboratory and store them at 4°C until used.

3.1 Culture requirements

3.1.1 Oxidation–reduction potential

Removal of oxygen from growth media and from all environments to which the methanogens may be exposed is essential to maintain low oxidation–reduction (redox) potential during cultivation. Hungate (18) pointed out that it is impossible simply by removal of O_2 to obtain a redox potential as low as -330 mV, which is apparently required for methanogens. Although O_2 is removed as completely as possible, some will still be present because it remains dissolved in medium, rubber and plastic tubing, and stoppers. Therefore, addition of reducing agents to the medium removes traces of O_2 and helps achieve the low redox potentials needed for growth of many anaerobic bacteria including methanogens. Many anaerobes probably prefer to grow in environments having redox potentials of -50 mV or less. This is particularly true for methanogens which may require redox potentials as low as -330 mV to initiate growth. As calculated by Hungate (18), to attain a redox potential of -330 mV, the concentration of O_2 would have to be of the order of 1.5×10^{-56} molecules/l. Thus, reducing agents such as cysteine, sulphide, or titanium salts (26, 27) are included in cultivation media to obtain low redox potential. Some reducing substances which are added to lower the redox potential in anaerobic media are listed in *Table 1*. The most commonly used reducing agent solution is a mixture of sodium sulphide and cysteine hydrochloride to lower the redox potential. Resazurin, which turns red at Eh values higher than -42 mV, is normally added to the medium as an indicator.

Table 1. Commonly used reducing agents for lowering the redox potential of media

Compound	Concentrations in medium	Redox potential (mV)
Cysteine–HCl	0.02–0.08%	−210
Dithiothreitol	0.05%	−330
FeS, amorphous	0.05 mM	< −350
Titanium (III) citrate	0.5–2.0 mM	−480
Titanium (III) nitrilotriacetate	0.5–5.0 mM	−480
Sodium sulphide or hydrogen sulphide	0.01–0.025%	−571
Sodium sulphide + cysteine sulphide	0.025%	−571
Dithionite	0.001%	< −600

Protocol 2. Preparation of reducing agent solution

1. Dissolve 12.5 g cysteine–HCl in 500 ml boiled and N_2-gassed water contained in a 2-l Erlenmeyer flask.
2. Adjust pH to 9.5 with NaOH.

Protocol 2. *Continued*

3. Wash 12.5 g crystals of $Na_2S \cdot 9H_2O$ in a plastic tray and then transfer these into the flask containing cysteine–HCl.
4. Bring volume up to 1000 ml with boiled and N_2-gassed water.
5. Bring the whole solution to boil under gassing with N_2.
6. Cool the solution under continued N_2 gas flow and dispense 25 ml in 58 ml capacity vials which were pre-flushed with N_2.
7. Seal the vials using butyl rubber stoppers and aluminium seals and autoclave for 15 min at 121°C.

3.1.2 Nutrient media

Media of different compositions have been used by various research groups to isolate, grow, and study methanogens (21, 28–33). The purpose here is to provide minimal nutrients to support the growth of methanogens. These media are either phosphate-buffered (21) or carbonate-buffered (30). The composition of general media for growing methanogens is given in *Table 2* (34, 35). The basal medium is supplemented with solutions of trace minerals and vitamins. Ammonium is the preferred source of nitrogen (28) but organic sources such as glutamine and urea can also replace it in certain cases (36). Sulphide, which is commonly added to the medium as a reducing agent, also serves as sulphur source but high concentrations have been reported to be inhibitory to the growth of methanogens (37–39). Mercapto-2-ethanol has also been used as a reducing agent to grow methanogens (40). Organic sulphur sources (cysteine, methionine) are also utilized by some methanogens, replacing the need for sulphide (36, 37). Sulphate should be omitted from enrichment medium to eliminate selection of sulphate reducers, particularly when common substrates such as acetate and H_2 are used.

Protocol 3. Preparation of a phosphate-buffered medium for methanogens

1. Dissolve medium components (except substrate, vitamins, reducing agent) into double distilled water and adjust pH to 7.2–7.4 (see *Table 2*).
2. Add 0.2% resazurin solution (1 ml/1000 ml of medium).
3. Remove dissolved O_2 from the medium by boiling it under a stream of O_2-free (passed over a heated copper catalyst) N_2 or $N_2:CO_2$ (95:5) gas.
4. Preflush anaerobic tubes/vials with O_2-free N_2 or $N_2:CO_2$ gas. Dispense appropriate volumes (10 or 50 ml) of medium anaerobically in the preflushed tubes/vials. Dispense medium either using a New Brunswick media dispenser or a Cornwall syringe.
5. Seal the tubes/vials with butyl rubber stoppers followed by aluminium crimps, and autoclave for 25 min at 121°C.

Table 2. General media for cultivation of methanogenic bacteria

Phosphate-buffered basal medium (PBBM)

PBBM composition (per 1 litre medium)

NaCl	0.9 g
$MgCl_2 \cdot 6H_2O$	0.2 g
$CaCl_2 \cdot 2H_2O$	0.1 g
NH_4Cl	1.0 g
Trace mineral solution II	10 ml
Resazurin solution, 0.2%	1 ml
Distilled H_2O to	1000 ml

Additions to autoclaved PBBM (per 10 ml)

Vitamin solution	0.1 ml
Phosphate buffer	0.1 ml
Reductant	0.25 ml
Substrate	As required

Composition of vitamin solution (per litre)

Dissolve in order:

Biotin	0.002 g
Folic acid	0.002 g
B_6 (pyridoxine) HCl	0.010 g
B_1 (thiamine) HCl	0.005 g
B_2 (riboflavin)	0.005 g
Nicotinic acid (niacin)	0.005 g
Pantothenic acid	0.005 g
B_{12}crystalline (cyanocobalamin)	0.001 g
PABA (*p*-aminobenzoic acid)	0.005 g
Lipoic acid (thioctic)	0.005 g
Distilled H_2O to	1000 ml

Composition of trace mineral solution II (per litre)

Nitrilotriacetic acid	12.80 g
$FeSO_4 \cdot 7H_2O$	0.10 g
$MnCl_2 \cdot 4H_2O$	0.10 g
$CoCl_2 \cdot 6H_2O$	0.17 g
$CaCl_2 \cdot 2H_2O$	0.10 g
$ZnCl_2$	0.10 g
$CuCl_2 \cdot 2H_2O$	0.02 g
H_3BO_3	0.01 g
Na molybdate	0.01 g
NaCl	1.00 g
Na_2SeO_3	0.017 g
$NiSO_4 \cdot 6H_2O$	0.026 g
$SnCl_2$	0.02 g

Procedure for making trace mineral II

(a) Add nitrilotriacetic acid to 200 ml H_2O and adjust pH to 6.5 with KOH.

(b) Add this solution to about 600 ml H_2O, dissolve components in order, add H_2O to 1 litre, and store the solution under N_2 gas phase.

(c) If a brown residue of $Fe(OH)_2$ forms, remove by filtration (immediately after adding $FeSO_4$ and before adding remaining reagents).

Bicarbonate-buffered MS-medium

Composition (per 1 l medium)

Yeast extract	2.0 g	$CoCl_2 \cdot 6H_2O$	1.5 mg
Trypticase peptones	2.0 g	$MnCl_2 \cdot 4H_2O$	1.0 mg
Mercaptoethanesulphonic acid	0.5 g	$FeSO_4 \cdot 7H_2O$	1.0 mg
NaOH	4.0 g	$ZnCl_2$	1.0 mg
$Na_2S \cdot 9H_2O$	0.25 g	$AlCl_3 \cdot 6H_2O$	0.4 mg
NH_4Cl	1.0 g	$Na_2WO_4 \cdot 2H_2O$	0.3 mg
$K_2HPO_4 \cdot 3H_2O$	0.4 g	$CuCl_2 \cdot 2H_2O$	0.2 mg
$MgCl_2 \cdot 6H_2O$	1.0 g	$NiSO_4 \cdot 6H_2O$	0.2 mg
$CaCl_2 \cdot 2H_2O$	0.4 g	H_2SeO_3	0.1 mg
Resazurin	1.0 mg	H_3BO_3	0.1 mg
$Na_2EDTA \cdot 2H_2O$	5.0 mg	$NaMoO_4 \cdot 2H_2O$	0.1 mg

Procedure

(a) Dissolve NaOH and equilibrate with a gas phase of N_2 and CO_2 (7:3), add all constituents except the sulphide, and dispense into culture vessels with a N_2 and CO_2 atmosphere.

(b) Add sulphide and soluble catabolic substrates from sterile, anoxic stock solutions, and gaseous catabolic substrates as an over-pressure after inoculation.

(c) The final pH is 7.2 to 7.3.

Protocol 3. *Continued*

6. Upon cooling add, per 10 ml of medium, filter-sterilized anaerobic vitamin solution (0.1 ml), 0.1 ml phosphate buffer (15% KH_2PO_4 + 29% K_2HPO_4, pH 7.0–7.1), and 0.1–0.2 ml of 2.5% reducing agent solution (see *Protocol 2*).
7. Add substrate according to the requirement to grow a specific methanogenic species.

3.2 Enrichment

Methanogens which utilize different substrates as energy sources are obtained from nature by selective enrichments. Enrichment procedures aid in isolation since some methanogens represent only a small fraction of the total microbial community in the natural environment and can be overgrown by other organisms. Selective pressure can be applied in the isolation of methanogens by the use of substrates, antibiotics which inhibit specifically eubacteria but not archaebacteria (41, 42), omission of sulphate, addition of organic components (yeast extract and trypticase, etc.) in very low concentrations, and, at times, supplementation with metal ions such as cobalt, nickel, tungsten, or molybdenum. In addition, salt concentration, pH, and temperature can also aid in having a selective pressure in isolation of a specific methanogenic species.

With the enrichment culture technique, the selective medium is inoculated with varying amounts of inoculum source and incubated under selective conditions until growth of organisms appears. Usually, several enrichment cultures are set up with variations in nutrient compositions, substrate, environmental factors such as pH, temperatures, etc. Enrichments showing production of methane in head-space gas are further selected for transfers until a reasonably good population of methanogens develops in the enrichment tube/vial. Different enrichments may provide different methanogenic strains/species. These enrichments are then plated on to a selective medium in plates, incubated anaerobically in a glove box. Roll tubes can also be prepared to obtain an isolated colony of a methanogenic bacterial culture.

Protocol 4. Enrichment of hydrogenotrophic methanogens

1. Prepare the reduced medium in tubes as described in *Protocol 3*.
2. Add to the medium penicillin, cycloserine, or vancomycin at 100 mg/ml.
3. Inoculate the tubes with a 10% inoculum.
4. Pressurize tubes with $H_2:CO_2$ (80:20) gas mixture to 30 psi (202.65 kPa).
5. Incubate at mesophilic or thermophilic temperatures depending upon the enrichment conditions.
6. After 7–14 days withdraw a gas phase sample using a gas-tight syringe and

Protocol 4. *Continued*

examine for the presence of methane by injecting the sample in a gas–liquid chromatograph.

7. If need be, replace the gas phase and repressurize the tubes with $H_2:CO_2$ gas and incubate until the methane content in gas phase is nearly 10%.
8. Transfer the enrichment at 10% in a homologous medium and continue to monitor for methane in the enrichments.
9. After three subsequent transfers, examine the enrichments in a microscope for the dominant cell morphology and for fluorescence of cells.
10. If the methanogenic population is dominant in the enrichment, serially dilute the culture, and plate 0.1 ml samples on to solid medium (same medium supplemented with 2% Bacto-purified agar) anaerobically in the glove box.

Protocol 5. Enrichment of methylotrophic methanogens

1. Prepare the medium as described in *Protocol 3*.
2. Add 40 mM methanol to the medium as substrate in place of $H_2:CO_2$.
3. Follow the enrichment procedure described in *Protocol 4* for hydrogenotrophic methanogens.
4. Maintain a gas phase of $N_2:CO_2$ (95:5) in tubes/vials.

Protocol 6. Enrichment of acetotrophic methanogens

1. Prepare the medium as described in *Protocol 3*.
2. Add 20–80 mM sodium acetate as substrate.
3. Maintain a gas phase of $N_2:CO_2$ (95:5) in tubes/vials.
4. Follow the enrichment procedure described in *Protocol 4* for hydrogenotrophic methanogens.
5. Incubate the tubes/vials for growth of acetate utilizing methanogens for 4 weeks or longer, since their growth rate is generally slow.

3.2.1 Variations on enrichment methods

(a) *Methanobacterium* species can be enriched on $H_2:CO_2$ or sodium formate (in a highly buffered medium), but in some enrichments add acetate or yeast extract and trypticase.

(b) *Methanospirillum* species can be enriched if enrichments are repeatedly incubated at lower temperatures (30–35°C).

(c) Use antibiotics (penicillin, cycloserine, or vancomycin) and methylamines as substrates to inhibit the growth of methylotrophic acetogens such as *Eubacterium limosum* and *Butyribacterium methylotrophicum*.

(d) Use high salt concentrations ($\geq$5% NaCl) to favour enrichment of halophilic methanogens.

(e) Use low concentrations of acetate and sulphide and long incubation times to favour and enrich *Methanosaeta*-type methanogens.

(f) Use methanol and H_2 to favour and enrich *Methanosphaera*-type of methanogens.

3.3 Gas chromatographic analysis

Since all methanogens produce methane, analysis of gas phase for production of CH_4 will confirm the growth of methanogens in tubes, vials, or any other inoculated sealed container.

Protocol 7. Analysis of methane by gas–liquid chromatography (GLC)

You will need the following materials for gas chromatographic analysis of methane in the gas phase

- Gas chromatograph (GLC)
- Chromosorb 101 column with a flame ionization detector (FID); or Poropak Q column with a thermal conductivity detector (TCD)
- H_2, oxygen-free N_2, and He gas tanks
- Gas-tight Mininert syringe valve (Alltech Assoc., Inc, Deerfield, IL)
- 23- and 25-gauge needles
- 1.0-ml glass syringe (Becton Dickinson and Co, Rutherford, NJ)
- Teflon tape

1. Prepare 1 to 10% CH_4 standards in sealed vials by mixing appropriate volumes of CH_4 with oxygen-free N_2 gas.
2. Flush the gas-tight glass syringe fitted with Mininert valve with N_2 gas.
3. Inject 0.4 ml CH_4 standard into a GLC equipped with an FID using Chromosorb 101 and N_2 as carrier gas (10 ml/min) at 75°C. Keep the injector temperature at 100°C and detector temperature at 200°C.
4. Withdraw 0.4-ml gas-phase sample from vials or tubes and inject into the GLC.
5. Calculate % CH_4 in the sample based on the reference peak from standard CH_4.
6. Alternatively, analyse CH_4 by injecting 0.4 ml standard methane and the gas phase sample into a GLC equipped with TCD using a Poropak Q column and N_2 or He as carrier gas at room temperature.

3.4 Microscopic analysis

The methanogenic bacteria are structurally diverse and display no unique features by which all species can be characterized. All basic morphological types found among bacteria, including cocci and packets of cocci, rods of different shape and size, spirillum, and filamentous forms, are represented in methanogenic bacteria. However, cells of methanogens can be recognized by their strong autofluorescence under oxidizing conditions. The major contributors to this phenomenon are coenzyme F_{420} and the methanopterin derivatives. Coenzyme (or factor) F_{420} has been found in almost all the methanogen cells examined at levels ranging from 1.2 (*Methanobrevibacter ruminantium*) to 65 (*Methanobacterium thermoautotrophicum*) mg of coenzyme per kg of cell dry weight (11, 43, 44). The compound has an absorption maximum at 420 nm. The absorbance is lost upon reduction. The fluorescence of methanogen cells can be examined using a microscope with a fluorescence attachment.

Protocol 8. Examination of fluorescence

You will need the following material for examination of cell morphology and its autofluorescence

- Microscope with fluorescence attachment
- Power supply unit for mercury lamp
- Glass slides
- Cover slips
- Non-fluorescing immersion oil
- Xylene

1. Put a drop of broth culture on a glass slide and cover it with a cover slip as for normal microscopic examination.
2. Put the slide on the microscope stage, bring the area of the specimen to be observed into the field of view, and focus with transmitted light emitted from the microscope's tungsten filament bulb.
3. Put a drop of non-fluorescing oil on the cover slip and view the cells using an oil-immersion objective for transmitted light phase contrast and reflected light fluorescence observation.
4. Switch off the tungsten bulb, and insert your selection of dichroic mirror, exciter, and barrier filters into the light path.
5. Observe the microbial cells for fluorescence.
6. If the fluorescence observation is to be interrupted briefly, cut off the beam of light by means of the shutter slider rather than turning off the mercury lamp since repeated on–off switching considerably shortens the useful life of the lamp.

Protocol 8. *Continued*

7. After use carefully wipe off the immersion oil deposited on the lens surfaces with gauze moistened with xylene.

8. At the end turn off the main switch of the power supply unit.

3.4.1 Precautions when using the fluorescence microscope

(a) Make it a practice to use the UV protective shade provided to protect your eyes from fluorescent light.

(b) Do not use alcohol or ether to wipe off the immersion liquid from the lens surfaces.

(c) Do not switch off the mercury lamp within 15 min after ignition.

(d) After the mercury lamp is switched off, do not re-ignite it for 3 min or more, in order to give it time to cool.

(e) Each time a lamp is replaced, the new lamp should be centred.

(f) For safety's sake, do not replace the mercury lamp for about 10 min after switching off.

(g) After lamp replacement, zero the life meter to correctly record the time period the lamp has been used.

3.5 Colony isolation

After the enrichment of a methanogenic bacterial population in a liquid medium is obtained, serial dilutions of the broth can be prepared and plated on to an agar medium to isolate single colonies. This can be done either by the roll-tube technique or by the streak-plate or pour-plate method. Methanogens can be plated with high plating efficiencies (45, 46).

Protocol 9. Isolation using roll-tube method

You will need the following materials for preparation of roll tubes

- Tube spinner (Bellco Glass Inc, Vineland, NJ)
- Water bath at 50°C
- Roll tubes
- Rubber stoppers
- Gassing system
- N_2 gas tank
- Sterile syringes (1 ml)
- Sterile needles (22- and 23-gauge)

1. Prepare sterile anaerobic agar medium (containing 15 g/l Difco purified agar)

Protocol 9. *Continued*

as described in *Protocol 3* and distribute 2–5 ml media per tube depending upon the size of tube.

2. Keep the tubes containing molten agar medium at 50°C in a water bath.
3. Prepare 10-fold serial dilutions of the enrichment culture in anaerobic sterile medium up to 10^{-9}.
4. Add 0.2–0.5 ml of the dilution from an appropriate dilution tube to the tubes containing molten agar medium. Use 10^{-3} to 10^{-9} dilutions to inoculate the roll tubes.
5. Mix the contents of the tubes avoiding formation and entrapment of gas bubbles.
6. Spin the tubes using the tube spinner under an ice cover, to have a uniform layer of solid medium on the inner wall of the tube. Alternatively, the tubes can be rolled under running cold tap water.
7. Incubate the roll tubes at appropriate temperatures. Make sure that the condensed water does not smear colonies during handling of roll tubes.

Protocol 10. Isolation by plating on solid media

You will need the following materials for plating methanogens

- Anaerobic chamber
- N_2, N_2:H_2 (95:5), and H_2:CO_2 (80:20) gas mixtures
- Sterile Petri dishes
- Sterile syringes (1 ml)
- Needles (22-gauge and 23-gauge)
- 15% Na_2S solution
- Anaerobic jar or modified paint-tank (47)

1. Prepare agar plates using the appropriate medium (see *Table 2*) containing 1.5% agar (Difco purified) in an anaerobic chamber. Allow these plates to dry for 6–8 h.
2. Take the anaerobic jar or modified paint-tank (W. R. Brown, Corp., Chicago, IL) inside the chamber.
3. Prepare dilutions of enrichment cultures, as described in *Protocol 9*, in the anaerobic chamber.
4. Spread 0.1 ml of appropriate dilutions (10^{-4} to 10^{-9}) on to the plates using sterile glass spreaders.
5. Put the plates in the anaerobic jar or paint-tank with a paper towel soaked with 5 ml of 15% Na_2S solution and immediately tighten the lid of the anaerobic jar or paint-tank.

Protocol 10. *Continued*

6. Take the anaerobic jar or paint-tank out of the chamber, pressurize it to 20 p.s.i. with $N_2:CO_2$ (95:5) gas, and incubate at the desired temperature.
7. Alternatively, for isolation of hydrogenotrophic methanogens pressurize the anaerobic jar or paint-tank to about 30 p.s.i. (202.65 kPa) with $H_2:CO_2$ (80:20) gas mixture.
8. Monitor the growth of methanogens by analysing production of methane gas during the incubation period by withdrawing gas samples from anaerobic jars or paint-tanks.
9. For hydrogenotrophic methanogens, repressurize the anaerobic jar with $H_2:CO_2$ when necessary.
10. When the gas phase contains $> 10\%$ methane gas, open the anaerobic jars or paint-tanks in the anaerobic chamber and examine the plates for the appearance of colonies on the plates.
11. Pick the isolated colonies into homologous liquid medium either by sterile tooth picks or by sterile syringe and needle.
12. Grow the isolated colonies as described earlier in *Protocols 4–6* and examine methanogenic cultures for cell growth, morphology, and fluorescence (see *Protocol 8*).
13. Streak the culture on to solid agar medium again to check a pure culture has been obtained.

3.6 Enumeration

The most common method used to enumerate specific types of methanogens is the most probable number (MPN) technique, where the sample is serially diluted and inoculated into a suitable broth medium with a specific substrate. Tubes positive for methane production are used to calculate the most probable numbers of methanogens in the sample. The roll-tube technique and plating of the sample in the presence of some antibiotics (penicillin, cycloserine, and vancomycin) can also help in enumeration but these will give numbers of viable cells of bacteria capable of utilizing specific substrates or able to grow under specific conditions (e.g. pH, temperature). Since methanogens contain unique coenzymes such as F_{420} and F_{350}, they can be enumerated based on their fluorescence properties. Immunological techniques has also been used recently to enumerate a specific methanogen in different environments or different methanogens in the same environment (48).

Protocol 11. Most probable number method for enumerating methanogens

1. Prepare several pressure tubes each with 9 ml of anaerobic medium, as described in *Protocol 3*.

Protocol 11. *Continued*

2. Prepare 10-fold serial dilutions of the sample from 10^{-1} to 10^{-10}.
3. Inoculate five tubes with each dilution from 10^{-3} to 10^{-10}, at 1 ml/tube.[a]
4. Keep five uninoculated tubes as controls.
5. Replace the gas phase of the tubes with $H_2:CO_2$ (80:20)[b] and pressurize tubes to 30 psi (202.65 kPa).
6. Incubate the uninoculated as well as inoculated tubes under shaking conditions at 37°C for 30 days.
7. Check presence of CH_4 in the gas phase of each tube.
8. Record number of positive tubes (on methane production basis) for all the dilutions.
9. Refer to a standard five-tube MPN table to calculate the value that can be expressed as the most probable number of hydrogen-utilizing methanogens per ml.

[a] As an alternative to inoculation of five tubes/dilution, you can also use three tubes/dilution but then you should refer to a standard three-tube MPN table for finding a most probable number.
[b] For enumerating acetate-utilizing methanogens, keep the gas phase of $N_2:CO_2$ (95:5) in place of $H_2:CO_2$ (80:20) and add 60 mM of sodium acetate as substrate.

4. Taxonomic characterization

After isolation of a methanogen in pure culture it needs to be characterized taxonomically. There have been a number of methanogens isolated during the last 10 years with the developments and improvements in the techniques and tools for cultivation and isolation of these strict anaerobes (1, 31). They have been largely classified as belonging to orders *Methanobacteriales*, *Methanococcales*, and *Methanomicrobiales*, based on their morphological, physiological, biochemical, and some genetic characteristics, which include:

(a) *Morphological characterization.* Cell shape, size, arrangement of cells as tetrads, bunches, chains, etc.; colony morphology; motility, Gram-staining; ultrastructure (scanning and transmission electron microscopy).

(b) *Physiological characterization.* Organic and inorganic carbon, nitrogen, and sulphur nutritional requirements, growth substrate(s), growth enhancers and growth rate; optimum and range of growth temperature; optimum and range of growth pH; resistance to antibiotics; requirement for NaCl, minerals, sulphide, and vitamins; toxicity and tolerance levels for sulphide and salt.

(c) *Biochemical characterization.* Total cellular protein profile, i.e. presence or absence of certain peptide band(s) in SDS/PAGE or native gels; DNA G+C ratios; immunological fingerprinting; cell envelope composition, membrane lipids, and key enzyme subunit profile, e.g. methyl reductase.

(d) *Molecular characterization*. DNA–DNA and DNA–RNA hybridization and 16S rRNA analysis.

Development of a systematic key based on some essential characteristics of the methanogens can help in their identification and classification. Based on the current data a simple schematic key is provided in *Figures 1* and *2* for identification, characterization, and classification of methanogens. To speciate a new methanogenic isolate it would be necessary to take into account the biochemical and molecular characteristics of the isolate.

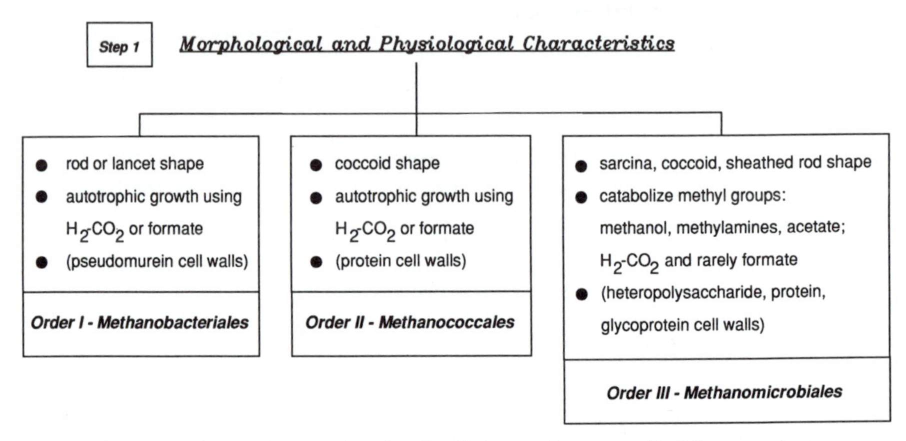

Figure 1. Comprehensive key for classifying methanogens in different orders.

5. Large-scale cultivation

Methanogens can be grown anaerobically at large scale to obtain cells to conduct basic studies involving biochemical characterization of enzymes and cofactors or genetics (e.g. isolation of DNA, etc.). Techniques described for growth of methanogens in 5–100 ml culture tubes and bottles (31) can be modified and applied to grow methanogens in fermentors of 10–100 l capacity. We will describe here the mass culture of methanogenic bacteria for obtaining large quantities of cells.

Protocol 12. Mass cultivation

You will need the following materials for mass culture of methanogens

- Fermenter
- pH controller
- Acid/base pump
- pH probe
- Anaerobic serum vials/plasma bottles

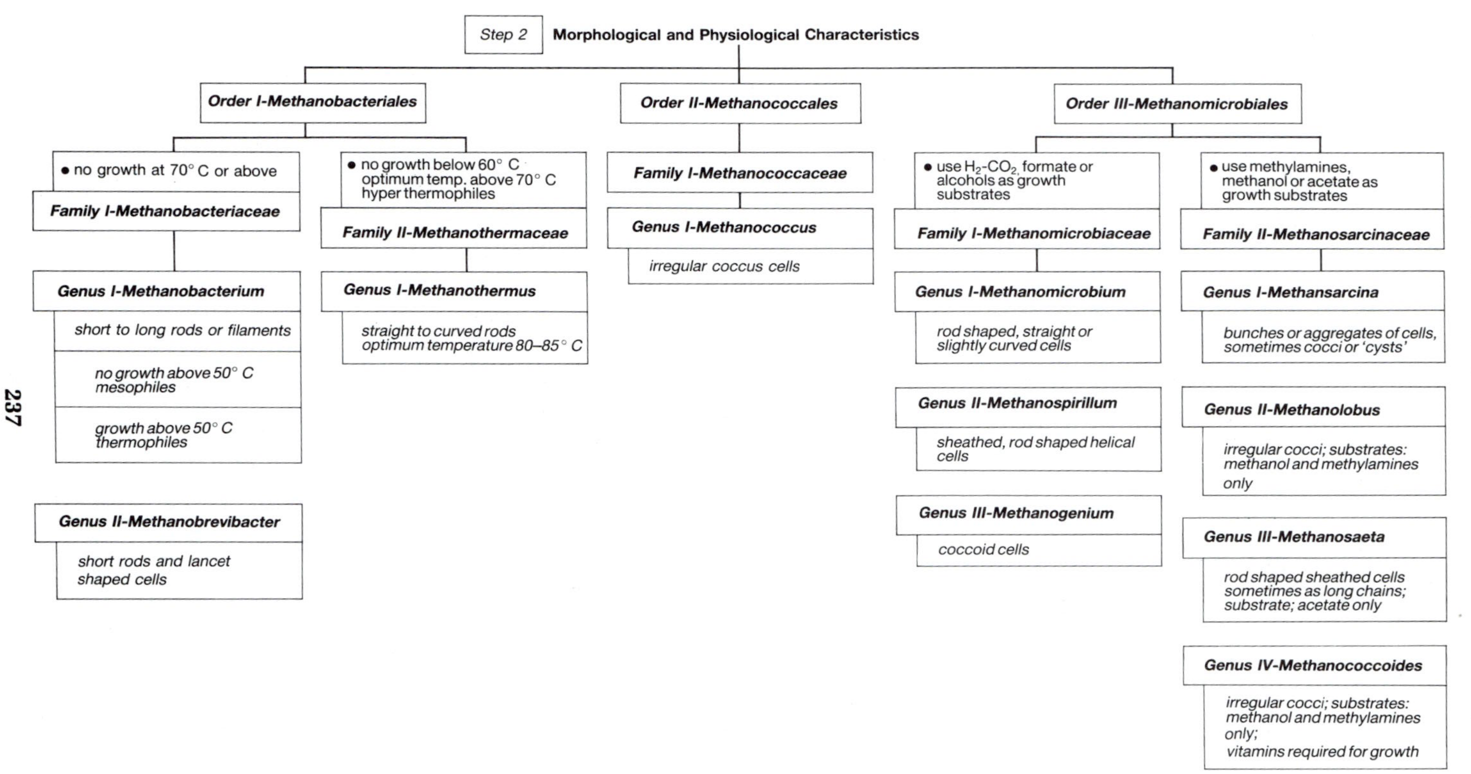

Figure 2. Comprehensive key for classifying methanogens in different families and genera.

Protocol 12. *Continued*

- Gas tanks ($N_2:CO_2$, $H_2:CO_2$, argon)
- Rubber tubing
- Sterile syringes
- Sterile needles (18-, 22-, and 23-gauge)
- Media constituents

1. Use multiple anaerobic bottles (Bellco) or plasma bottles (49) with suitable medium (50–200 ml) to grow the organism, as a pre-culture.
2. Obtain 0.5 to 1.0 l of the preculture which will be sufficient to inoculate a 10–15 l fermenter or carboy (working volume).
3. Use a fermenter with a capacity of 10–15 l working volume (Chemapac SARL, Switzerland, or New Brunswick, Microferm) and modify it to grow methanogens (50, 51).
4. Dissolve all the sterilizable components of the medium into an appropriate volume of distilled water in the fermenter.
5. Bubble $N_2:CO_2$ gas (150–200 ml/min) through the medium for 30–60 min while heating the fermenter for sterilization. Adjust pH of the medium.
6. Stop $N_2:CO_2$ (95:5) gas bubbling and sterilize the medium in place in the fermenter according to the given specifications (for the kind of fermenter used).
7. Prepare other components of the medium which are sterilized separately by autoclaving (e.g. phosphates and reductant Na_2S in case of PBBM medium) or filter sterilization (e.g. vitamins, etc.).
8. After sterilization, bubble the fermenter with $N_2:CO_2$ again and wait until the fermenter cools down to desired temperature.
9. Sterilize the fermenter ports by spraying alcohol and flaming locally. Add the remaining components of the medium through the sterilized ports.
10. For hydrogenotrophic methanogens, gas the medium with $H_2:CO_2$ (80:20) at 100–200 ml/min during the entire time course of growth. To maintain the optimal sulphide concentration in solution as reductant and sulphur source, either add Cys–S or add 1–2 ml/l of Na_2S (2.5%) daily. In place of Na_2S addition, H_2S (0.2%) with incoming $H_2:CO_2$ can also be used (52).
11. Inoculate the completed medium in the fermenter with an appropriate amount of pre-culture (~5–10% v/v).
12. Control and maintain growth temperature and pH known for the methanogen, using 6M NaOH or 4M HCl.
13. Control the agitation of the medium according to the substrate used (e.g. 500–1000 r.p.m. when $H_2:CO_2$ is used or less when soluble substrate such as acetate is used) or the methanogen.

Protocol 12. *Continued*

14. Monitor growth of the organism during the run by aseptically withdrawing liquid samples from the fermenter. Also, check gas phase for production of methane. Once desired growth is reached, start harvesting cells as described in *Protocol 13*.

Protocol 13. Harvesting and storing cells

The following materials will be needed to harvest methanogenic bacterial cell mass from fermentation broth.

- Sharples centrifuge or membrane-based cell concentrator
- Preparatory centrifuge
- Centrifuge bottles (large size)
- Gas tanks (nitrogen, argon)
- Spatula
- Anaerobic tubes/vials
- −80°C freezer
- Liquid nitrogen

1. Prepare Sharples centrifuge (enclosed in wooden unit) for harvesting cells. Cool the Sharples unit to 4°C. Flush argon through the unit to replace air/oxygen. Alternatively, concentrate cells using a Millipore Pellicon system (50) and then centrifuge to collect cells.
2. Connect the fermenter drain to the Sharples centrifuge and start harvesting.
3. Monitor liquid culture volume in the fermenter. Stop centrifugation leaving some liquid culture (~100–200 ml) in the fermenter.
4. Take out the centrifuge drum containing cell paste while flushing with nitrogen or argon.
5. Remove cell paste using a long spatula to appropriate tubes and freeze the small chunks of cell paste by dropping directly into liquid nitrogen.
6. Store the frozen cell paste either in liquid nitrogen or at −80°C.

6. Preparation of cell extracts

Cell extracts are prepared from the cell paste using a French Pressure cell under strict anaerobic conditions (53). Special centrifuge tubes with rubber bung tops are used. The frozen cell paste can be thawed with continuous flushing of N_2 gas. Hydrogen can be added to keep the extract reduced. Anaerobic conditions are maintained throughout the preparation of the cell extracts by flushing the French Pressure cell and the tubes. All operations are carried out at 4°C.

Protocol 14. Cell extract preparation

You will need the following materials for preparation of cell-free extracts

- French Press with cell
- N_2 and H_2 gas tanks
- Spinco tubes with bung
- Regular Spinco tube caps (with a hole drilled in centre)
- Anaerobic tubes/vials to store extract
- Cell paste
- Required buffer (anaerobic)
- Ultracentrifuge
- Preparatory centrifuge
- Glass syringes (10 ml)
- Needles (18-, 22-, and 23-gauge)
- Ice bucket
- Microscope

1. Take the frozen cell paste (~5 g) and transfer into the special centrifuge tube (Spinco) with anaerobic distilled water (20 ml) or appropriate buffer. Keep flushing N_2 gas until the cell paste is completely thawed.
2. Wash cells thrice under anaerobic conditions by centrifuging at 10 000–20 000 *g* for 10 min and resuspend the cells in buffer to a 10-ml final volume.
3. Use a cold French Pressure cell kept at 4°C overnight for disrupting cells. Fill the pressure cell with the cell suspension (under N_2 gas flushing). Use pressure cell with capacity of 5 or 50 ml depending upon volume of cell material to be French-pressed.
4. Make sure that the N_2 gas flushing system is a closed circuit with exhaust only through a syringe needle stuck in the tube rubber bung (see *Figure 3*).
5. Start applying the pressure and carefully monitor the pressure increase to 10 000–18 000 lb/in^2 (depends on the pressure cell used).
6. Once the desired pressure is reached, make sure that the disrupted cells are collected out of the pressure cell outlet drop by drop.
7. Collect the disrupted cells into another centrifuge tube (kept in ice bucket and flushed with N_2 gas). Check the efficiency of breaking the cells under the microscope. If less than 30% of cells are broken, run cell suspension again through the French Pressure cell.
8. Centrifuge the broken cell suspension at 30 000–48 000 *g* for 30 min at 4°C. The supernatant is designated as crude extract.

Protocol 14. *Continued*

9. Remove the supernatant into another tube and store at 4°C for immediate use (enzyme assay; cofactor analysis, protein gels, etc.).
10. For long-term storage of crude extracts, drop directly into liquid N_2 using a syringe and store the resulting beads in liquid N_2, or freeze the crude extract at −80°C in a closed glass tube/vial with H_2 head space.

Methods for assay of F_{420}-dependent hydrogenase and methyl-coenzyme M reductase are described in *Protocols 12* and *13* of Chapter 7.

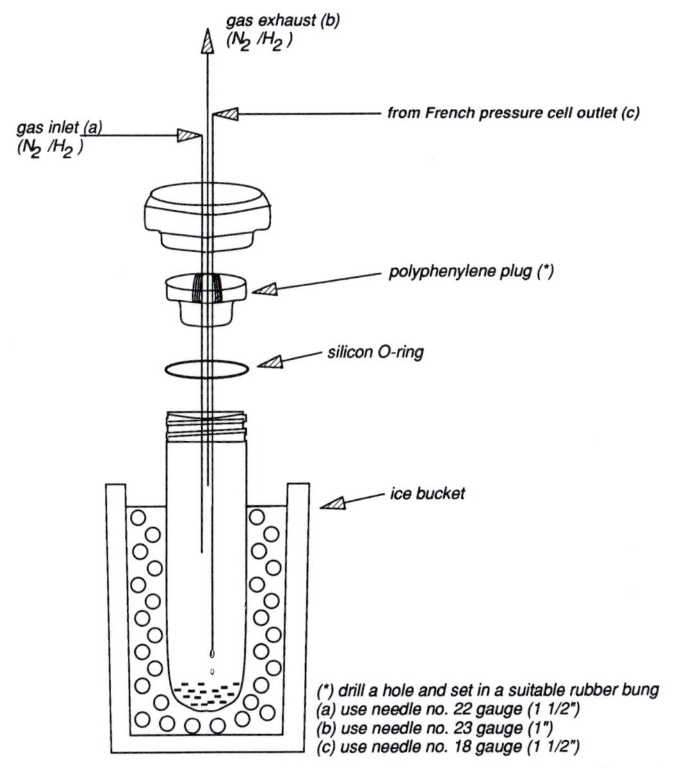

Figure 3. Schematic diagram for collection of French-pressed (broken) cells under anaerobic conditions.

7. Maintenance and preservation

Maintaining the viability of cultures is still one of the major problems to be overcome when working with methanogenic bacteria. These bacteria are obligate anaerobes, most of which are very sensitive to oxygen even when briefly exposed. Methanogenic bacteria can be maintained by subculturing, freezing, and freeze-drying (54, 55). The choice of method will depend upon many considerations that include the type of strain, the reason for preservation, availability of equipment, and the time required for initial preparation of culture for preservation and for maintenance of the preservation systems. For long-term storage, freezing and storing them in liquid nitrogen is a very satisfactory method. Excellent results have been obtained by using this technique (54).

Protocol 15. Freezing in glass ampoules

You will need the following materials for freezing in glass ampoules

- Liquid nitrogen gas
- Glass ampoules
- Sterile glycerol
- Sterile dimethyl sulphoxide (DMSO)
- Calibrated Pasteur pipettes
- Sterile syringes
- Needles

1. Grow the methanogenic bacteria in an appropriate medium in 158-ml serum bottles (containing 20 ml medium for hydrogen utilizers or 50 ml for acetate utilizers) until mid-logarithmic growth.
2. Centrifuge the cells in the same bottles and remove the supernatant aseptically using syringe and needle. Centrifuge the cultures that use methanol before active gas production ceases.
3. Suspend the cell pellet in 2–5 ml of fresh medium that contains either 10% (v/v) glycerol or 5% (v/v) DMSO.
4. Pool cell suspensions from several bottles into one bottle if a large number of ampoules are to be prepared.
5. Clean and sterilize glass ampoules such as the 2-ml Wheaton gold-band cryule (Wheaton Scientific Division, 651486).
6. Flush the ampoules with oxygen-free $N_2:CO_2$ using a gassing cannula and, during gassing, transfer 0.2 ml of concentrated cell suspension to the bottom of each ampoule using syringe and needle, or a calibrated Pasteur pipette.
7. Move the gassing cannula near the top of the ampoule neck and heat the ampoule in the middle of the long neck with a thin, hot flame from a gas burner, drawing it out and finally sealing it in the process.
8. Immediately place these ampoules in an ice bath to prevent further warming of the contents from the heated upper part of the ampoule; then remove and dry from outside.
9. Freeze the cell suspension in the ampoules by placing them in the cold gas phase of liquid nitrogen using hanging aluminium dippers.

Protocol 15. *Continued*

10. Store the glass ampoules in the vapour phase and not immersed in the liquid nitrogen.

11. For recovery, remove an ampoule from the liquid nitrogen (using protective gloves) and quickly thaw by immediately placing it in warm water at 30–37°C; then dry it from outside.

12. Break the ampoule by scratching with a diamond pencil and transfer the contents into a sealed tube containing 5 ml of medium using a syringe and needle.

8. Culture sources

Cultures of methanogenic bacteria may be obtained from German Collection of Microorganisms and Cell Cultures (DSM), American Type Culture Collection (ATCC), and Oregon Collection of Methanogens (OCM).

References

1. Jain, M. K., Bhatnagar, L., and Zeikus, J. G. (1988). *Indian J. Microbiol.* **28**, 143.
2. Thiele, J. T. and Zeikus, J. G. (1988). *Appl. Environ. Microbiol.* **54**, 20.
3. Kandler, O. and Hippe, H. (1977). *Microbiology* **11**, 357.
4. Kandler, O. and Konig, H. (1985). In *The Bacteria*, Vol. 8 (ed. C. R. Woese and R. S. Wolfe), pp. 413–57. Academic Press, New York.
5. Tornabene, T. G. and Langworthy, T. A. (1973). *Science* **203**, 51.
6. Tornabene, T. G., Wolfe, R. S., Balch, W. E., Holzer, G., Fox, G. E., and Oro, J. (1978). *J. Mol. Evol.* **11**, 256.
7. Langworthy, T. A. (1985). In *The Bacteria*, Vol. 8 (ed. C. R. Woese and R. S. Wolfe), pp. 459–97. Academic Press, New York.
8. Bryant, M. P., Tzeng, S. F., Robinson, I. M., and Joyner, A. E. (1971). In *Anaerobic Biological Treatment Processes*, Advances in Chemistry Series 105 (ed. R. F. Gould), pp. 23–40, American Chemical Society, Washington, DC.
9. Taylor, C. D., McBride, B. C., Wolfe, R. S., and Bryant, M. P. (1974). *J. Bacteriol.* **120**, 974.
10. Cheeseman, P., Toms-wood, A., and Wolfe, R. S. (1972). *J. Bacteriol.* **112**, 527.
11. Eirich, L. D., Vogels, G. D., and Wolfe, R. S. (1979). *J. Bacteriol.* **140**, 20.
12. Gunsalus, R. P. and Wolfe, R. S. (1978). *FEMS Microbiol. Lett.* **3**, 191.
13. Vogels, G. D., Keltjens, J. T., Hutten, T. J., and van der Drift, C. (1982). *Zentralbl. Bakteriol. Parasitenkd. Infektionskr. Hyg.* Abt. 1, **Orig C3**, 258.
14. Leigh, J. A. and Wolfe, R. S. (1983). *J. Biol. Chem.* **258**, 7536.
15. Fox, G. E., Stackebrandt, E., Hespell, R. B., Gibson, J., Maniloff, J., Dyer, T. A., Wolfe, R. S., Balch, W. E., Tanner, R. S., Magrum, L. J., Zablen, L. B., Blakemore, R., Gupta, R., Bonen, L., Lewis, B. J., Stahl, D. A., Kuehrsen, K. R., Chen, K. N., and Woese, C. R. (1980). *Science* **209**, 457.
16. Jones, W. J., Whitman, W. B., Fields, R. D., and Wolfe, R. S. (1983). *Appl. Environ. Microbiol.* **46**, 220.

17. Konig, H. and Stetter, K. O. (1989). *Archaeobacteria* (ed. J. T. Staley, M. P. Bryant, N. Pfennig, and J. G. Holt), pp. 2171–253. Williams & Wilkins, Baltimore.
18. Hungate, R. E. (1969). In *Methods in Microbiology*, Vol. 3B (ed. J. R. Norris and D. W. Ribbons), pp. 117–32. Academic Press, New York.
19. Wolfe, R. S. (1971). *Adv. Microbiol. Physiol.* **6**, 107.
20. Bryant, M. P. (1972). *Am. J. Clin. Nutr.* **25**, 1324.
21. Zeikus, J. G. (1977). *Bacteriol. Rev.* **41**, 514.
22. Wolfe, R. S. (1979). *Antonie von Leeuwenhoek* **45**, 353.
23. Hespell, R. B. and Bryant, M. P. (1981). In *The Prokaryotes: A Handbook on Habitats, Isolation and Identification of Bacteria* (ed. M. P. Starr, H. Stolp, H. G. Truper, A. Balows, and H. G. Schlegel), pp. 1479–94. Springer-Verlag, New York.
24. Mah, R. A., Ward, D. M., Baresi, L., and Glass, T. L. (1977). *Ann. Rev. Microbiol.* **31**, 309.
25. Keiner, A., and Leisinger, T. (1983). *Syst. Appl. Microbiol.* **4**, 305.
26. Zehnder, A. J. B. and Wührman, K. (1976). *Science* **194**, 1165.
27. Moench, T. T. and Zeikus, J. G. (1983). *J. Microbiol. Meth.* **1**, 199.
28. Mah, R. A. and Smith, M. R. (1981). In *The Prokaryotes: A Handbook on Habitats, Isolation and Identification of Bacteria* (ed. M. P. Starr, H. Stolp, H. G. Truper, A. Balows, and H. G. Schlegel), pp. 948–77. Springer-Verlag, New York.
29. Huser, B. A., Wuhrmann, K., and Zehnder, A. J. B. (1982). *Arch. Microbiol.* **132**, 1.
30. Widdel, F. (1986). *Appl. Environ. Microbiol.* **51**, 1056.
31. Balch, W. E., Fox, G. E., Magrum, L. J., Woese, C. R., and Wolfe, R. S. (1979). *Microbiol. Rev.* **43**, 260.
32. Lynd, L., Kerby, R., and Zeikus, J. G. (1982). *J. Bacteriol.* **149**, 255.
33. Widdel, F. and Pfennig, N. (1981). *Arch. Microbiol.* **129**, 395.
34. Kenealy, W. and Zeikus, J. G. (1981). *J. Bacteriol.* **144**, 133.
35. Boone, D. R., Johnson, R. L., and Liu, Y. (1989). *Appl. Environ. Microbiol.* **55**, 1735.
36. Bhatnagar, L., Jain, M. K., Aubert, J.-P., and Zeikus, J. G. (1984). *Appl. Environ. Microbiol.* **48**, 785.
37. Scherer, P. and Sahm, H. (1981). *Acta Biotechnol.* **1**, 57.
38. Ronnow, P. H. and Gunnarsson, L. A. H. (1981). *Appl. Environ. Microbiol.* **42**, 580.
39. Ronnow, P. H. and Gunnarsson, L. A. H. (1982). *FEMS Microbiol. Lett.* **14**, 311.
40. Bhatnagar, L., Henriquet, M., Zeikus, J. G., and Aubert, J.-P. (1984). *FEMS Microbiol. Lett.* **22**, 155.
41. Hilpert, R., Winter, J., Hammes, W., and Kandler, O. (1981). *Zentralblatt Bakteriol., Mikrobiol. Hyg.* Abt. 1, **Orig C2**, 11.
42. Bock, A. and Kandler, O. (1985). In *The Bacteria*, Vol. 8 (ed. C. R. Woese and R. S. Wolfe), pp. 525–44. Academic Press, New York.
43. Eirich, L. D., Vogels, G. D., and Wolfe, R. S. (1978). *Biochemistry* **17**, 4583.
44. Eirich, L. D., Vogels, G. D., and Wolfe, R. S. (1982). In *Flavins and Flavoproteins* (ed. V. Massey and C. H. Williams), pp. 435–41. Elsevier/North Holland Publishing Co, Amsterdam.
45. Jones, W. J., Whitman, W. B., Fields, R. D., and Wolfe, R. S. (1983). *Appl. Environ. Microbiol.* **46**, 220.
46. Jain, M. K. and Zeikus, J. G. (1987). *Appl. Environ. Microbiol.* **53**, 1387.
47. Belyaev, S. S., Wolkin, R., Kennealy, W. R., DeNiro, M. J., Epstein, S., and Zeikus, J. G. (1983). *Appl. Environ. Microbiol.* **45**, 691.
48. Macario, A. J. L. and Conway de Macario, E. (1988). *Appl. Environ. Microbiol.* **54**, 79.

49. Bhatnagar, L., Henriquet, M., and Longin, R. (1983). *Biotechnol. Lett.* **5**, 39.
50. Daniels, L., Belay, N., and Mukhopadhyay, B. (1984). *Biotech. Bioeng. Symp.* **14**, 199.
51. Bhatnagar, L., Zeikus, J. G., and Aubert, J.-P. (1986). *J. Bacteriol.* **165**, 638.
52. Schonheit, P., Moll, J., and Thauer, R. K. (1980). *Arch. Microbiol.* **127**, 59.
53. Zeikus, J. G., Fuchs, A., Kenealy, W., and Thauer, R. K. (1977). *J. Bacteriol.* **132**, 604.
54. Hippe, H. (1984). In *Maintenance of Microorganisms—A Manual of Laboratory Methods* (ed. B. E. Kirsop and J. J. S. Snell), pp. 69–81. Academic Press, London.
55. Impey, C. S. (1986). In *Anaerobic Bacteria in Habitats Other Than Man* (ed. E. M. Barnes and G. C. Mead), p. 411–19. Blackwell Scientific Publications, Oxford.

13

Isolation and cultivation of anoxygenic phototrophic bacteria

B. J. TINDALL

1. Introduction

Photosynthetic or phototrophic bacteria include two groups of organisms, the anoxygenic phototrophic bacteria and the oxygenic cyanobacteria (often referred to as the blue-green algae). Although it has been demonstrated that certain cyanobacteria are capable of anaerobic growth, the subjects of the present chapter are the anoxygenic red, purple, and green phototrophic bacteria.

Phototrophic bacteria have been the subject of scientific study for over 100 years. Many of the first detailed observations on such organisms were made by Bavendamm (1), Buder (2), Cohn (3), Ehrenberg (4), Lauterborn (5), Molisch (6), van Niel (7, 8), and Winogradsky (9). During the course of such work the concept was developed that all such organisms could be grouped into one of three major families, the 'Athiorhodaceae' (Rhodospirillaceae or purple non-sulphur bacteria), the 'Thiorhodaceae' (Chromatiaceae or purple sulphur bacteria), and the 'Chlorobacteriaceae' (Chlorobiaceae or green sulphur bacteria), and that there was a unifying taxonomic character in their ability to grow phototrophically under anaerobic conditions (10). Studies on the amino-acid sequences of the cytochromes (11–13) and cataloguing of the 16S rRNA molecules (14) demonstrated an unexpected heterogeneity in the evolutionary tree of members of the family Rhodospirillaceae and led to the discovery that members of the family Rhodospirillaceae were distributed throughout three subdivisions of one of the major branches of the prokaryotic evolutionary tree (15). While we can no longer refer to the non-sulphur purple bacteria as the family Rhodospirillaceae in the classical sense, the other families seem to be less heterogeneous, although they are scattered throughout the prokaryotic evolutionary tree, which indicates that the property of phototrophic growth can no longer be used as a unifying taxonomic feature of these organisms. The currently described genera and species, together with some of their properties, are listed in *Tables 1–5*.

Table 1. Characteristics of members of the family Chromatiaceae[a]

Species[b]	**Pigments**[c]	**G:C ratio (mol %)**	**Shape**	**Size (μm)**	**Gas vacuoles/ motility**	**Colour**
Amoebobacter pedioformis	a/1a	65.5	Sphere/ovoid	2.0 × 2.0–3.0	g.v./–	Pink-red
Amoebobacter pendens	a/1a	65.3	Sphere/ovoid	1.5 × 2.0	g.v./–	Pink-red
Amoebobacter purpureus	a/1a	63.4–64.1	Sphere/ovoid	1.9 × 2.3/2.0 × 3.2	g.v./–	Purple-red
Amoebobacter roseus	a/1a	64.3	Sphere/ovoid	2.0–3.0	g.v./–	Pink-red
Chromatium buderi	a/2	62.2–62.8	Rod	3.5–4.5 × 4.5–9.0	+	Purple-violet
Chromatium gracile	a/1a	68.9–70.4	Rod	1.0–1.3 × 2.0–6.0	+	Brown-red
Chromatium minus	a/4a	52.0–62.2	Rod	2.0 × 2.5–6.0	+	Purple-red
Chromatium minutissimum	a/1a	63.7	Rod	1.0–1.2 × 2.0	+	Brown-red
Chromatium okenii	a/4a	48.0–50.0	Rod	4.5–6.0 × 8.0–15.0	+	Purple-red
Chromatium purpuratum	a/4a	68.9	Rod	1.2–1.7 × 3.0–4.0	+	Purple-red
Chromatium salexigens	a/2	64.6	Rod	2.0–2.5 × 4.0–7.5	+	Purple-violet
Chromatium tepidum	a/1a	61.5	Rod	1.0–2.0 × 2.8–3.2	+	Red
Chromatium vinosum	a/1a	61.3–66.3	Rod	2.0 × 2.5–6.0	+	Brown-red
Chromatium violascens	a/2	61.8–64.3	Rod	2.0 × 2.5–6.0	+	Purple-violet
Chromatium warmingii	a/2	55.1–60.2	Rod	3.5–4.0 × 5.0–11.0	+	Purple-violet
Chromatium weissei	a/4a	48.0–50.0	Rod	4.5 × 7.0–9.0	+	Purple-red

Lamprobacter modestohalophilus	a/4a	64.0	Rod	2.0–2.5 × 4.0–5.0	g.v./+	Purple-red
Lamprocystis roseopersicina	a/2	63.8	Sphere/ovoid	3.0–3.5	g.v./+	Purple-violet
Thiocapsa pfennigii	b/1a	69.4–69.9	Spherical	1.2–1.5	–	Orange-brown
Thiocapsa roseopersicina	a/1a	63.3–66.3	Spherical	1.2–3.0	–	Pink-red
Thiocystis gelatinosa	a/4b	69.4–69.9	Spherical	3.0	+	Purple-red
Thiocystis violacea	a/2	63.3–66.3	Spherical	2.5–3.0	+	Purple-violet
Thiodictyon bacillosum	a/2	66.3	Rod	1.5–2.0 × 3.0–6.0	g.v./–	Purple-violet
Thiodictyon elegans	a/2	65.3–66.3	Rod	1.5–2.0 × 3.0–8.0	g.v./–	Purple-violet
Thiopedia rosea	a/4a	62.5–63.6	Spherical/in sheets	1.0–2.0 × 1.2–2.5	g.v./–	Purple-red
Thiospirillum jenense	a/1a	45.5	Spirillum	2.5–4.5 × 30.0–40.0	+	Orange-brown

[a] Data from references 26 and 62.
[b] Names not included in the Approved Lists of Bacterial Names, or subsequently validly published, are enclosed in quotation marks.
[c] Pigment type as defined in references 10 and 61; bacteriochlorophyll/carotenoid composition.

Table 2. Characteristics of members of the family Ectothiorhodospiraceae[a]

Species[b]	Pigments[c]	G:C ratio (mol %)	Shape	Size (μm)	Gas vacuoles/ motility	Colour
Ectothiorhodospira mobilis	a/1a	67.3–69.9	Vibrio	0.7–1.0 × 2.0–2.6	+	Red
Ectothiorhodospira shaposhnikovii	a/1a	62.3–64.0	Vibrio	0.8–0.9 × 1.5–2.5	+	Red
Ectothiorhodospira vacuolata	a/1a	61.4–63.6	Rod	1.5 × 2.0–2.4	g.v./+	Red/pink
Ectothiorhodospira marismortui	a/1a	65.0	Rod	0.9–1.3 × 1.5–3.3	+	Red
Ectothiorhodospira abdelmalekii	b/1a	63.3–63.8	Spirillum	0.9–1.2 × 4.0–6.0	+	Green
Ectothiorhodospira halochloris	b/1a	52.9	Spirillum	0.5–0.6 × 2.5–8.0	+	Green
Ectothiorhodospira halophila	a/1a	68.4	Spirilium	0.8 × 5.0	+	Red

[a] Data from references 26 and 30.
[b] Names not included in the Approved Lists of Bacterial Names, or subsequently validly published, are enclosed in quotation marks.
[c] Pigment type as defined in references 10 and 61; bacteriochlorophyll/carotenoid composition.

2. Enrichment methods

2.1 General considerations

A number of methods have been developed for the enrichment and isolation of strains of anoxygenic phototrophic bacteria. The two commonly used methods are inoculation directly into synthetic medium, or the use of pre-enrichment in mud and water-filled glass cylinders. Irrespective of the method used a number of factors have to be taken into consideration which affect the type of strains isolated, and these include salinity, temperature, and pH.

2.1.1 Temperature

It is obvious that the incubation temperature will affect the isolation of mesophilic or thermophilic strains of anoxygenic phototrophic bacteria. To date only a limited number of thermophilic strains have been isolated, including *Chloroflexus aurantiacus* (16), *Chromatium tepidarius* (17), and *Heliothrix oregonensis* (18). While such organisms require temperatures of about 45°C and above, the majority of anoxygenic phototrophic bacteria require temperatures not in excess of 30°C. Temperatures between 25°C and 30°C generally select for the more rapid growing strains (19, 20). Lower temperatures of between 15°C and 20°C have been used to enrich for gas vacuolate strains, for large-celled members of the family Chromatiaceae, and members of the family Chlorobiaceae (19, 20). When incubating cultures in the laboratory it should be remembered that a standard light bulb will also generate heat. This can be used to advantage by placing glass columns of hot spring mat material close to a light source, which has the effect of generating a temperature gradient in the column (21). The heating effect of a light bulb can be critical in the cultivation of large-scale cultures, and the heating effect can be minimized by placing a flat bottle(s) filled with water between the light source and the culture.

2.1.2 pH

van Niel (7) was one of the first to demonstrate the effect of pH on the development of green and purple sulphur photosynthetic bacteria. Under alkaline conditions it has been shown that a number of different species may be isolated, although the organisms that are favoured under these conditions and that may give rise to blooms in the natural environment are members of the family Ectothiorhodospiraceae (22–24). It has been shown in some strains within the family Ectothiorhodospiraceae that there may be an optimum at pH 8.0–8.5 under photoautotrophic conditions, while under photoheterotrophic conditions the optimum is higher, at pH 9.5–10.5 (24). There is, in addition, evidence to show that members of the genus *Chloroflexus* (16) also have an alkaline pH optimum, of pH 8.0–8.5. At the other end of the extreme there are few acidophiles, although *Rhodopseudomonas acidophila* (25) is a notable exception. Enrichment at pH values of between 5.0 and 5.5 may be used to select for *Rhodomicrobium*

Table 3. Characteristics of 'purple non-sulphur' phototrophic bacteria (formerly family Rhodospirillaceae)[a]

Species[b]	Pigments[c]	G:C ratio (mol %)	Shape	Size (µm)	Motility	Colour
Rhodobacter adriaticus	a/3	66.7	Ovoid/rod	0.5–0.8 × 1.3–1.8	–	Yellow-brown/brown
Rhodobacter capsulatus	a/3	65.5–66.8	Ovoid/rod	0.5–1.2 × 2.0–2.5	+	Yellow-brown/brown
Rhodobacter euryhalinus	a/3	62.1–68.6	Ovoid/rod	0.7–1.0 × 1.5–3.0	+	Yellow-brown/brown
Rhodobacter sphaeroides	a/3	68.4–69.9	Sphere/ovoid	0.7–4.0 × 2.5–4.0	+	Green-brown/brown
Rhodobacter sulfidophilus	a/3	68.9–73.2	Ovoid/rod	0.5–0.8 × 1.3–1.8	+	Yellow-green/brown
Rhodobacter veldkampii	a/3	64.4–67.5	Ovoid/rod	0.6–0.8 × 1.0–1.3	–	Yellow-brown/red
Rhodocyclus gelatinosus	a/3	70.5–72.4	Curved rod	0.4–0.7 × 1.0–3.0	+	Peach
Rhodocyclus purpureus	a/2	65.3	Half circle	0.6–0.7 × 2.7	–	Purple-violet
Rhodocyclus tenuis	a/1a,2	64.8	Curved rod	0.3–0.5 × 1.5–6.0	+	Brown-red
Rhodomicrobium vannielii	a/1a	61.8–63.8	Ovoid	1.0–1.2 × 2.0–2.8	+	Pink/brown/red

Rhodopila globiformis	a/4b	66.3	Sphere/ovoid	1.6–1.8	+	Purple-red
Rhodopseudomonas acidophila	a/1a	62.2–66.8	Rod	1.0–1.3 × 2.0–5.0	+	Orange-brown/red
Rhodopseudomonas blastica	a/3	65.3	Rod	0.6–0.8 × 1.0–2.5	−	Brown
'Rhodopseudomonas julia'	a/1a	63.5	Rod	1.5 × 2.5	+	Pink/Purple
Rhodopseudomonas marina	a/1a	61.5–63.8	Rod	0.7–0.8 × 1.0–2.5	+	Pink/red
Rhodopseudomonas palustris	a/1a	64.8–66.3	Rod	0.6–0.9 × 1.2–2.0	+	Red-brown/red
Rhodopseudomonas rutila	a/?	67.6–69.4	Rod	0.4–1.0 × 1.5–3.0	+	Red
Rhodopseudomonas sulfoviridis	b/3	67.8–68.4	Rod	0.5–0.9 × 1.2–2.0	+	Olive green
Rhodopseudomonas viridis	b/1c	66.3–71.4	Rod	0.6–0.9 × 1.2–2.0	+	Olive green
Rhodospirillum fulvum	a/1a	64.3–65.3	Spirillum	0.5–0.7 × 3.5	+	Brown
'Rhodospirillum mediosalinum'	a/1a	66.6	Spirillum	0.8–1.0 × 2.2–6.0	+	Pink/brown-red
Rhodospirillum molischianum	a/1a	60.5–64.8	Spirillum	0.7–1.0 × 5.0–8.0	+	Brown
Rhodospirillum photometricum	a/1a	64.8–65.8	Spirillum	1.2–1.5 × 7.0–10.0	+	Brown
Rhodospirillum rubrum	a/1a	63.8–65.8	Spirillum	0.8–1.0 × 7.0–10.0	+	Red
Rhodospirillum salexigens	a/1a	64.0	Spirillum	0.6–0.7 × 1.0–6.0	+	Red
Rhodospirillum salinarum	a/1a	67.4	Spirillum	0.8–0.9 × 2.0–3.5	+	Red

[a] Data from references 26, 38, 42, and 64.
[b] Names not included in the Approved Lists of Bacterial Names, or subsequently validly published, are enclosed in quotation marks.
[c] Pigment type as defined in references 10 and 61; bacteriochlorophyll/carotenoid composition.

Table 4. Characteristics of members of the family Chlorobiaceae[a]

Species[b]	Pigments[c]	G:C ratio (mol %)	Shape	Size (μm)	Gas vacuoles/ motility	Colour
Ancalochloris perfilievi[d]	n.d./n.d.	n.d.	Prosthecate	0.5–1.0	g.v./–	Green/yellow-green
'Clathrochloris sulfurica'[d]	n.d./n.d.	n.d.	Spherical	0.5–1.5	g.v./–	Green
Chlorobium chlorovirioides	d/5b	54.0	Rod to vibrio	0.3–0.4 × 0.4–0.8	–	Green/green-yellow
Chlorobium limicola	c(d)/5b	51.0–52.0	Rod, in chains	0.7–1.1 × 0.9–1.5	–	Green
Chlorobium phaeobacteroides	e/5a	49.0–50.0	Rod	0.6–0.8 × 1.3–2.7	–	Yellow- or red-brown
Chlorobium phaeovibrioides	e/5b	52.0–53.0	Rod to vibrio	0.3–0.4 × 0.4–0.8	–	Yellow- or red-brown
Chlorobium vibrioforme	d(c)/5b	52.0–57.0	Rod to vibrio	0.5–0.7 × 1.0–2.0	–	Green
Chloroherpeton thalassium	c/5c	47.8	Filament	1.0 × 8.0–20.0	g.v./gliding	Green
Pelodictyon clathratiforme[d]	c/5b	n.d.	Rod, network	0.7–1.2 × 1.5–2.5	g.v./–	Green
Pelodictyon luteolum	c or d/5b	58.1	Rod or ovoid	0.6–0.9 × 1.2 – 2.0	g.v./–	Green
Pelodictyon phaeum	e/5a	n.d.	Rod	0.6–0.9 × 1.0–2.0	g.v./–	Brown
Pelodictyon phaeoclathratiforme	e/5a	47.9	Rod, network	0.7–1.1 × 1.5–3.0	g.v./–	Brown
Prosthecochloris aestuarii	c/5b	52.0–56.1	Prosthecate	0.5–0.7 × 1.0–1.2	–	Green
Prosthecochloris phaeoasteroidea	e/5a	52.2	Prosthecate	0.5–0.6 × 0.5–1.2	–	Brown

[a] Data from references 10, 26, and 63.
[b] Names not included in the Approved Lists of Bacterial Names, or subsequently validly published, are enclosed in quotation marks.
[c] Pigment type as defined in references 10 and 61; bacteriochlorophyll/carotenoid composition.
[d] Not in pure culture.

Table 5. Characteristics of green anoxygenic phototrophic bacteria not in the family Chlorobiaceae[a]

Species[b]	Pigments[c]	G:C ratio (mol %)	Shape	Size (μm)	Gas vacuoles/ motility	Colour
Family Chloroflexaceae						
Chloroflexus aurantiacus	c/5c	53.0–55.0	Filament	0.6–0.7 × 2.0–6.0	Gliding	Green
Genera of uncertain affiliation						
Chloronema giganteum	d/n.d.	n.d.	Filament	2.0–2.5 × 3.5–4.5	g.v./gliding	Green
'Chloronema spiroideum'[d]	c/n.d.	n.d.	Filament	1.5–2.0 × 2.5–3.0	g.v./gliding	Green
Heliothrix oregonensis	a/n.d.	n.d.	Filament	1.5 × 10.0–20.0	Gliding	Orange
Oscillochloris chrysea	c/n.d.	n.d.	Filament	4.5–5.5 × 3.5 – 7.0	Gliding	Yellow-green
Oscillochloris trichoides	c/5c	n.d.	Filament	1.0–1.4 × 2.3–3.8	Gliding	Yellow-green
Gram-positive anoxygenic phototrophs						
Heliobacterium chlorum	g/n.d.	52.0	Rods	1.0 × 4.0–10.0	Gliding	Green
'Heliobacillus mobilis'	g/n.d.	50.3	Rods	1.0 × 7.0–10.0	Flagellated	Green

[a] Data from references 26 and 50.
[b] Names not included in the Approved Lists of Bacterial Names, or subsequently validly published, are enclosed in quotation marks.
[c] Pigment type as defined in references 10 and 61; bacteriochlorophyll/carotenoid composition.
[d] The status of *'Chloronema spiroideum'* is uncertain (26).

vannielii and *Rhodopseudomonas globiformis* (26). The majority of the members of the family Chromatiaceae are enriched and grow at pH values between 7.0 and 8.5, although van Niel (7) presented evidence that under autotrophic conditions the optimum for some strains is about pH 8.0. In contrast, members of the family Chlorobiaceae generally have optima in the region pH 6.5–7.0. Due to the overlap in pH optima between members of the families Chlorobiaceae and Chromatiaceae it is not usually possible to use pH alone to enrich selectively for one group to the exclusion of the other. The remaining anoxygenic phototrophic bacteria have optima at neutral pH.

2.1.3 Salinity

Anoxygenic phototrophic bacteria have been isolated from a number of saline habitats. In many cases the strains isolated from sea water have been shown to be closely related to similar organisms from fresh-water habitats (27). A number of other strains have been described which have been isolated only from saline environments, typically highly saline bodies of water. Such moderate or extreme halophiles include *Ectothiorhodospira vacuolata* (28), *E. mobilis* (29), *E. marismortui* (30), *E. halochloris* (31), *E. abdelmalekii* (32), *E. halophila* (33), *Chromatium salexigens* (34), *C. buderi* (35), *Rhodospirillum salexigens* (36), *Rhsp. salinarum* (37), '*Rhsp. mediosalinum*' (38), *Rhodopseudomonas sulfidophila* (39), *Rhps. marina* (40), *Rhodobacter adriatica* (41), *Rhb. euryhalinus* (42), and *Lamprobacter modestohalophilus* (43). Under certain conditions these organisms may form blooms in the water column (e.g. members of the genus *Ectothiorhodospira* in alkaline lakes) or may be found in distinct zones in the upper layers of anoxic, illuminated sediment (e.g. the gypsum crust or trona crust of neutral or alkaline saline lakes).

2.1.4 Sulphide concentration

Early research suggested that there was a clear distinction between members of the purple non-sulphur bacteria, which were inhibited by and could not utilize sulphide, and members of the green and purple sulphur bacteria which grew on sulphide. This has been shown not to be the case, and certain strains of purple non-sulphur bacteria have been shown to metabolize sulphide (44). The ability of some strains to tolerate sulphide is evident in enrichments from saline environments, where media designed for the isolation of members of the family Chromatiaceae at low salinity often give rise to populations of sulphide-tolerant purple non-sulphur bacteria.

Sulphide, above a certain concentration, is toxic to all anoxygenic phototrophic bacteria and this should be taken into consideration when preparing media. Green sulphur bacteria are generally enriched or grown in media containing 3.0 mM sulphide, while purple sulphur bacteria tolerate lower concentrations and 1.5 mM sulphide is used for such organisms (20). Higher cell densities are achieved by feeding these organisms at regular intervals with sulphide. To a certain extent the level of sulphide utilized in enrichment cultures may influence the type of organisms isolated.

2.1.5 Light intensity

Light intensity plays a significant role in the ecology and isolation of anoxygenic phototrophic bacteria (19, 20). Members of the family Chlorobiaceae are favoured under comparatively low light intensity 5–50 lux, and may be enriched from samples where purple sulphur bacteria normally predominate. Under conditions where members of the family Chlorobiaceae predominate, gas vacuolate strains may be enriched at 20–100 lux, while higher light intensities (500–2000 lux) favour members of the genera *Chlorobium* or *Prosthecochloris*.

Light intensity may also be used to select for certain strains of purple sulphur bacteria, small-celled non-gas-vacuolate members of this family being enriched at 1000–2000 lux, while large-celled strains and gas-vacuolate strains are often favoured at 50–300 lux, usually with a light–dark regime of 12 h light and 12 h dark, 6 h light and 6 h dark, or 4 h light and 8 h dark. In the absence of a suitable light incubator, placing the enrichment cultures in a window which does not receive direct sunlight (a north-facing window is ideal) often gives good results. As a guide to the light intensity produced by incandescent lamps Pfennig (19) gave the following values

- 100–300 lux, 45–25 cm from a 25-watt bulb
- 300–700 lux, 25–15 cm from a 25-watt bulb
- 700–2000 lux, 25–15 cm from a 60-watt bulb

Apart from light intensity, the use of selected wavelengths may be advantageous in the enrichment and isolation of certain strains (20). Under conditions where members of the family Chlorobiaceae would normally be enriched, their development may be suppressed by illuminating cultures at wavelengths above their bacteriochlorophyll absorption maximum (i.e. above 800 nm). Using wavelengths of about 1000 nm and greater, strains containing bacteriochlorophyll b may be selectively enriched (e.g. *E. abdelmalekii*, *E. halochloris*, *Thiocapsa pfennigii*, *Rhodopseudomonas sulfoviridis*, or *Rhps. viridis*). Details of suitable filters may be obtained from the *CRC Handbook of Chemistry and Physics* (46). The use of tungsten filament lamps as opposed to fluorescent lights is usually recommended, although a number of species have been successfully grown using Philips 'daylight' fluorescent tubes (22).

2.1.6 Oxygen concentration

When anoxygenic phototrophic bacteria are isolated or enriched from natural samples under phototrophic conditions, the medium is prepared anaerobically and oxygen concentration is not an important factor. One exception has been the methods used for the aerobic, heterotrophic isolation of *Chloroflexus aurantiacus* (16). However, many strains within the families Chromatiaceae, Ectothiorhodospiraceae, and the purple non-sulphur bacteria are tolerant of oxygen to some degree (26, 47, 48). Several species within the purple non-sulphur bacteria will grow aerobically, while some members of the family Chromatiaceae and Ectothiorhodospiraceae also grow chemoauto- or chemoheterotrophically in the dark under aerobic or microaerophilic conditions (26, 47, 48). Members of the family Chlorobiaceae (26) or members of the genera *Heliobacterium* (49) and '*Heliobacillus*' (50) do not appear to be capable of growth in the presence of oxygen.

2.1.7 Organic substrates

The use of an organic substrate for the isolation of members of the anoxygenic phototrophic bacteria is a factor in determining which strains will be enriched. In the case of members of the purple non-sulphur bacteria the range of substrates utilized for phototrophic growth has been well studied, and it is possible to use certain substrates in conjunction with other factors to selectively enrich for certain strains (26, 45). Benzoate may be used to enrich for *Rhodocyclus purpureus* or *Rhodopseudomonas palustris* (26, 45), while fatty acids are suitable for the enrichment of a restricted range of strains, such as brown-coloured *Rhodospirillum* spp. (45). In enrichment cultures for members of the green and sulphur bacteria, the inclusion of organic substrates is not advisable when sulphide-tolerant strains of purple non-sulphur bacteria are present (20). The use of organic substrates will often encourage the growth of faster growing species. However, the inclusion of acetate in pure cultures of green and sulphur bacteria will give higher cell yields, and 0.05% (w/v) or 0.03% (w/v) ammonium acetate (or a mixture of ammonium and magnesium acetate) may be added to cultures of purple and green sulphur bacteria, respectively. In certain members of the family Chromatiaceae (*Thiodictyon elegans* and *Lamprocystis roseopersicina*) 0.05% (w/v) acetate may be too high, resulting in rupture of the cell due to the overproduction of storage material (polyhydroxybutyrate), and 0.025% (w/v) acetate is recommended for these strains (51).

2.1.8 Diazotrophic growth

It has been shown that many strains of anoxygenic phototrophic bacteria are capable of fixing nitrogen. By selecting a medium in which ammonium salts are omitted and nitrogen is used in the gas phase, the different rates at which various strains grow under nitrogen-fixing conditions may be used for selective enrichment (52).

2.1.9 Combination of factors

In considering the factors influencing the enrichment of various strains the use of a combination of conditions should not be forgotten. Thus it is often useful when isolating different strains of green and purple sulphur bacteria to use a combination of light intensity, temperature, pH, etc. In this way the selective pressure exerted by each parameter narrows the range of strains which may be expected.

3. Enrichment in natural samples

In many cases the presence of anoxygenic phototrophic bacteria may be detected by the pigmentation of natural environments due to the presence of blooms of these organisms. In the absence of such blooms it is possible to remove material from the natural environment and incubate it under conditions which are suitable for the enrichment of certain strains. This is the type of method which was used by Winogradsky and subsequent workers and is a valuable method for preliminary enrichment. Pfennig (19) reviewed the use of this method, which is often referred to as the 'Winogradsky column'.

3.1 Preparation of Winogradsky columns

Winogradsky columns are prepared in tall glass cylinders. In practice either simple glass columns may be used, with a tightly fitting lower rubber bung, or glass measuring cylinders (500-ml, 1-l, or 2-l sizes). In many cases a source of sulphate for sulphate-reducing bacteria is added to provide a long-term supply of sulphide. Since these columns are designed to be small-scale ecosystems it is not important to completely fill the column, nor does air have to be excluded from the head space. However, the mud and water layer should be sufficiently deep to allow anaerobic conditions to develop in the lower layers.

Protocol 1. Preparation of a Winogradsky column for green and purple sulphur bacteria

1. Fill the column with a mixture of mud and water from the natural environment under study.
2. Add to this mixture chopped plant material and small pieces of gypsum.
3. Cover the cylinder and incubate in a window without direct sunlight (a north-facing window is ideal).
4. Green sulphur bacteria develop in the mud on the illuminated side, while purple sulphur bacteria develop at the mud surface and in the water (19).

Protocol 2. Alkaline enrichment in Winogradsky columns

1. In the bottom of the column place a layer of proteinaceous material (such as egg white or dried meat).
2. Overlay this with several cm of garden soil, in which is embedded a pea-sized piece of potassium sulphide, surrounded by small pieces of gypsum.
3. On top of this layer, place about 2 cm of sand.
4. Add the sample of mud and water and cover the column.
5. Incubate as described in *Protocol 1*.

Protocol 2. *Continued*

6. Due to the higher pH, this type of column usually enriches for small-celled *Chromatium* spp. (19).

Protocol 3. Enrichment in the presence of sulphate-reducing bacteria

1. Mix together two parts of fresh sewage sludge, two parts of garden soil, and one part of precipitated calcium sulphate.
2. Fill the bottom quarter of the cylinder with this mixture.
3. Overlay with rain water (or tap water).
4. Incubate in the dark at 20–28°C for 14 days, until sulphide production commences.
5. Inoculate the column with mud or water containing the anoxygenic phototrophs.
6. Green sulphur bacteria often develop in this type of column to the exclusion of purple sulphur bacteria.
7. Purple sulphur bacteria can be enriched by using infra-red filters to suppress the development of green sulphur bacteria.
8. The temperature and light intensity influence which species of purple sulphur bacteria will be enriched (19).

Protocol 4. Enrichment of sulphide-intolerant strains

1. Mix mud or soil with a suitable organic material (decomposing plant matter, dried meat, or eggs) and place in the cylinder.
2. Overlay with water, either from the natural environment or tap water.
3. Place the cylinder in indirect sunlight.
4. In the absence of significant sulphide production this method is ideal for enrichment of strains that are inhibited by sulphide or are not dependent on sulphide (usually non-sulphur purple bacteria).

---

4. Isolation of pure cultures

A number of methods has been used for the isolation of pure cultures of anoxygenic phototrophic bacteria. Although the majority of strains can be cultivated on media solidified with agar there are some isolates, particularly the large-celled members of the family Chromatiaceae, that have only been cultivated successfully in liquid culture. The most commonly used methods can be divided into two categories, purification using agar-solidified medium and purification in liquid culture. When using agar media it should be remembered that certain strains will grow heterotrophically in the dark under aerobic conditions, and thus may be isolated using methods suitable for normal aerobes, although this method is not commonly used.

4.1 Purification in agar deeps

Purification of anoxygenic phototrophic bacteria is usually carried out using the well tested method of agar deeps or agar shakes (20, 45, 53). Using this method the medium is solidified with soft agar (0.9% w/v) in glass tubes, and the anoxygenic phototrophic bacteria grow as distinct colonies in the agar. In the method described by Pfennig (53) normal test tubes are used with tight-fitting cotton bungs, but the use of Hungate-type tubes fitted with a rubber septum appears to be equally suitable and may even have certain advantages. In order to prepare agar deeps a suitable medium is prepared in the usual manner (see *Tables 6–14* and Section 5), but agar is added to give a final concentration of 0.9% (w/v), resulting in a semi-solid agar. To free the agar of soluble impurities the powder should be washed several times with distilled water before use, either on a filter or by suspending in a flask.

Protocol 5. Purification of anoxygenic phototrophs in semi-solid agar using media not containing bicarbonate, carbonate, or high salt concentrations

You will need

- Water bath at 50°C
- Cold water bath
- Sterile mixture of liquid paraffin: paraffin wax, usually in ratio of 3:1 (v/v)[a]

1. Prepare medium as described in *Table 8*.
2. Add agar to a final concentration of 0.9% (w/v). Adjust pH to the desired value.
3. Heat the medium to dissolve the agar; then dispense in 10-ml aliquots prior to autoclaving. Alternatively, the medium may be dispensed aseptically after autoclaving; this method is preferable if heat-sensitive compounds are to be added to the medium after autoclaving.
4. Keep the tubes of sterile medium in a water bath at 50°C, and have a bath of cold water ready for solidifying the inoculated medium.
5. Inoculate the first tube with the culture to be purified. Mix by gentle inversion. Do not shake or use a vortex mixer.

Protocol 5. *Continued*

6. Transfer 0.5–1.0 ml of the inoculated medium to a second tube; then place the first tube in the cold water bath to solidify the medium.
7. Mix the contents of the second tube; then inoculate a third tube in the same way. Continue until a dilution series of six to 10 tubes is obtained.
8. It is important to prevent drying out of the medium or the entry of oxygen. If Hungate tubes are used (see Chapter 1), flush out each tube with oxygen-free nitrogen and then seal tightly. If test tubes sealed with a cotton-wool plug are used, add a 1-cm layer of the sterile paraffin wax mixture, which solidifies at 51–54°C. Some contraction of the wax plug will occur as it sets. Gently rewarm the plug before incubating to ensure a good seal.
9. If cotton-wool plugs and a wax seal are used, the medium will have become slightly aerated. Incubate the tubes for several hours in the dark to allow for the removal of oxygen. This is a wise routine to adopt even if tubes are flushed out with nitrogen.
10. Incubate the tubes under suitable conditions in the light until colonies form.

[a] For incubation temperatures of 30°C or higher, a higher proportion of wax is required.

Protocol 6. Purification of anoxygenic phototrophs in semi-solid agar using media containing bicarbonate, carbonate, or high salt concentrations[a]

1. Prepare 3% (w/v) agar and dispense in 3-ml aliquots in test tubes before autoclaving for 15 min at 121°C.
2. After autoclaving place the tubes in a water bath at 50°C[b] and add to each tube 6 ml of sterile growth medium.
3. Inoculate and mix the tubes as described in *Protocol 5*.

[a] For isolation of members of the families Ectothiorhodospiraceae, Chromatiaceae, and Chlorobiaceae (see *Tables 6* and *11* for suitable media) and for strains requiring high salt concentrations.

[b] At high salinities the water bath temperature should be increased to 55–60°C.

4.1.1 Removing single colonies from agar deeps

Single colonies may be obtained from agar deeps by several methods, each of which has its own advantages and disadvantages. Once isolated, the colonies may be broken up by resuspending in a small volume of sterile growth medium, and this suspension can then be used for inoculation of fresh media and for microscopic examination.

Table 6. Media for the cultivation of green and purple sulphur bacteria (20, 56, 57)

Constituents	Medium 1 (g/l)	Medium 2 (g/l)	*Chloroherpeton* (g/l)	*Heliothrix* (g/l)
Solution 1				
KH_2PO_4	0.34	1.0	0.01	0.3
NH_4Cl	0.34	0.5	0.42	—
$(NH_4)_2SO_4$	—	—	—	0.5
KCl	0.34	—	0.38	0.5
(NaCl)[a]	(20.0–30.0)	(20.0–30.0)	9.5–12.0	1.0
$MgSO_4 \cdot 7H_2O$[a]	0.5	0.4	—	0.4
$MgCl_2 \cdot 6H_2O$	—	—	1.02	—
$CaCl_2 \cdot 2H_2O$	0.25	0.05	0.15	0.32
$FeCl_3$ solution (0.29 g/l)	—	—	—	1.0 ml
Trace element solution	—	—	—	10 ml (SL7)
Distilled water	800 ml	950 ml	950 ml	1000 ml
pH	—	—	—	7.6
Solution 2				
Distilled water	172 ml	—	—	—
Solution 3				
Vitamin B_{12}[b]	1.0 ml	1.0 ml	2.5 ml	625 μl (medium IM)
Vitamin solution (IMC)[c]	—	—	—	0.5 ml (medium IMC)
Solution 4				
Trace element solution[d]	1.0 ml (SL7)	1.0 ml (SL8)	1.0 ml (SL8)	—
Solution 5				
Bicarbonate solution	20 ml (7.5%)	40 ml (5.0%)	50 ml (5.0%)	100 ml (0.42%)
Solution 6				
10% $Na_2S \cdot 9H_2O$ solution	4 or 6 ml	—	1.5 ml	—
5% $Na_2S \cdot 9H_2O$ solution	—	6 or 12 ml	—	—

[a] For strains from marine habitats, 2.0–3.0% (w/v) NaCl is added, and the $MgSO_4 \cdot 7H_2O$ concentration may be raised to 3.0% (w/v). *Chromatium salexigens* is cultivated in medium containing 10.0% (w/v) NaCl, 2.0% (w/v) $MgSO_4 \cdot 7H_2O$ and 1.0% $MgCl_2 \cdot 6H_2O$. *C. buderi* is cultivated in media containing 5% NaCl.

[b] 2 mg/100 ml, filter-sterilized.

[c] See *Table 14*.

[d] A number of formulations have been used, including trace element solutions SL7, SL8, and SL11 (see *Table 13*).

[e] Prepare the sodium bicarbonate solution in water, saturate it with CO_2 by passing CO_2 through it and filter-sterilize it with CO_2 overpressure (the CO_2 need not be sterile) using a suitable apparatus (e.g. Sartorius SM 165 79, SM 162 63, or similar unit) and store in sterile bottles.

Table 7. Medium for *Chloroflexus* (21)

Constituents	(g/l)
Nitrilotriacetic acid	0.1
Micronutrient solution[a]	0.5 ml
$FeCl_3$ solution (0.29 g/l)	1.0 ml
$CaSO_4 \cdot 2H_2O$	0.06
$MgSO_4 \cdot 7H_2O$	0.10
NaCl	0.008
KNO_3	0.10
$NaNO_3$	0.70
Na_2HPO_4	0.11

[a] See *Table 13.*

Table 8. Medium for the cultivation and isolation of purple non-sulphur bacteria (45)

Constituents	(g/l)
KH_2PO_4	0.5
NH_4Cl	0.4
NaCl	0.4
$MgSO_4 \cdot 7H_2O$	0.2
$CaCl_2 \cdot 2H_2O$	0.05
Yeast extract	0.2
Fe–citrate solution (0.1 g/100 ml)	5 ml
Organic substrate (e.g. acetate)	1.0
Trace element solution (SL7a)[a]	1 ml
Vitamin B_{12} solution (1 mg/100 ml)[b]	1 ml
Distilled water	993 ml

[a] See *Table 13.*
[b] Filter-sterilized; add after autoclaving.

Table 9. Medium for the cultivation of *Heliobacterium* and *Heliobacillus* (49, 50)

Constituents	(g/l)
K_2HPO_4	1.0
$MgSO_4 \cdot 7H_2O$	0.5
Yeast extract	10.0
Na pyruvate	1.1
Distilled water	1000 ml
pH	7.0

Table 10. Medium for the isolation of *Heliobacterium* and *Heliobacillus* (50)

Constituents	(g/l)
Na pyruvate	2.2
L-methionine	0.15
EDTA di-sodium salt	0.005
$MgSO_4 \cdot 7H_2O$	0.2
$CaCl_2 \cdot 2H_2O$	0.075
$(NH_4)_2SO_4$	1.0
KH_2PO_4	0.6
K_2HPO_4	0.9
Trace element solution[a]	1 ml *(Heliobacterium)*
Chelated iron solution[b]	1 ml
Biotin (1.5 mg/100 ml)[c]	1 ml
Distilled water	1000 ml
pH	6.8

[a] See *Table 13*.
[b] Chelated iron solution contains (per litre) 500 mg $FeCl_2 \cdot 4H_2O$, 1 g EDTA di-sodium salt, and 1.5 ml concentrated HCl.
[c] Filter-sterilized; add after autoclaving.

Table 11. Media for the cultivation of members of the family Ectothiorhodospiraceae

Constituents	***E. mobilis/ E. vacuolata*** (g/l)	***E. marismortui*** (g/l)	***E. halophila/ E. halochloris*** (g/l)	***E. abdelmalekii*** (g/l)
KH_2PO_4	1.0	0.33[c]	0.5	0.8
NH_4Cl	1.0	0.33[c]	0.8	0.8
KCl	—	0.33[c]	—	—
$MgCl_2 \cdot 6H_2O$	—	0.33[c]	0.1	0.1
$MgSO_4 \cdot 7H_2O$	0.1	—	—	—
$CaCl_2 \cdot 6H_2O$	0.05	0.33[c]	0.05	0.05
NaCl	30.0	100.0	180.0	120.0
Na_2SO_4	—	0.5	20.0	15.0
Yeast extract	1.0	0.1	0.5	0.5
Na acetate	1.0	0.5	—	—
Na succinate	—	—	1.0	1.0
Na_2CO_3	1.1	1.5	6.0	4.0–5.0
$NaHCO_3$	—	—	14.0	10.0
$Na_2S \cdot 9H_2O$	0.01	—	1.0	0.5
Vitamin solution VA[a]	—	—	1 ml	1 ml
Trace elements[b]	1 ml (SLA)	1 ml (SL7)	1 ml (SLA)	1 ml (SLA)
pH	9.0–9.5	6.5–6.8	8.5	8.5

[a] See *Table 14*.
[b] See *Table 13*.
[c] Prepare as stock solutions (3.3% w/v) and add to other ingredients.

Table 12. Media for the enrichment and cultivation of *Rhsp. salexigens* (36) and *E. halophila* (33)

Constituents	***Rhsp. salexigens* (g/l)**	***E. halophila* (g/l)**
Solution C		
Nitrilotriacetic acid	10	10
$MgCl_2 \cdot 6H_2O$	30	24
$CaCl_2 \cdot 2H_2O$	3.34	3.3
$FeCl_3 \cdot 4H_2O$	—	1.1
$(NH_4)_6Mo_7O_{24} \cdot 4H_2O$	—	0.1
Rhsp. salexigens trace soln.[a]	100 ml	—
E. halophila trace soln[a]	—	50 ml
Distilled water	900 ml	950 ml
Growth medium		
Solution C	20 ml	20 ml
$(NH_4)_2SO_4$	0.5	0.5
K succinate	—	1.0
Na acetate	1.0	—
Glutamic acid	1.0	—
NaCl	80	220
$Na_2S \cdot 9H_2O$	—	0.1
1.0 M potassium phosphate	20 ml	20 ml
$NaHCO_3$	—	2.0
Vitamins[b]	1 ml *Rhsp. salexigens*	0.5 ml *E. halophila*
Distilled water	960 ml	960 ml
pH	6.9	7.4–8.0

[a] See *Table 13.*
[b] See *Table 14.*

Protocol 7. Removing single colonies from deep agar cultures

1. Crack the tube, remove the glass fragments, and place the intact agar in a sterile Petri dish.
2. Section the agar rod using a sterile scalpel and remove isolated colonies with a loop or needle.
3. Alternatively, using a Pasteur pipette drawn into a fine capillary, draw up isolated colonies from the agar.
4. Use the material obtained to incubate fresh cultures.

Protocol 8. Obtaining single colonies by forcing the agar out of the tube[a]

1. Insert a sterile Pasteur pipette down the side of the tube between the tube wall and the agar until it touches the bottom of the tube.

Protocol 8. *Continued*

2. Attach the pipette to a gas line (either air or nitrogen) at a pressure of about 2–5 bar.
3. Carefully force the agar out of the tube into a sterile Petri dish.
4. Pick sterile colonies as described in *Protocol* 7.

[a] This method is not suitable for use with screw-cap Hungate tubes, due to the constriction at the neck of the tube.

Protocol 9. Removing single colonies from deep agar cultures

1. Draw out a Pasteur pipette into a fine capillary.
2. Melt the paraffin wax overlay and decant from the culture tube.
3. Attach the Pasteur pipette to a controllable vacuum line; then carefully push the pipette into the agar until the tip is in the centre of the chosen colony.
4. Apply a gentle vacuum until the colony has been drawn into the pipette; then transfer to fresh medium.
5. Reseal the original tube as described in *Protocol* 5; then re-incubate.

4.2 Use of agar plates or roll tubes

Agar plates may be used in one of two ways. They may be streaked in the normal way or be poured from a bacterial dilution series (54). In both cases the plates are incubated in the light either in an illuminated anaerobic chamber, or in transparent polycarbonate anaerobic jars. A number of problems may be encountered with the latter method, although it has been successfully used in a number of laboratories.

(a) One problem is obtaining sufficient illumination in the centre of an anaerobic jar. One solution has been to use smaller plates and, using an inner transparent sleeve, to place them at the periphery of the jar (22).

(b) Another problem is the generation of excess water, which condenses in the bottom of the jars, when gas-generating envelopes are used. This may be countered to some extent by using anhydrous calcium chloride to absorb excess moisture.

(c) A third problem relates to the use in certain media of sulphide, which poisons the palladium catalyst used in anaerobic jars and chambers. An alternative is to flush the jars with an anaerobic gas mixture to remove the majority of the oxygen present, and to use sulphide in the gas phase to remove traces of oxygen, omitting the catalyst. Irgens (54) suggested the use of thioacetamide as a source of sulphide in such a system.

Table 13. Composition of trace element solutions

Constituents	**SL7 (mg/l)**	**SL7a (mg/l)**	**SL8 (mg/l)**	**SL11 (mg/l)**	**SLA (mg/l)**	**Micronutrient solution (mg/l)**	***Rhsp. salexigens* (mg/100 ml)**	***E. halophila* (mg/100 ml)**	***Heliobacterium* (mg/l)**
HCl (25%)	6.5 ml	1.0 ml	—	—	—	—	—	—	—
H_2SO_4	—	—	—	—	—	0.5 ml	—	—	—
EDTA di-Na salt	—	—	5200	5200	—	—	125	250	10 000
$FeCl_3 \cdot 4H_2O$	1500	—	1500	1500	1800	—	251	36	—
H_3BO_3	62	60	62	6	500	500	5.7	11	400
$MnCl_2 \cdot 4H_2O$	100	100	100	100	70	—	90	83	800
$MnSO_4 \cdot H_2O$	—	—	—	—	—	2280	—	—	—
$CoCl_2 \cdot 6H_2O$	190	200	190	190	250	45	—	—	80
$Co(NO_3)_2 \cdot 6H_2O$	—	—	—	—	—	—	12.4	25	—
$ZnCl_2$	70	70	70	70	100	—	261	—	200
$ZnSO_4 \cdot 7H_2O$	—	—	—	—	—	500	—	—	—
$NiCl_2 \cdot 6H_2O$	24	20	24	24	10	—	—	—	200
$CuCl_2 \cdot 2H_2O$	17	20	17	2	10	—	7.8	17	400
$CuSO_4 \cdot 5H_2O$	—	—	—	—	—	25	—	—	—
$Na_2MoO_4 \cdot 2H_2O$	36	40	36	36	30	25	—	—	400
$(NH_4)_6Mo_7O_{24} \cdot 4H_2O$	—	—	—	—	—	—	9.25	—	—
$Na_2SeO_3 \cdot 5H_2O$	—	—	—	—	10	—	—	—	20
$NaVO_3$	—	—	—	—	—	—	—	—	20
Distilled water	993 ml	999 ml	1000 ml	1000 ml	1000 ml	1000 ml	100 ml	100 ml	1000 ml

Table 14. Vitamin solutions

Constituents	VA (mg)	*Rhsp. salexigens* (mg)	*E. halophila* (mg)	IMC (mg)
Biotin	10	2	2	2
Nicotinamide	35	—	—	—
Nicotinic acid	—	200	200	200
Thiamine dihydrochloride	30	100	100	100
p-aminobenzoic acid	20	20	20	20
Pyridoxal hydrochloride	10	—	—	—
Calcium pantothenate	10	—	—	—
Vitamin B_{12}	5	—	0.1	—
Distilled water	100 ml	100 ml	100 ml	100 ml

(d) An additional problem may be encountered when media containing bicarbonate or carbonate are used in a gas phase with a high CO_2 content. Diffusion of CO_2 into the agar will result in a drop in pH of finely buffered media, and may inhibit growth.

Roll tubes have not been used extensively, although this method is similar in principle to the use of agar plates.

4.3 Purification in liquid culture

The only successful method for the isolation of pure cultures of some strains is the use of a dilution series in a mineral medium and growth of the strain under photoautotrophic conditions (i.e. on sulphide and CO_2).

Protocol 10. Purification of gas-vacuolate strains in liquid cultures

1. Incubate the culture until gas vacuole production is pronounced.
2. Store the culture at 10–15°C overnight. Under these conditions gas-vacuolate strains float to the top of liquid cultures.
3. Remove material from the dense pellicle and inoculate fresh cultures.

Using appropriate equipment it may be possible to transfer individual cells to small volumes of growth medium using fine flattened capillaries. The problems associated with this method are directly related to the probability of obtaining a pure culture. A logical alternative which does not appear to have been tried is the use of density gradients prepared from sterile, anaerobic Percoll mixtures.

4.4 Checking for purity

In the case of autotrophically grown strains, contamination may be limited to sulphur- or sulphate-reducers, and media suitable for the growth of such organisms should be used to check the purity of isolated strains of phototrophs.

The medium selected should take factors such as salinity, temperature, and pH into account. It is worth noting that many of the motile green sulphur bacteria have been shown to be consortia of non-motile green sulphur phototrophic bacteria and motile colourless bacteria.

Where organic compounds are added to the medium (yeast extract, organic substrates) or the strains produce slime, contamination with heterotrophic bacteria may be a problem. In particular, the presence of both anaerobic and facultatively anaerobic strains should be checked.

5. Media suitable for enrichment and cultivation

A number of media have been formulated over the years for the cultivation of anoxygenic phototrophic bacteria. The media used for the cultivation of members of the purple non-sulphur bacteria generally are variations and modifications of the medium used by van Niel (8), while the media used for members of the families Chromatiaceae and Chlorobiaceae are variations and modifications of the medium described by Pfennig (19). The media described here are those most commonly used, or special modifications needed for a particular strain. It should be noted that, in the original description by Pfennig (19), anaerobiosis of the medium was obtained by adding sulphide or another reducing agent to remove oxygen from the medium, which was usually filtered (under CO_2 or N_2 overpressure). This method is ideal for a laboratory not equipped for work with anaerobes, although the modification of this method by using the Hungate technique considerably reduces the oxygen concentration in the medium prior to addition of a reducing agent. When filling culture vessels for the cultivation of anoxygenic phototrophic bacteria it is important to leave a small air space to allow for expansion; otherwise the bottles may burst when placed in front of the light source (see Section 2.1.1). When cultures need to be fed with sulphide, the equivalent amount of medium should be removed from filled bottles before adding the feeding solution.

5.1 Special equipment

Media for the cultivation of anoxygenic phototrophic bacteria usually require that several components are prepared separately and then mixed. It is often convenient to prepare media in large quantities and to dispense it into suitable growth vessels (tubes or narrow-necked bottles). To assist in the preparation and dispensing of media two types of container have been described (20).

(a) The first is a conical flask or bottle, fitted at the bottom with an outlet which is connected to a length of soft plastic tubing, and terminating in an inverted glass bell (Bellco type 5610 or 5611, or similar design). The flow of medium is controlled with a simple pinch clamp on the tubing. This vessel is suitable for preparing and dispensing most media.

(b) The second type of vessel is a bottle, or ideally an inverted conical flask, with the mouth sealed by a glass blower. The flask is inverted and four outlets are then made in the top (formerly the bottom), one as a gas inlet, the second as a gas outlet, the third to allow additions to the medium, and the fourth which has a tube reaching to the bottom for the distribution of media. To the medium outlet is attached a soft rubber tube and a glass bell for dispensing media (as described for the first vessel). The medium is forced out by N_2 (or N_2/CO_2) under gentle pressure. This apparatus is used in the preparation of medium 1 (see *Table 6*).

Media prepared by filtration may be filtered using pressure filtration units, with either a built-in reservoir or a separate reservoir (such as those available from Sartorius).

5.2 Preparation of media

In this section media necessary for enrichment and cultivation of pure cultures of anoxygenic phototrophs are described. Detailed recipes for each of the media are given in *Tables 6–11* and the preparation of several media is described in *Protocols 11–16*.

5.2.1 Media for green and purple sulphur bacteria

Protocol 11. Preparation of medium 1

1. Prepare solution 1 as described in *Table 6* and autoclave at 121°C for 15 min.
2. Allow to cool to room temperature under an atmosphere of nitrogen.
3. Saturate the cold medium with CO_2 by bubbling sterile CO_2 (at 0.05–0.01 atm overpressure) through it.
4. Aseptically add solutions 2–6 and adjust the pH with 2 M Na_2CO_3 or 2 M HCl (the final pH is 7.2–7.4 for members of the Chromatiaceae and pH 6.5–6.8 for members of the Chlorobiaceae).
5. Stir and dispense the sterile medium into screw-cap bottles or tubes.

Protocol 12. Preparation of medium 2

1. Prepare solution 1 as described in *Table 6* and autoclave at 121°C for 15 min.
2. Allow to cool to room temperature under an atmosphere of nitrogen.
3. Aseptically add solutions 3–6 (see *Table 6*).
4. Adjust pH as described in step 4 of *Protocol 11*.
5. Stir and dispense the sterile medium into screw-cap bottles or tubes.

Protocol 13. Preparation of *Chloroherpeton* medium

1. Prepare solution 1 as described in *Table 6*.
2. Add 0.8% (w/v) washed agar and autoclave at 121°C for 15 min.
3. Cool to 50°C in a water bath.
4. Aseptically add solutions 3–6.
5. Adjust pH to 6.8–7.0 as described in step 4 of *Protocol 11*.
6. Dispense the sterile medium while still molten.

5.2.2. Media for *Heliothrix*

Heliothrix oregonensis grows only in co-culture with *Isosphaera pallida*. It was originally isolated by growth of mat material on IM medium (see *Protocol 14*). Filaments of *H. oregonensis* which grew out from the inoculum were cultivated with *I. pallida* in medium IMC (see *Protocol 15*). It is thus possible to 'isolate' *H. oregonensis*, but it will not grow further without *I. pallida*.

Protocol 14. Preparation of *Heliothrix* isolation medium (IM)

1. Add 650 ml distilled water to 250 ml of solution 1 (see *Table 6*).
2. Autoclave at 121°C for 15 min and allow to cool.
3. Aseptically add solutions 3 and 5.
4. Stir and dispense the sterile medium into screw-cap bottles or tubes.

Protocol 15. Preparation of medium IMC for co-culture of *H. oregonensis* and *I. pallida*

1. To 250 ml solution 1 (see *Table 6*) add
   - Glucose 0.225 g
   - Vitamin-free casamino acids 0.225 g
   - Distilled water 650 ml
2. Autoclave at 121°C for 15 min and allow to cool.
3. Asptically add solutions 3 and 5.
4. Stir and dispense the sterile medium into screw-cap bottles or tubes.

5.2.3 Media for *Chloroflexus*

Protocol 16. Preparation of *Chloroflexus* medium

1. Prepare the inorganic salts solution shown in *Table 7*, as a 20-fold concentrated stock solution. Store this stock solution unautoclaved at 4°C.
2. To prepare the medium, dilute the stock solution and add
   - Glycylglycine 0.8 g/l
   - NH_4Cl 0.2 g/l
   - Yeast extract 1.0–2.0 g/l
   - Vitamin-free casamino acids 2 g/l
3. Adjust the pH to 8.2 with 2 M NaOH.
4. Autoclave and dispense the sterile medium into bottles or tubes.

5.2.4 Media for purple non-sulphur bacteria

The composition of the basic medium for purple non-sulphur bacteria is shown in *Table 8*. The concentration of NaCl can be adjusted for halophilic strains. This may range from 3% (w/v) NaCl for marine strains to 10–20% (w/v) for extreme halophiles. The pH is usually 6.8–7.2, although it can be lowered for acidophilic strains. A reducing agent, such as ascorbate (0.2% w/v), may be added if desired.

5.2.5 Media for the cultivation of *Heliobacterium* and *Heliobacillus* spp.

Media for the cultivation of members of these two genera (49, 50) are prepared using the components shown in *Tables 9* and *10*.

5.2.6 Media for the family Ectothiorhodospiraceae

Several media have been described for the Ectothiorhodospiraceae (22, 30–32). The composition of these media is shown in *Tables 11* and *12*. They may be sterilized either by autoclaving or by filtration using nitrogen under pressure; in the latter case the sterile trace-elements solution should be added aseptically after filtration of the remainder of the medium. If the medium is sterilized by autoclaving, it should be noted that at higher pH ammonia will be lost from the medium (although this is compensated for by the addition of yeast extract). In addition, $NaHCO_3$ and the vitamin solution cannot be autoclaved, and must be added from filter-sterilized stock solutions.

When preparing the medium for *Ectothiorhodospira marismortui*, the NaCl, Na_2SO_4, and yeast extract should be autoclaved in 900 ml of medium, the remaining components being added aseptically from sterile stock solutions.

5.2.7 Media for enrichment and cultivation of *Rhodospirillum salexigens* and *Ectothiorhodospira halophila*

These media (see *Table 12*) may be filter-sterilized or autoclaved. In the latter case, the vitamins, $NaHCO_3$, and potassium phosphate should be added after autoclaving. *E. halophila* can also be enriched and cultivated in the medium shown in *Table 11*, while *Rhsp. salexigens* also grows in the medium described for purple non-sulphur bacteria (see *Table 8*), with the addition of 8% (w/v) NaCl.

5.2.8 Vitamin solutions

Vitamin solutions (see *Table 14*) are filter-sterilized and stored in the dark at 4°C. If the medium is autoclaved, the vitamin solution is added to the cooled medium.

5.2.9 Sulphide/bicarbonate stock-feeding solution

When strains are grown photoautotrophically the toxicity of sulphide limits the initial concentration which may be added to the medium, and it is advantageous to add sulphide and bicarbonate at regular intervals to increase the cell yield. The feeding solution contains 28.0 g/l $Na_2S \cdot 9H_2O$ and 10.6 g/l Na_2CO_3. After sterilization the pH of the solution is adjusted to about pH 7.2 by gassing with sterile CO_2. A special glass bottle suitable for the preparation of this solution was described by Siefert and Pfennig (58).

5.2.10 Acetate feeding solution

Increased cell yields in medium 1 and medium 2 (see *Protocols 11* and *12*) can be obtained by feeding the cultures with a mixture of 2.5 g ammonium acetate and 2.5 g magnesium acetate in 100 ml water. The acetate solution is added to give a final concentration of 0.05 or 0.03%.

6. Storage of pure cultures

The usual method of storing pure cultures is either in liquid culture or in agar stabs in the cold. Cultures should be stored before growth has reached the stationary phase and, in the case of members of the families Chromatiaceae, Ectothiorhodospiraceae, and Chlorobiaceae, they should be fed with a small amount of sulphide before storing (20). An alternative is the use of freeze-drying, although this can be time-consuming and problematic for use in the routine laboratory (59). The best alternative for long-term storage appears to be the use of liquid nitrogen using 5% (w/v) dimethylsulphoxide as cryoprotectant. Although glass or plastic vials (usually 1.0–1.5 ml in volume) are generally recommended, the glass capillary method for methanogenic bacteria (60) also gives good results. When using the latter method for anaerobically grown strains it is important to carry out manipulations in an anaerobic atmosphere, either in a chamber or by gassing with nitrogen. The advantages of the capillary method over vials are that up to 40 capillaries can be stored in the same space, and that loss of a valuable culture through explosion of the vial on removal from the storage tank is eliminated.

7. Summary

It is almost impossible to cover all aspects of the isolation and maintenance of all anoxygenic phototrophic bacteria. While the general methods and principles employed in the isolation and cultivation of these organisms are based on many years of experience the reader should not forget that a chance error in the preparation of media or the use of a new combination of parameters during isolation may result in the isolation of a new strain. The most recent example of such serendipity is the isolation of anoxygenic phototrophic bacteria related to the Gram-positive bacteria (*Heliobacterium* and '*Heliobacillus*'). For the reader wishing to obtain more information, *Bergey's Manual of Systematic Bacteriology* (26) and various chapters in *The Prokaryotes* (20, 21, 45) are a good starting point.

References

1. Bavendamm, W. (1924). In *Pflanzenforschung* (ed. R. Kolkwitz), pp. 7–156. Gustav Fischer Verlag, Jena.
2. Buder, J. (1919). *Jahrb. Wissensch. Botan.* **58**, 525.
3. Cohn, F. (1875). *Beitr. Biologie Pflanz.* **1**, 141.
4. Ehrenberg, C. G. (1838). *Die Infusionsthierchen als vollkommene Organismen.* Voss, Leipzig.
5. Lauterborn, R. (1915). *Ver. Naturkundl. Med. Ver. Heidelb.* **13**, 395.
6. Molisch, H. (1907). *Die purpurbakterien nach neuen Untersuchungen.* Gustav Fischer Verlag, Jena.
7. van Niel, C. B. (1932). *Arch. Mikrobiol.* **3**, 1.
8. van Niel, C. B. (1944). *Bact. Rev.* **8**, 1.
9. Winogradsky, S. N. (1888). *Beiträge zur Morphologie und Physiologie der Bakterien.* Heft 1. *Zur Morphologie und Physiologie der Schwefelbakterien.* Felix, Leipzig.
10. Pfennig, N. and Trüper, H. G. (1974). In *Bergey's Manual of Determinative Bacteriology* (8th edn) (ed. R. E. Buchanan and N. E. Gibbons), pp. 24–64. Williams and Wilkins, Baltimore.
11. Ambler, R. P., Meyer, T. E., and Kamen, M. D. (1979). *Nature* **278**, 661.
12. Ambler, R. P., Daniel, N., Hermoso, J., Meyer, T. E., Bartsch, R. G., and Kamen, M. D. (1979). *Nature* **278**, 659.
13. Dickerson, R. E. (1980). *Nature* **283**, 210.
14. Gibson, J., Stackebrandt, E., Zablen, L. B., Gupta, R., and Woese, C. R. (1979). *Current Microbiol.* **3**, 59.
15. Woese, C. R. (1987). *Microbiol. Rev.* **51**, 221.
16. Pierson, B. K. and Castenholz, R. W. (1974). *Arch. Microbiol.* **100**, 4.
17. Madigan, M. T. (1986). *Int. J. System. Bacteriol.* **36**, 222.
18. Pierson, B. K., Giovannoni, S. J., Stahl, D. A., and Castenholz, R. W. (1985). *Arch. Microbiol.* **142**, 164.

19. Pfennig, N. (1965). *Zentralbl. Bakteriol. Parasitenk.* **1** (1), 179.
20. Pfennig, N. and Trüper, H. G. (1981). In *The Prokaryotes, A Handbook on Habitats, Isolation, and Identification of Bacteria* (ed. M. P. Starr, H. Stolp, H. G. Trüper, A. Balows, and H. G. Schlegel), pp. 279–89. Springer-Verlag, Berlin.
21. Castenholz, R. W. and Pierson, B. K. (1981). In *The Prokaryotes, A Handbook on Habitats, Isolation, and Identification of Bacteria* (ed. M. P. Starr, H. Stolp, H. G. Trüper, A. Balows, and H. G. Schlegel), pp. 290–8. Springer-Verlag, Berlin.
22. Tindall, B. J. (1980). Unpublished D.Phil. thesis, University of Leicester.
23. Imhoff, J. F., Sahl, H. G., Soliman, G. S. H., and Trüper, H. G. (1978). *Geomicrobiol. J.* **1**, 183.
24. Tindall, B. J. (1988). In *Halophilic Bacteria* (ed. F. Rodriguez-Valera), pp. 31–67. CRC Press, Boca Raton, FL.
25. Pfennig, N. (1969). *J. Bacteriol.* **99**, 597.
26. Pfennig, N. (ed.) (1984). *Bergey's Manual of Systematic Bacteriology* Vol. 3. Williams and Wilkins, Baltimore.
27. Trüper, H. G. (1970). *Helgol. wissensch. Meeres.* **20**, 6.
28. Imhoff, J. F., Tindall, B. J., Grant, W. D., and Trüper, H. G. (1981). *Arch. Microbiol.* **130**, 238.
29. Trüper, H. G. (1968). *J. Bacteriol.* **95**, 1910.
30. Oren, A., Kessel, M., and Stackebrandt, E. (1989). *Arch. Microbiol.* **151**, 524.
31. Imhoff, J. F. and Trüper, H. G. (1977). *Arch. Microbiol.* **114**, 114.
32. Imhoff, J. F. and Trüper, H. G. (1981). *Zentralbl. Bakteriol. Hyg. I Abt. Orig. C* **2**, 228.
33. Raymond, J. C. and Sistrom, W. R. (1969). *Arch. Mikrobiol.* **69**, 121.
34. Caumette, P., Baulaigue, R., and Matheron, R. (1988). *System. Appl. Microbiol.*, **10**, 284.
35. Trüper, H. G. and Jannasch, H. (1968). *Arch. Mikrobiol.* **61**, 363.
36. Drews, G. (1981). *Arch. Microbiol.* **130**, 325.
37. Nissen, H. and Dundas, I. D. (1984). *Arch. Microbiol.* **138**, 251.
38. Kompantseva, E. I. and Gorlenko, V. M. (1984). *Microbiology* **53**, 775.
39. Hansen, T. A. and Veldkamp, H. (1973). *Arch. Mikrobiol.* **92**, 45.
40. Imhoff, J. F. (1983). *System. Appl. Microbiol.* **4**, 512.
41. Neutzling, O., Imhoff, J. F., and Trüper, H. G. (1984). *Arch. Microbiol.* **137**, 256.
42. Kompantseva, E. I. (1985). *Microbiology* **54**, 771.
43. Gorlenko, V. M., Krasil'nikova, E. N., Kikina, O. G., and Tatarinova, N. Y. (1979). *Izv. Akad. Nauk. SSSR Ser. Biol.* **5**, 755.
44. Hansen, T. A. and van Germenden, H. (1972). *Arch. Mikrobiol.* **86**, 49.
45. Biebl, H. and Pfennig, N. (1981). In *The Prokaryotes, a Handbook on Habitats, Isolation, and Identification of Bacteria* (ed. M. P. Starr, H. Stolp, H. G. Trüper, A. Balows, and H. G. Schlegel), pp. 267–73. Springer-Verlag, Berlin.
46. Weast, R. C. (1986). *CRC Handbook of Chemistry and Physics.* CRC Press, Boca Raton, FL.
47. Kämpf, B. and Pfennig, N. (1986). *J. Basic Microbiol.* **9**, 507.
48. Kämpf, B. and Pfennig, N. (1986). *J. Basic Microbiol.* **9**, 517.
49. Gest, H. and Favinger, J. L. (1983). *Arch. Microbiol.* **136**, 11.
50. Beer-Romero, P. and Gest, H. (1987). *FEMS Microbiol. Lett.* **41**, 109.
51. Pfennig, N., Markham, M. C., and Liaaen-Jensen, S. (1968). *Arch. Mikrobiol.* **62**, 178.
52. Gest, H., Favinger, J. L., and Madigan, M. T. (1985). *FEMS Microbiol. Ecol.* **31**, 317.
53. Pfennig, N. (1965). *Zentralbl. Bakteriol. Parasitenk.* **1** (1), 503.

54. van Niel, C. B. (1971). In *Methods in Microbiology* Vol. 23a (ed. S. P. Colwick and N. V. Kaplan), pp. 3–28. Academic Press, New York.
55. Irgens, R. L. (1983). *Curr. Microbiol.* **8**, 183.
56. Gibson, J., Pfennig, N., and Waterbury, J. B. (1984). *Arch. Microbiol.* **138**, 96.
57. Pierson, B. K., Giovannoni, S. J., and Castenholz, R. W. (1984). *Appl. Env. Microbiol.* **47**, 576.
58. Siefert, E. and Pfennig, N. (1986). *Arch. Microbiol.* **139**, 100.
59. Malik, K. A. (1990). *J. Microbiol. Methods* **12**, 117.
60. Hippe, H. (1984). In *Maintenance of Microorganisms* (ed. B. E. Kirsop and J. J. S. Snell), pp. 69–81. Academic Press, London.
61. Tindall, B. J. and Grant, W. D. (1986). In *Anaerobic Bacteria in Habitats Other Than Man* (ed. E. M. Barnes and G. C. Mead), pp. 115–55. Academic Press, London.
62. Eichler, B. and Pfennig, N. (1988). *Arch. Microbiol.* **149**, 395.
63. Overmann, J. and Pfennig, N. (1989). *Arch. Microbiol.* **152**, 401.
64. Kompantseva, E. I. (1989). *Microbiology* **58**, 254.

14

Anaerobic protozoa and fungi

G. S. COLEMAN

1. Introduction

There are many different species of 'anaerobic' amoebae, flagellates, and ciliates although they differ considerably in their degree of aerotolerance. For example, *Entamoeba histolytica* grows in semi-solid media containing a reducing agent such as cysteine or ascorbic acid but without any other anaerobic precautions, whereas the rumen ciliates (e.g. *Entodinium caudatum*) will only grow in media from which all dissolved oxygen has been removed, which contain a reducing agent, and which have O_2-free gases in the headspace above the medium. Anaerobic protozoa are found in a variety of habitats such as the sapropel (the ecosystem at the bottom of fresh or marine waters that contains decomposing organic material and which is rich in sulphide), the hind gut of termites and other xylophagous insects, the reticulorumen of ruminants such as sheep, cows, deer, antelope, giraffe, etc., the hind gut and caecum of many herbivorous animals, and the intestines and genitourinary tract of many animals. Some, such as the pond amoebae, will grow only at 16–24°C, whereas others, such as the rumen ciliates, will grow only at 37–42°C. Some, such as *Trichomonas vaginalis*, require very complex organic media for growth axenically, whereas the rumen entodiniomorphid protozoa, which have only been grown in the presence of bacteria, will only grow on very simple media. It should also be noted that some, such as *Entamoeba* and *Giardia* spp., are pathogens, whereas the protozoa from the rumen and termite hind gut are beneficial or essential to the host.

Thus it will be apparent that it is not possible to give simple instructions for the cultivation of anaerobic protozoa. The problems experienced in trying to grow the pathogenic protozoa *in vitro* are associated with their requirements for complex organic compounds rather than their need for anaerobiosis, although the presence of materials such as serum and casein digest will also help to keep the redox potential down. The reader who wishes to grow the pathogenic *Entamoeba*, *Giardia*, and *Trichomonas* spp. is referred to Diamond (1).

In this chapter it is proposed, as far as the protozoa are concerned, to describe techniques involved in the isolation, growth, and handling of rumen entodiniomorphs and flagellates, termite flagellates, and the sapropelic protozoa. Although 'anaerobic' fungi occur in a number of habitats, most of the recent work has been carried out with rumen fungi and these will be considered in greatest detail.

2. Cultivation of anaerobic protozoa

In any attempt to grow protozoa *in vitro* it is essential to try and reproduce, as far as possible, the conditions under which the organisms grow in the wild. As many ciliates utilize particulate food, this should be provided as, for example, starch grains, cellulose or plant fibres, bacteria, etc. The use of particulate food also discourages the growth of bacteria, which can be troublesome. As already mentioned, these protozoa require not just the absence of free oxygen, but also a low redox potential in the medium. An inert gas such as nitrogen must therefore be bubbled through the medium in order to remove dissolved oxygen, and a reducing agent such as cysteine or ascorbic acid must be added to the medium. Most anaerobic organisms also require CO_2 for growth and it is essential to add at least 5% (v/v) CO_2 to the gas phase. Some organisms require 100% CO_2 in the gas phase and in these circumstances it is essential to add $NaHCO_3$ to the medium in order to maintain a constant pH.

2.1 Cultivation of rumen entodiniomorphid protozoa

To cultivate these protozoa you will need the equipment and gases in *Table 1* and the media and other materials given in *Protocols 1–5*.

Table 1. Equipment and gases needed for cultivation of rumen entodiniomorphid protozoa

- An incubator at 39°C
- 50-ml heavy-walled centrifuge tubes
- 180 mm × 12 mm test tubes
- Size 11, 25, and 27 rubber bungs to fit the tubes
- CO_2—medical quality[a]
- O_2-free N_2
- 95% N_2+5% CO_2 (v/v)[a]

[a] Can be used without treatments to remove any contaminating oxygen.

Protocol 1. Preparation of double-strength Coleman salt solution

1. Dissolve 127 g K_2HPO_4, 100 g KH_2PO_4, and 13 g NaCl separately in 700 ml water with warming and place in an aspirator of 11 l capacity. Dilute to 6–7 l with distilled water.
2. Dissolve 0.9 g $CaCl_2$ (dried) in about 50 ml water and dilute to about 1 l. Add to aspirator.[a]

Protocol 1. *Continued*

3. Dissolve 1.8 g $MgSO_4 \cdot 7H_2O$ in about 50 ml water and dilute to about 1 l. Add to the aspirator.[a]
4. Add distilled water to a total volume of 10 l and mix thoroughly.

[a] If the calcium and magnesium salts are added before the phosphates have been dissolved in the bulk of the water a precipitate will form.

Protocol 2. Preparation of dried grass

1. Air-dry lawn mowings on the bench with frequent turning for 3 days or until dry.
2. Grind in an attrition or other suitable mill to produce a coarse powder containing particles up to 5 mm long (coarse dried grass).
3. Grind for 16 h in a ball mill to produce a fine powder (powdered dried grass).
4. Dried grass is added to cultures from the end of a small spatula. The amount added is not critical and with practice can soon be judged by eye.
5. Store in the dark at room temperature.

Protocol 3. Preparation of rumen fluid for culture media

1. Remove rumen fluid (about 500 ml) from a sheep or other ruminant via a rumen cannula (or obtain it from a slaughter house) and strain the fluid through one layer of muslin.
2. Centrifuge the strained fluid at 500 *g* for 5 min to sediment the protozoa.
3. Remove the supernatant fluid and store overnight at 4°C to kill any residual protozoa. This 'fresh rumen fluid' can be stored for up to 2 weeks at 4°C.
4. 'Autoclaved rumen fluid' is prepared by autoclaving (115°C, 20 min) fresh rumen fluid in sealed bottles under CO_2 (*Table 1*).

Protocol 4. Preparation of caudatum-type medium

1. Immediately before use prepare 2% cysteine solution by dissolving 1 g L-cysteine hydrochloride in 5 ml water, neutralizing it with 6 ml 1 M NaOH and diluting to 50 ml.
2. Mix 100 ml double-strength Coleman salt solution (see *Protocol 1*), 100 ml distilled water, 1.0 ml 15% (w/v) sodium acetate, and 2.0 ml cysteine solution. If required (see *Table 2*), add 20 ml prepared fresh rumen fluid (see

Protocol 4. *Continued*

Protocol 3), mix, and three-quarters fill 50-ml centrifuge tubes (for established cultures) or place 2 ml in 180 mm × 12 mm tubes.

3. Bubble 95% N_2 + 5% CO_2 (v/v) gas mixture (*Table 1*) through the media for $1\frac{1}{2}$–2 min and seal with a rubber bung.

Table 2. Conditions for the cultivation of rumen entodiniomorphid protozoa

Protozoon	Salts medium preferred	Stimulation of growth of established cultures by rumen fluid	Preferred food source[a]
Diplodinium pentacanthum	Caudatum	+	DG
Diploplastron affine	Caudatum	+	WF
Enoploplastron triloricatum	Simplex	−	DG
Entodinium caudatum	Caudatum	+	WF
Entodinium simplex	Simplex	+	WF
Epidinium caudatum	Caudatum	−	WF
Eremoplastron bovis	Caudatum	+	DG
Eudiplodinium maggii	Caudatum	−	DG
Ophryoscolex caudatus	Caudatum	+	WF
Ostracodinium dilobum	Simplex	+	DG
Polyplastron multivesiculatum	Caudatum	+	WF

[a] DG, powdered dried grass; WF, wholemeal flour + coarse dried grass.

Protocol 5. Preparation of simplex-type medium

1. Prepare 2% cysteine solution as described in step 1 of *Protocol 4*.
2. Mix 80 ml double-strength Coleman salt solution (see *Protocol 1*), 120 ml distilled water, 30 ml 5% (w/v) sodium hydrogen carbonate, and 2.2 ml 2% cysteine. If required (see *Table 2*), add 20 ml prepared fresh rumen fluid (see *Protocol 3*), mix, and three-quarters fill 50-ml centrifuge tubes (for established cultures) or place 2 ml in 180 mm × 12 mm tubes (for isolation of protozoa).
3. Bubble CO_2 (*Table 1*) through the media for $1\frac{1}{2}$–2 min and seal with a rubber bung.

2.1.1 Maintenance of established cultures

Daily treatment

Remove the 50-ml tubes containing the cultures from the incubator, remove the bung, and add one drop (for media without rumen fluid) or three drops (for media with rumen fluid) 1.5% (w/v) aqueous suspension of wholemeal flour or a few mg powdered dried grass. The type of food added depends upon the protozoal species (see *Table 2*). The gas space is then filled with CO_2 (regardless of the gas used in the preparation of the medium), the tubes sealed with a rubber bung, and the cultures are re-incubated at 39°C.

Dilution of the cultures

Once or twice a week the cultures are mixed, half poured into another 50 ml tube, and each made up to the original volume with fresh medium of the same type as used previously. The cultures are then fed as under 'daily treatment' except that a few mg of coarse dried grass are added to those tubes that are given wholemeal flour. The cultures are then gassed, sealed, and re-incubated at 39°C.

2.1.2 Initial isolation of protozoal species

Before attempting to isolate a single protozoal species it is advisable to determine the medium on which that protozoon grows best. This is done by inoculating a few drops of crude rumen contents containing the protozoon into two 50-ml tubes each of caudatum and simplex-type media containing prepared fresh rumen fluid (see *Protocol 3*). Feed one of each pair daily with wholemeal flour and coarse dried grass and the other pair with powdered dried grass only. Incubate and dilute the cultures twice a week with fresh medium as described in Section 2.1.1. After 1–3 weeks it should be apparent by microscopical examination, which is the preferred growth medium. The procedure used to pick out individual protozoal species from crude rumen contents or an enrichment culture is given in *Protocol 6*.

Protocol 6. Removal and culture of individual protozoa from a crude mixture of species

In addition to the equipment and gases in *Table 1*, you will need

- An inverted microscope with a ×4 objective or a normal microscope with a long working distance between the objective and the slide.
- A micropipette prepared by drawing out a thin-walled glass tube.
- A micromanipulator for use in guiding the tip of the pipette over the slide.
- A source of gentle suction or pressure so that liquid can be sucked up into, or expelled from, the pipette; a purpose-made pump is ideal although a standard automatic pipette can be adapted for use

1. Prepare 180 × 12 mm culture tubes with 2 ml caudatum or simplex-type media containing prepared fresh rumen fluid (see *Protocols 4* and *5*).
2. Place 2–3 drops of caudatum-type medium containing prepared fresh rumen fluid on a slide and add a small volume of the crude protozoal mixture. Experience will determine how much will give a suitable protozoal density. Pick out at least six cells of the desired protozoon, inoculate the medium, feed with a few mg powdered dried grass or, if starch is required, with a small drop

Protocol 6. *Continued*

of the supernatant fluid obtained by allowing a 1.5% (w/v) suspension of wholemeal flour to stand for a few min, plus a few mg coarse dried grass. Gas with CO_2 and incubate at 39°C.

3. If it is not possible to easily pick out the desired protozoon, do not worry if other protozoa are removed as well. Then expel this enriched protozoal mixture into a fresh few drops of caudatum-type medium on the slide and repeat the picking process until only one species is removed. It is essential to work rapidly as the protozoa are killed by exposure to air.
4. Feed the cultures each day (as described in Section 2.1.1) with a minimal amount of wholemeal flour (if required) or powdered dried grass. As the protozoal population density increases, the amount of food given each day can be increased.
5. After a week, if the protozoa are growing well, the volume of medium can be doubled by adding fresh medium; this process can be continued until the culture is in standard 50-ml tubes.

The following are possible reasons for failure to culture any protozoa or for production of contaminated cultures.

(a) Too long is taken in the picking process so that the protozoa are killed by air. With experience the worker will become quicker and therefore more successful. If an anaerobic chamber is available this will facilitate the operation, although it is not essential.

(b) Too much starch is added during the first few days of culture. Try adding less. Starch that is not engulfed by the protozoa is fermented by bacteria with the production of conditions that are not favourable to protozoal growth.

(c) It is very easy to inadvertently pick up small transparent entodinia when trying to remove larger protozoa. It is therefore advisable to inoculate at least 10 tubes when isolating one species.

If more information is required on the cultivation of these ciliates consult Coleman (2).

2.1.3 Preparation of monoxenic and axenic cultures

As all entodiniomorphid protozoa contain bacteria in vesicles in their endoplasm, they can only be rendered bacteria-free by treatment with antibiotics. Although 'bacteria-free' protozoa have been prepared, they will not grow under axenic conditions. Limited success has been obtained in attempts to grow these ciliates in the presence of one bacterial species (3, 4).

2.1.4 Continuous culture of rumen ciliate protozoa

It is not possible to grow entodiniomorphid protozoa in a conventional continuous-culture apparatus, owing, in part, to the inability of the protozoa to engulf starch grains in continuously stirred medium. The problem has been overcome by providing a large dead space where the protozoa can sequester and feed. In the most successful methods the rate of passage of liquid and solid phases are different. The reader is not advised to attempt the continuous cultivation of these protozoa until he is familiar with the cultivation methods described in Sections 2.1.1 and 2.1.2. Information on continuous culture can be found in references 5–9.

2.2 Cultivation of rumen flagellate protozoa

Seven species of flagellate protozoa of the genera *Chilomastix*, *Monocercomonas*, *Monocercomonoides*, *Pentatrichomonas*, *Tetratrichomonas*, and *Tritrichomonas* occur in the rumen at low population density. For many years there were believed to be two types of flagellate protozoa in the rumen, namely, the above-mentioned and those of the genera *Neocallimastix*, *Piromonas*, and *Sphaeromonas*. The former are genuine flagellates, whereas the latter are the zoospores of anaerobic chytrid fungi, the cultivation of which is described in Section 3.1.

There has been very little work done on the cultivation of the true flagellates although Jensen and Hammond (10) claimed to have easily cultured some species in media similar to those used for the pathogenic trichomonads and in conventional bacteriological culture media containing antibiotics to suppress bacterial growth. Unfortunately, attempts by others to repeat these studies have been unsuccessful.

2.3 Cultivation of flagellates from the termite hindgut

Flagellates from the hind gut of the termite *Zootermopsis* can be grown both in the presence of bacteria (*Protocol 8*) and axenically (*Protocol 9*), although the population densities are smaller under the latter conditions (11, 12). The limited evidence available suggests that flagellates from other termites can also be grown under the same conditions.

Protocol 7. Preparation of double-strength salt solution for cultivation of termite flagellates

1. Dissolve 37.6 g K_2HPO_4, 18.8 g KH_2PO_4, 32.0 g KCl, and 28.6 g NaCl separately in about 1 l distilled water with warming and place in an aspirator of 11-l capacity. Dilute to 6–7 l.
2. Dissolve 26.0 g $MgSO_4 \cdot 7H_2O$ and 1.3 g $CaCl_2$ (dried) separately in 1 l distilled water with warming and add to the aspirator.
3. Add the two solutions and distilled water to make a total volume of 10 l and mix thoroughly.

2.3.1 Cultivation of *Trichomitopsis termopsidis*

The medium and procedure for the isolation and cultivation of *T. termopsidis* in the presence of bacteria is given in *Protocol 8*.

Protocol 8. Medium preparation and cultivation of *Trichomitopsis termopsidis* in the presence of bacteria

You will need

- An incubator which can be maintained at temperatures 20–27°C
- 150 mm × 16 mm roll tubes and butyl rubber bungs to fit, screw-top tubes of a similar size with pierced caps sealed with butyl rubber membranes, or any tubes available for the Hungate technique for the cultivation of anaerobic bacteria
- Sterile plastic syringes fitted with 21-gauge 25-mm needles
- O_2-free N_2 and a means of passing it through a sterile filter

1. Take 50 ml double-strength salt solution (*Protocol* 7), 43.5 ml distilled water, and 100 mg cellulose powder.[a] Boil the mixture and bubble O_2-free N_2 through the solution while allowing it to cool.[b]
2. Dispense 8.35-ml quantities into the roll or alternative tubes and either plug with cotton wool or screw down the caps. Autoclave and allow to cool. If cotton-plugged tubes are used, step 3 must be carried out as soon as the medium has cooled.
3. When the medium has cooled, add 0.4 ml 2.1% (w/v) $NaHCO_3$ (filter-sterilized by applying positive pressure to the liquid being filtered rather than negative pressure to the underside of the filter) and 0.25 ml heat-inactivated filter-sterilized fetal calf serum.
4. Bubble sterile O_2-free N_2 through each tube for 30 sec and seal with a rubber bung or the method appropriate to the tubes in use.
5. Obtain the flagellates from *Zootermopsis* spp. by washing out the hind gut with medium.
6. Using a sterile plastic syringe, inoculate the suspension obtained in step 5 into a tube of the medium prepared in steps 1–4. Before any transfer of material the air in the barrel of the sterile plastic syringe must be replaced by O_2-free N_2.
7. Incubate the inoculated tubes at 27°C and subculture every 2–4 weeks. After 7–8 weeks a thriving population of *Trichomitopsis termopsidis* should develop. *Trichonympha* spp. originally present in the termites will also survive steps 5–7, but will not multiply.
8. Remove the *Trichonympha* spp., which are heavier than *T. termopsidis*, by

Protocol 8. *Continued*

mixing the culture tubes before subculturing, allowing the tubes to stand, and then removing liquid from the top.

[a] The cellulose particles must be small enough for the protozoa to engulf; Sigmacell Type 20, Sigma Chemical Co, is recommended.

[b] Although the original method recommends boiling the mixture, dissolved oxygen is removed effectively just by bubbling O_2-free N_2 through the medium.

It is possible, with care, to produce a pure culture of *T. termopsidis*. The cultures produced using *Protocol 8* contain a mixed bacterial flora but axenic protozoal cultures can be prepared by subculture in the medium in *Protocol 9*.

Protocol 9. Medium preparation and cultivation of *Trichomitopsis termopsidis* in the absence of bacteria

You will need the same equipment and gases as in *Protocol 8*.

1. Take 50 ml double-strength salt solution (*Protocol* 7) and 8.5 ml distilled water and dissolve 100 mg reduced glutathione in this solution.
2. Dispense 5.85-ml quantities into tubes and autoclave as in step 2 of *Protocol 8*.
3. When the medium has cooled, add
   - 0.4 ml 2.1% (w/v) $NaHCO_3$, filter-sterilized as in step 3 of *Protocol 8*
   - 0.25 ml heat-inactivated, filter-sterilized fetal calf serum
   - 1.0 ml 10% (w/v) yeast extract, sterilized by autoclaving
   - 1.0 ml autoclaved rumen fluid (*Protocol 3*)
   - 0.5 ml filter-sterilized penicillin (10^4 units/ml)
   - 0.5 ml 2% (w/v) filter-sterilized streptomycin
4. Bubble sterile O_2-free N_2 through each tube for 30 sec and seal with a rubber bung or the method appropriate to the tubes in use.
5. Follow the inoculation procedure in steps 5 and 6 of *Protocol 8*, but using the medium described in steps 1–4 of this protocol.
6. Incubate the inoculated tubes at 27°C and subculture every 3 weeks. After two to three transfers the cultures should be bacteria-free.
7. To test for the presence of bacteria, omit the antibiotics from the medium. Once the antibiotic concentration has dropped below the minimum inhibitory concentration, any bacteria will grow vigorously and their presence can be determined by visual or microscopic inspection.

If all the bacteria have been killed, the cultures can be maintained axenic for long periods, provided that aseptic technique is followed when subculturing.

2.3.2 Cultivation of *Trichonympha sphaerica*

In Section 2.3.1. it was shown that *Trichonympha* spp. remains alive, but does not grow, when placed in the medium described in *Protocol 8* and incubated at 27°C. The growth of *Trichonympha sphaerica* can be encouraged by dropping the incubation temperature to 20°C. *Trichonympha sphaerica* together with *Trichomitopsis termopsidis* can be grown at 20°C in the absence of bacteria in the medium described in *Protocol 9*, modified by the use of streptomycin and penicillin at quarter concentration (i.e. 0.5% (w/v) and 2500 units/ml, respectively). Pure cultures of *Trichonympha sphaerica* are prepared by picking out individual cells with a micropipette (as described in *Protocol 6*) and inoculating them into fresh medium. After 8 weeks there should be a flourishing protozoal culture and subcultures can be made. Thereafter the flagellates should be subcultured every 4 weeks. After several subcultures the flagellates should be bacteria-free and the antibiotics can be omitted from the medium.

2.4 Cultivation of *Metopus striatus*

Metopus striatus is a rhomboid or triangular, brown ciliate found in the sapropel of fresh water ponds. The habitat is characterized not only in being highly anaerobic and rich in decaying organic matter, but in having a comparatively high concentration of sulphide. For convenience this protozoon can be grown easily and successfully in laboratory aquaria, although it can now be grown in monoculture in the presence of bacteria (13).

Protocol 10. Basal mineral salts solution for cultivation of *Metopus striatus*

1. Dissolve 0.06 g NH_4Cl, 0.06 g K_2HPO_4, and 0.04 g KCl in about 6 l water.
2. Dissolve separately in about 1 l water, 0.04 g $CaCl_2 \cdot 2H_2O$ and 0.05 g $MgSO_4 \cdot 7H_2O$, and add to the other salts solution.
3. Dilute to 10 l and mix.

Protocol 11. Medium preparation and cultivation of *Metopus striatus*

You will need

- Carrel (tissue culture) flasks and butyl rubber stoppers to fit
- 95% N_2 + 5% CO_2 (v/v) gas mixture
- An incubator at 20°C (the temperature must not be allowed to exceed 30°C or the protozoa die)

Protocol 11. *Continued*

1. Place 20 ml basal mineral salts (*Protocol 10*) in a Carrel flask and add 0.2 ml 5% (w/v) $NaHCO_3$ and 0.10 ml 0.48% (w/v) $Na_2S \cdot 9H_2O$ (freshly prepared).
2. Bubble 95% N_2+5% CO_2 through the medium for 2 min to remove dissolved oxygen and bring the pH to 6.5–7.2.
3. Flush out air from the space above the liquid and seal with a rubber bung.
4. To initiate the culture pick individual protozoa from the sediment in a laboratory aquarium using the technique described in *Protocol 6*. Inoculate them into the medium.
5. Feed the cultures initially and every 2 days with a suspension of waste water sediment (containing 10^9 bacteria/ml) prepared as follows.
   (a) Centrifuge the pond water at 12000 *g* for 30 min.
   (b) Remove the bulk of the supernatant fluid.
   (c) Resuspend the sediment in a small volume of the original supernatant fluid.
6. Incubate the cultures at 20°C and subculture every 3 weeks.

2.5 Cultivation of *Metopus contortus*

M. contortus is found under very similar conditions to those of *M. striatus*, differing only in that it has a marine habitat. The basal salts medium used must therefore contain a high concentration of NaCl. It is recommended that artificial sea salts (obtainable from most manufacturers of laboratory chemicals) be used.

Protocol 12. Cultivation of *Metopus contortus*

1. Fill 250-ml measuring cylinders with a 3.25% (w/v) artificial sea salt solution (13), supplemented with 0.1% (w/v) yeast extract.
2. Inoculate the cylinders with sea mud, e.g. from salt marshes, and some decomposing sea grass. Seal with a rubber bung.
3. Incubate the cylinders at 20°C. When the original sea grass disappears, feed the cylinders with ground dried grass (1% (w/v); see *Protocol 2*).
4. When, after some weeks, the numbers of *M. contortus* are sufficient, pick out the individual protozoa (see *Protocol 6*) and inoculate them into the following medium (25 ml contained in 100-ml serum bottles).
   - Artificial sea salt 32 g/l
   - Na_2CO_3 0.5 g/l
   - $Na_2S \cdot 9H_2O$ 1.2 g/l

Protocol 12. *Continued*

5. Remove dissolved oxygen by bubbling with O_2-free N_2, flush the head space with N_2, and seal the bottles with a butyl rubber bung.
6. Incubate in the dark at 20°C.
7. Initially, and at intervals when required, feed the cultures with ground dried grass (400 mg initially and 10 mg subsequently).
8. Twice a week remove half of the medium through a 7-μm filter (to prevent loss of protozoa) while flushing the gas space in the bottles with O_2-free N_2. Then add an equal quantity of fresh medium from which oxygen has been removed.
9. Subculture every 2–4 weeks.

2.6 Cultivation of *Pelomyxa palustris*

The giant sapropelic amoeba, *Pelomyxa palustris*, which is found in fresh-water lakes, has proved difficult to cultivate. However, Kudo (14), using a technique similar to that described in *Protocol 11* for *Metopus striatus*, with lake water as medium and chopped *Spirogyra* as food, successfully cultured this protozoon for an undefined period.

3. Cultivation of anaerobic fungi

3.1 Cultivation of rumen fungi

As mentioned in Section 2.2, organisms of the genera *Neocallimastix*, *Piromonas*, and *Sphaeromonas*, which were originally believed to be flagellate protozoa, are now known to be chytrid fungi, the zoospores of which resemble flagellates. These fungi undergo a distinct life cycle both *in vitro* and *in vivo*. The motile zoospores germinate on plant material and form a short mycelium on which sporangia form. These sporangia eventually burst and liberate more zoospores.

As bacteria are much more numerous than fungi in the rumen, all methods for the isolation of the fungi are based initially on enrichment with the use of antibiotics to inhibit bacterial growth. In the most successful methods (15) this enriched material was used to inoculate plates on which comparatively large (15-mm diameter) colonies form. These colonies appear to be formed of concentric alternating regions of 'mycelium' and 'sporangia'. By picking material from the edge of one of these colonies one has the best chance of obtaining fungi free from contaminating bacteria.

There are two types of media available for the growth of the fungi, one of which contains rumen fluid, while the other does not. Neither are completely defined, as even the one without rumen fluid contains materials such as yeast extract. The media both contain many different compounds, but preparation can be simplified by making up a number of mixtures, aliquots of which can be frozen and stored indefinitely (see *Protocol 13*). When solid media are required, 2% (w/v) agar is added to the liquid medium and the whole is steamed for 40 min.

Protocol 13. Preparation of medium components

1. To prepare growth factor solution, dissolve the following in 1 l of 5 mM Hepes buffer
   - 1,4 naphthoquinone 0.25 g
   - Calcium D-pantothenate 0.2 g
   - Nicotinamide 0.2 g
   - Riboflavin 0.2 g
   - Thiamin 0.2 g
   - Pyridoxine HCl 0.2 g
   - Biotin 0.025 g
   - Folic acid 0.025 g
   - Cyanocobalamin 0.025 g
   - *p*-aminobenzoic acid 0.025 g
2. Sterilize by filtration through a 0.2-μm membrane when required.
3. Dissolve the following antibiotics in distilled water
   - Streptomycin sulphate 2 g
   - Penicillin G 8 g
   - Chloramphenicol 6 g
   - Oxytetracycline 5 g
   - Neomycin sulphate 6 g
4. Sterilize by membrane filtration and use immediately.
5. Add the following fatty acids to 700 ml 0.2 M NaOH
   - Acetic acid 6.85 ml
   - Propionic acid 3.00 ml
   - Butyric acid 1.84 ml
   - 2-methylbutyric acid 0.55 ml
   - Isobutyric acid 0.47 ml
   - *n*-valeric acid 0.55 ml
   - Isovaleric acid 0.55 ml
6. Adjust the pH to 7.5 with 1 M NaOH, then dilute to 1 l with distilled water.
7. Prepare trace element solution by dissolving the following salts in 1 l 0.2-M HCl
   - $MnCl_2 \cdot 4H_2O$ 0.25 g

Protocol 13. *Continued*

- $NiCl_2 \cdot 6H_2O$ 0.25 g
- $NaMoO_4 \cdot 2H_2O$ 0.25 g
- H_3BO_4 0.25 g
- $FeSO_4 \cdot 7H_2O$ 0.20 g
- $CoCl_2 \cdot 6H_2O$ 0.05 g
- SeO_2 0.05 g
- $NaVO_3 \cdot 4H_2O$ 0.05 g
- $ZnCl_2$ 0.025 g
- $CuCl_2 \cdot 2H_2O$ 0.025 g

8. Dissolve the sodium salt of 2-mercaptoethanesulphonic acid (MESNA) in distilled water to a final concentration of 0.4% (w/v).
9. Dissolve 0.1 g haemin in 10 ml ethanol, then dilute to 1 l with 0.05 M NaOH.
10. Prepare a solution containing 0.4% (w/v) lysozyme and 0.3% (w/v) disodium–EDTA. Filter-sterilize and use immediately.
11. Prepare a reducing solution containing 2.5% (w/v) $Na_2S \cdot 9H_2O$ and 2.5% (w/v) L-cysteine hydrochloride. Flush with oxygen-free N_2; then autoclave in sealed bottles under nitrogen.
12. Prepare a solution containing 3.75% (w/v) glucose. Flush with CO_2; then autoclave in sealed bottles at 115°C for 10 min only, to minimize caramelization.
13. Prepare mineral salt solution 1, containing 0.3% (w/v) K_2HPO_4.
14. Prepare mineral salt solution 2 by dissolving the following in 150 ml distilled water
    - KH_2PO_4 0.45 g
    - $(NH_4)_2SO_4$ 0.9 g
    - NaCl 0.9 g
    - $MgSO_4 \cdot 7H_2O$ 0.09 g
    - $CaCl_2$ (dried) 0.09 g
15. Prepare rumen fluid as described in *Protocol 3*, but after step 2 centrifuge the fluid at 12 000 *g* for 30 min to remove bacteria.

Protocol 14. Preparation of basal solution A

1. Dissolve 2 g yeast extract and 2 g trypticase peptone in 100 ml water.

Protocol 14. *Continued*

2. To this add

• Clarified rumen fluid (*Protocol 13*, step 15)	150 ml
• Mineral salt solution I (*Protocol 13*, step 13)	150 ml
• Mineral salt solution II (*Protocol 13*, step 14)	150 ml
• Haemin solution (*Protocol 13*, step 9)	10 ml
• Fatty acid solution (*Protocol 13*, step 5)	10 ml
• 0.1% (w/v) resazurin	1.0 ml
• 10% (w/v) cellulose	100 ml (if required in medium)

3. Adjust pH to 6.8 with 1 M KOH and dilute to 830 ml.
4. Boil the mixture and pass CO_2 through the solution as it cools.
5. Autoclave at 115°C for 20 min in sealed vessels. The experimenter has a choice of preparing the complete medium (*Protocol 15*) in bulk and dispensing small quantities into tubes, or of dispensing basal solution A in small quantities in tubes and adding proportionally smaller amounts of the other solutions given in *Protocol 15*.

Protocol 15. Preparation of medium A

1. Mix together under sterile conditions the following solutions, all of which have been sterilized by the appropriate method

• Basal solution A (*Protocol 14*)	830 ml
• 8% (w/v) Na_2CO_3 (sterilized by autoclaving under N_2 after bubbling with N_2)	50 ml
• Growth factor solution (*Protocol 13*, step 1)	10 ml
• Reducing solution (*Protocol 13*, step 11)	10 ml
• Distilled water	100 ml

If required, the distilled water can be replaced by 3.75% (w/v) glucose (*Protocol 13*, step 12) or by the following mixture

• Antibiotic solution (*Protocol 13*, step 3)	50 ml
• Lysozyme solution (*Protocol 13*, step 10)	10 ml
• Distilled water	40 ml

2. Bubble sterile CO_2 through the complete mixture for 5–10 min.
3. Dispense into tubes which should be flushed with CO_2. Seal with butyl rubber bungs or, if screw caps are used, screw down the caps. Any tubes showing pink colouration after 30 min should be rejected.

Protocol 16. Preparation of basal solution B

1. Dissolve the following in 200 ml distilled water
- KCl 0.6 g
- NaCl 0.6 g
- $MgSO_4 \cdot 7H_2O$ 0.5 g
- $CaCl_2 \cdot 2H_2O$ 0.2 g
- NH_4Cl 0.54 g
- Trypticase peptone 1 g
- Pipes buffer 1.5 g

2. Add
- Coenzyme M solution (*Protocol 13*, step 8) 10 ml
- Fatty acid solution (*Protocol 13*, step 5) 10 ml
- Trace element solution (*Protocol 13*, step 7) 10 ml
- Haemin solution (*Protocol 13*, step 9) 10 ml
- 0.1% (w/v) resazurin 1.0 ml
- 10% (w/v) cellulose suspension 100 ml (if required)

3. Adjust the pH to 6.8 with 1M KOH and dilute to 810 ml.

4. Boil the mixture and pass CO_2 through the solution as it cools.

5. Autoclave at 115°C for 20 min in sealed vessels under CO_2.

Protocol 17. Preparation of medium B (no rumen fluid)

1. Mix together under sterile conditions the following solutions, all of which have been sterilized by the appropriate method
- Basal solution B (see *Protocol 16)* 810 ml
- 6.8% (w/v) KH_2PO_4 (sterilized by autoclaving) 10 ml
- 5.0% (w/v) yeast extract (sterilized by autoclaving) 10 ml
- 8.0% (w/v) Na_2CO_3 (sterilized by autoclaving under N_2 after bubbling with N_2) 50 ml
- Growth factor solution (*Protocol 13*, step 1) 10 ml
- Reducing solution (*Protocol 13*, step 11) 10 ml
- Distilled water 100 ml

If required, the distilled water can be replaced by 3.75% (w/v) glucose (*Protocol 13*, step 12) or by the following mixture
- Antibiotic solution (*Protocol 13*, step 3) 50 ml

Protocol 17. *Continued*

- Lysozyme solution (*Protocol 13*, step 10) 10 ml
- Distilled water 40 ml

2. Bubble sterile CO_2 through the complete mixture for 5–10 min.
3. Dispense into tubes which should be flushed with CO_2 and sealed. Any tubes showing pink colouration after 30 min should be rejected.

Protocol 18. Enrichment and isolation of rumen fungi

1. Place 0.1 g milled barley straw (prepared as for dried grass; see *Protocol 3*) in a culture tube and autoclave.
2. Add 10 ml medium B (see *Protocol 17*), containing antibiotics and lysozyme, but no cellulose or glucose.
3. Inoculate with 1 ml rumen contents and incubate under CO_2 for 5 days at 39°C.
4. Prepare in an anaerobe chamber plates containing a lower layer of medium B agar without glucose, overlaid by medium B agar containing cellulose, antibiotics, and lysozyme.
5. Take the enrichment cultures into the anaerobe chamber and inoculate the plates with the straw particles, which should be colonized with fungi.
6. Incubate the plates at 39°C in anaerobe jars under CO_2.
7. If the method has been successful, fungal colonies of up to 20 mm diameter should have developed. If they have, transfer the cultures again into the anaerobe chamber. Remove small plugs of agar from the edges of the colonies with a sterile Pasteur pipette and inoculate into liquid medium B containing glucose, but no antibiotics or lysozyme.
8. Incubate the liquid medium under CO_2 at 39°C and examine daily for fungal and bacterial growth. Reject any tubes contaminated with bacteria.
9. Subculture the fungal cultures daily. Experience will show how long cultures can be safely left between transfers without dying.

The method for the isolation of rumen fungi given in *Protocol 18* obviously depends on the availability of an anaerobe chamber. It is possible to use the classical Hungate roll-tube technique (16) described in Chapter 1, but the fungal colonies are much smaller (1–2 mm) and the technique is more difficult to use.

Once the fungi have been isolated they grow well on both media A and B. The choice of medium will depend on the availability of rumen fluid for medium A, or on the reasons why the fungi have been isolated.

3.2 Cultivation of other anaerobic fungi

Fungi are normally regarded as obligate aerobes which are unable to survive for long periods in the absence of oxygen (17). However, some such as *Blastocladia* spp. (18) will grow equally well in the presence or absence of oxygen, while a very few such as the rumen fungi and *Aqualinderella fermentans* are unable to grow in the presence of air. However, in the latter case this is due to a requirement for a high concentration of CO_2 (19). *A. fermentans* therefore grows well in anaerobic habitats because these contain a high concentration of CO_2, rather than because they lack oxygen. *A. fermentans* grows well in air supplemented with 18% CO_2. Apart from this requirement for CO_2 by *A. fermentans*, both this organism and *Blastocladia* spp. grow well on conventional mycological culture media in air; their cultivation is therefore not discussed further.

References

1. Diamond, L. S. (1987). In *In Vitro Methods for Parasite Cultivation* (ed. A. E. R. Taylor and J. R. Baker), pp. 1–28. Academic Press, London.
2. Coleman, G. S. (1987). In *In Vitro Methods for Parasite Cultivation* (ed. A. E. R. Taylor and J. R. Baker), pp. 29–51. Academic Press, London.
3. Coleman, G. S. (1962). *J. Gen. Microbiol.* **28**, 271.
4. Hino, T. and Kametaka, M. (1977). *J. Gen. Appl. Microbiol.* **46**, 693.
5. Weller, R. A. and Pilgrim, A. F. (1974). *Br. J. Nutr.* **32**, 341.
6. Nakamura, F. and Kurihara, Y. (1978). *Appl. Environ. Microbiol.* **35**, 500.
7. Abe, M. and Kurihara, Y. (1984). *J. Appl. Microbiol.* **56**, 201.
8. Gijzen, H. J., Zwart, K. B., van Gelder, P. T., and Vogels, G. D. (1986). *Appl. Microbiol. Biotechnol.* **25**, 155.
9. Gijzen, H. J., Zwart, K. B., Teunissen, M. J., and Vogels, G. D. (1988). *Biotech. Bioeng.* **32**, 749.
10. Jensen, E. R. and Hammond, D. M. (1964). *J. Protozool.* **11**, 386.
11. Yamin, M. A. (1978). *J. Protozool.* **25**, 535.
12. Yamin, M. A. (1981). *Science* **211**, 58.
13. van Bruggen, H. (1986). Thesis, University of Nijmegen.
14. Kudo, R. R. (1957). *J. Protozool.* **4**, 154.
15. Lowe, S. E., Theodorou, M. K., Trinci, A. P. J., and Hespell, R. B. (1985). *J. Gen. Microbiol.* **131**, 2225.
16. Joblin, K. N. (1981). *Appl. Environ. Microbiol.* **42**, 1119.
17. Cochrane, V. W. (1958). *Physiology of Fungi.* Wiley, New York.
18. Emerson, R. and Cantino, E. C. (1948). *Am. J. Bot.* **35**, 157.
19. Emerson, R. and Held, A. A. (1969). *Am. J. Bot.* **56**, 1103.

A1

Manufacturers and suppliers of specialist items

Alltech Associates, 2051 Waukegan Road, Deerfield, IL 60015, USA.

American Type Culture Collection, 12301 Parklawn Drive, Rockville, MD 20852–1776, USA.

Amersham International plc, Lincoln Place, Green End, Aylesbury HP20 2TP, UK.

Anaerobe Systems, 3066 Scott Boulebard, Santa Clara, CA 95050, USA.

API-Analytab Products, 200 Express Street, Plainview, NY 11803, USA.

API-bioMérieux (UK) Ltd, Grafton Way, Basingstoke RG22 6HY, UK.

Baird & Tatlock Ltd, PO Box 1, Romford, Essex RM1 1HA, UK.

Baxter Scientific Products, 1430 Waukegan Road, McGaw Park, IL 60085–6787, USA.

BBL Microbiology Systems, PO Box 243, Cockeysville, MD 21030, USA.

Becton Dickinson UK Ltd, Between Towns Road, Cowley, Oxford OX4 3LY, UK.

Bio-Rad Laboratories Ltd, Bio-Rad House, Maylands Avenue, Hemel Hempstead HP2 7TD, UK.

Boehringer Mannheim Biochemicals, PO Box 50414, Indianapolis, IN 46250, USA; **Boehringer Corporation Ltd**, Bell Lane, Lewes, East Sussex BN7 1LG, UK.

W. R. Brown Division Intermatic Inc, Intermatic Plaza, Spring Grove, IL 60081, USA.

Calbiochem Brand Biochemicals: Behring Diagnostics, 10933 North Torrey Pines Road, La Jolla, CA 92037, USA; **Calbiochem-Behring, Cambridge BioScience**, 42 Devonshire Road, Cambridge CB1 2BL, UK.

Chemap AG, Alte Landstrasse-415, 8708 Mannedorf ZH, Switzerland.

Deutsche Sammlung von Mikroorganismen und Zellkulturen GmbH, Mascheroder Weg 1b, D-3300 Braunschweig, Germany.

Diagnostics Pasteur, 3 Boulevard Raymond Poincarré, B.P.3, 92430 Marnes-La-Coquette, France.

Difco Laboratories, PO Box 331058, Detroit, MI 48232-7058, USA; PO Box 14B, Central Avenue, East Molesey, Surrey KT8 0SE, UK.

Disposable Products Pty Ltd, PO Box 90, Ingle Farm 5095, Adelaide, South Australia.

DuPont (UK) Ltd, Wedgwood Way, Stevenage, Hertfordshire SG1 4QN, UK.
Gibco, 3175 Staley Road, Grand Island, NY 14072, USA; PO Box 35, Trident House, Renfrew Road, Paisley PA3 4EF, UK.
Hybaid Ltd, 111–113 Waldegrave Road, Teddington TW11 8LL, UK.
ICN Biomedicals Ltd, Cressex Industrial Estate, High Wycombe HP12 3XJ, UK.
LabM Ltd, Topley House, PO Box 19, Bury BL9 6AU, UK.
Massachusetts Public Health Biologic Laboratories, Boston, MA, USA.
Mast Diagnostics Ltd, Mast House, Derby Road, Bootle L20 1EA, UK.
MDH (Microflow Dent & Hellyer), Walworth Road, Andover, Hampshire SP10 5AA, UK.
Mercia Diagnostics Ltd, Mercia House, Broadford Park, Guildford GU4 8EW, UK.
Merck, Alltech, Kellet Road, Carnforth, Lancashire LA5 9XP, UK.
National Collection of Type Cultures, Central Public Health Laboratory, 61 Colindale Avenue, London NW9 5HT, UK.
Oregon Collection of Methanogens, Department of Environmental Science and Engineering, Oregon Graduate Institute of Science and Technology, 19600 NW von Neumann Drive, Beaverton, Oregon 97006-1999, USA.
Oxoid Ltd, Wade Road, Basingstoke, Hants. RG24 0PW, UK; **Oxoid U.S.A. Inc**, 9017 Red Branch Road, Columbia, MD 21045, USA; **Oxoid Canada Inc**, 217 Colonnade Road, Nepean, Ontario K2E 7K3, Canada; **Oxoid Australia Pty Ltd**, PO Box 220, West Heidelberg, Victoria 3081, Australia.
Pharmacia LKB, Biotechnology AB, Bjorkgatan 30, 751 82 Uppsala, Sweden; **Pharmacia Ltd**, Midsummer Boulevard, Milton Keynes MK9 3HP, UK; **Pharmacia LKB Biotechnology Inc**, 800 Centennial Avenue, PO Box 1327, Piscataway, NJ 08855-1327, USA.
Phase Sep, Phase Separations Ltd, Deeside Industrial Park, Queensferry, Clwyd CH5 2NU, UK; **Phase Sep Inc**, 140 Water Street, Norwalk, CT 06854, USA.
Sclavo Inc, 5 Mansard Court, Wayne, NJ 07470, USA.
Sigma Chemical Co, PO Box 14508, St. Louis, MO 63178–9916, USA; Fancy Road, Poole, Dorset BH17 7NH, UK.
Supelco Inc, Supelco Park, Bellefonte, PA 16823-0048, USA; **Radleys**, London Road, Sawbridgeworth, Hertfordshire CM21 9JH, UK.
TechLab Inc, 1861 Pratt Drive, Corporate Research Center, Blacksburg, VA 24060, USA.
Wako Chemicals USA Inc, 12300 Ford Road, Suite 130, Dallas, TX 75234, USA.
Wellcome Diagnostics, Temple Hill, Dartford, Kent DA1 5AH, UK.
Whatman Biochemicals Ltd, Springfield Mill, Maidstone, Kent ME14 2LE, UK.
Wheaton Instruments, 1301 North Tenth Street, Millville, NJ 08332, USA.
Don Whitley Scientific Ltd, 14 Otley Road, Shipley, West Yorks BD17 7SE, UK; **Tekmar Co**, PO Box 371856, Cincinnati, OH 45222–1856, USA.
Worthington Biochemical Corporation, Freehold, NJ 07728, USA.

Index

Actinomyces 34, 43, 163
alcohol shock 20, 22, 23, 194
p-aminobenzoic acid 19
Amoebobacter 248
anaerobic cabinets 6–8, 189–90, 295
anaerobic cocci 35, 196
anaerobic infections
  antibiotic-associated diarrhoea 20, 22–4
  botulism 20, 24–6, 163–8
  Clostridium perfringens food-poisoning 20, 21–2
  infant botulism 25
  pseudomembranous colitis 20, 22–4
anaerobic jars
  evacuation-replacement technique 3–4
  catalyst 3, 4, 7
  culture of
    anaerobic fungi 295
    anoxygenic protozoa 267–8
  disposable pouches 6
  indicators of anaerobiosis 5, 7, 9, 11
  internal gas-generation 3, 5–6
Ancalochloris perfilievi 254
antigens 102–4
antimicrobial sensitivity testing 51–62
  agar dilution methods 60–1
  broth dilution methods 61
antisera 109–11
Aqualinderella fermentans 296
Arabidopsis thaliana 161
Archaeoglobus 209
artificial electron carriers 130
Athiorhodaceae 247
axenic culture of anaerobic protozoa 279, 284, 287

Bacillus 30, 169
  B. subtilis 151, 153
bacterial microflora 183–99
  enumeration 186–7, 189–95, 210–13, 234–5
  metabolism 195–199
  sampling 184–6, 224
bacteriochlorophyll 248–55
bacteriocin typing 21
Bacteroides 15, 18, 30–1, 35, 40, 43, 52–62, 65, 84, 101, 121, 131, 145–52, 163, 194, 196
  'fragilis-group' 52–62, 102, 194, 196
  'melaninogenicus-group' 31, 32
  B. asaccharolyticus 31, 40
  B. bivius 40, 53
  B. capillosus 53
  B. disiens 40, 53
  B. distasonis 40, 53
  B. eggerthii 40
  B. endodontalis 31
  B. fragilis 32, 40, 53, 54, 58, 86, 131–2, 135
  B. gingivalis 31
  B. intermedius 40
  B. levii 40
  B. melaninogenicus 40, 53
  B. oralis 32, 40, 53
  B. ovatus 40, 53
  B. ruminicola 40, 53
  B. splanchnicus 40
  B. thetaiotaomicron 40, 53, 58
  B. uniformis 40
  B. ureolyticus 30, 32, 40
  B. vulgatus 40, 53
Bifidobacterium 34, 65, 194, 196
bile tolerance 41, 42
Blastocladia 296
blood cultures 18–20
botulism 13, 24–6, 163–8
Butyribacterium methylotrophicum 230

Capnocytophaga 88
capsules 108–9, 163
carotenoids 248–55
cell fractionation 124–6, 239–41

cell walls 74–80, 102–7
amino acids 77–9
peptidoglycans 74–6
polysaccharides 76–8
reducing sugars 80
cellular lipids 80–9, 102, 103, 107–8, 235
Fatty acids 84–6
Menaquinones 86–9
Polar lipids 80–4
Chilomastix 285
Chlorobacteriaceae 247
Chlorobiaceae 247, 254–8, 262, 270, 274
Chlorobium 254, 257
Chloroflexaceae 255
Chloroflexus 264, 272
C. aurantiacus 251, 255, 257
Chloroherpeton 263, 272
C. thalassium 254
Chloronema 255
Chromatiaceae 247–9, 256–60, 262, 270, 274
Chromatium 248, 256, 260
C. buderi 263
C. salexigens 263
C. tepidarius 251
chytrid fungi 285, 290–5
'*Clathrochloris sulfurica*' 254
Clostridium 35, 43, 54, 56, 65, 122, 145, 149, 151, 163, 196
C. acetobutylicum 151, 152
C. argentinense 164
C. barati 26, 164
C. bifermentans 24, 38, 39, 178
C. botulinum 24–6, 39, 101, 163–8, 171, 172
C. butyricum 25, 39, 164
C. carnis 30
C. chauvoei 164, 165
C. difficile 22–4, 39, 46, 48, 117, 153, 157–8, 163–5, 172–8, 179, 194, 196
C. fallax 178
C. histolyticum 30, 164, 169, 178–9
C. innocuum 39
C. novyi 2, 3, 39, 164, 178
C. paraputrificum 39
C. perfringens 20–2, 31, 38, 39, 41, 48, 56, 58, 101, 102, 149, 152, 163–5, 169–71, 172, 178, 179
C. ramosum 39, 194, 196
C. septicum 30, 39, 164, 169, 178
C. sordellii 26, 39, 164, 172, 178
C. spiroforme 39, 164, 165, 171, 172
C. sporogenes 30, 39, 178
C. tertium 30, 39
C. tetani 5, 30, 39, 101, 151, 163–5, 168–9
C. thermocellum 139, 140
C. thermohydrosulfuricum 139
coenzymes in methanogens 136, 223, 231, 234, 241
conjugation 153–4
Coprococcus 189
p-cresol 22, 23, 46, 48
crossed immunoelectrophoresis 113–14, 116
cyanobacteria 247

deoxyribonucleic acid
cloning 145, 179
determination of base composition 91
DNA–DNA hybridization 91–4, 236
isolation and purification 89–91, 145–50
polymerase chain reaction 145, 155–61
Desulfobacter 207, 214
Desulfobacterium 208–9
Desulfobulbus 207, 214
Desulfococcus 207
D. multivorans 214
Desulfomicrobium 209
Desulfomonas 207
D. pigra 214
Desulfomonile 209
Desulfonema 208
D. limicola 214
Desulfosarcina 208
Desulfotomaculum 206–7, 214
D. acetoxidans 202, 214
Desulfovibrio 206, 214
D. vulgaris 202
diaminopimelic acid 79
diazotrophic growth
of anoxygenic phototrophs 258

Index

p-dimethylaminocinnamaldehyde (DMACA) 41
Diplodinium pentacanthum 282
Diploplastron affine 282

Ectothiorhodospira 250, 256, 257, 265
  E. halophila 266, 273
  E. marismortui 273
Ectothiorhodospiraceae 250, 258, 265, 273, 274
Eikenella corrodens 30
electroporation 154–5
Enopoplastron triloricatum 282
enrichment methods 23–4, 228–30, 251–60, 290, 295
Entamoeba histolytica 279
entodiniomorphid protozoa 280–285
Entodinium caudatum 279, 282
  E. simplex 282
enzyme assays 126–40, 198–9, 214–17
enzyme electrophoresis 73–4, 130
enzyme-linked immunosorbent assay 21, 24, 111, 115–16
Epidinium caudatum 282
Eremoplastron bovis 282
Escherichia coli 54, 140, 151
Eubacterium 34, 43, 65, 194, 196
  E. lentum 197
  E. limosum 230
Eudiplodinium maggii 282

faeces 20–6, 158, 167, 176–8, 184–7, 189–96
fermentation tests 33–7, 39, 40
flagella 109
flagellate protozoa,
  in rumen 285
  in termite hindgut 285–8
fluorescence 15, 18, 22, 23, 31, 214, 231–2
food 21–6
French press 104, 124, 239–41
Fusobacterium 31, 32, 35, 43, 53–62, 65, 163, 194, 196
  F. mortiferum 39
  F. necrophorum 30, 39
  F. nucleatum 39, 75, 94–5

gas-liquid chromatography 14, 22, 33, 34, 43–9, 66, 84–6, 194, 230
gel diffusion 21
Gemmiger 189
Giardia 279
Gram-stain 15, 30, 31, 48, 186, 194, 235

β-haemolysis 30, 39
heat shock 20, 21, 25
'*Heliobacillus*' 258, 264, 265, 273, 275
  '*H. mobilis*' 255
Heliobacterium 258, 264, 265, 273, 275
  H. chlorum 255
Heliothrix 263, 272
H. oregonensis 251, 255, 272
high performance liquid chromatography 66–7, 87–8
hydrolysis reactions in identification 36, 37, 39

identification 196
  agar plate identification method 35–8
  by antibiotic resistance 31–2
  by dye resistance 32–3
  methanogens 235–7
  miniaturized methods 33, 41–2
  non-sporing anaerobes 194–5
  sulphate-reducing bacteria 213–14
immunoblotting 111, 116–18
immunoelectrophoresis 112–15, 116
immunogold electron microscopy 111, 118–19
indole 34, 39, 41
isoelectric focusing 68, 71–3
Isosphaera pallida 272
isotopes 141

β-lactamase 52–5, 62
lactic acid determination 67–8
lactobacilli 30
Lactobacillus 34, 43
Lamprobacter modestohalophilus 249, 256

Lamprocystis roseopersicina 249, 258
latex agglutination 21, 22, 24, 171
lecithinase 38, 39, 40
lipase 38, 39
Listeria 169

media
  anaerobic fungi 290–5
  anaerobic protozoa 280–90
  anoxygenic phototrophs 263–74
  clinical microbiology 15–43
  egg yolk agar 25, 35, 38, 41, 48, 101
  methanogens 225–8
  normal flora studies 188–9, 191–4
  PRAS 8–11, 33–5, 122–4, 189, 193, 197–8
  selective media 15–17, 21–5, 187–8, 191–4
  sulphate reducing bacteria 201–10
metabolic studies 141–3, 195–9, 214–21
Methanobacteriales 235
Methanobacterium 169
  M. thermoautotrophicum 138, 151, 231
Methanobrevibacter ruminantium 231
Methanococcales 235
Methanococcus thermolithotrophicus 160–1
methanogens 122, 136–8, 145, 146, 223–43, 274
  enumeration 234–5
  identification 235–7
  isolation 232–4
  large scale cultivation 236–9
  storage 241–3
Methanomicrobiales 235
Methanosaeta 230
Methanosphaera 230
Methanospirillum 229
Metopus contortus 289–90
  M. striatus 288–9
metronidazole 17, 18, 29
Monocercomonas 285
Monocercomonoides 285
monoxenic cultures,
  of anaerobic protozoa 284

nagler plate 38, 41
Neocallimastix 285, 290
Neurospora crassa 161
nitrate reduction test 41, 42
non-volatile fatty acids 43–9

Oscillochloris 255
Ostracodinium dilobum 282

Pelodictyon 254
Pelomyxa pelustris 290
penicillin-binding proteins 54
Pentatrichomonas 285
peptidoglycans 74–6
Peptococcus 32
Peptostreptococcus 32, 43, 65
  P. anaerobius 32, 48
phototrophic bacteria 247–75
Piromonas 285, 290
plasmids 54, 148–50, 151–4
polyacrylamide gel electrophoresis 68–71, 235
polymerase chain reaction 145, 155–61
  PCRMOP 159–61
Polyplastron multivesiculatum 282
Porphyromonas 15, 31, 88
  P. asaccharolytica 81, 86
  P. gingivalis 81, 88, 89
pre-reduced, anaerobically sterilized (PRAS) media 1, 8–11, 33–5, 122–4, 189, 193, 197–8
preservation
  anoxygenic phototrophs 274
  methanogens 241–3
Prevotella 88
  P. buccalis 86
propionibacteria 30
Propionibacterium 34, 43, 133
Prosthecochloris 254, 257
proton motive force 137
protozoa 279–90
Pseudomonas 5, 54
pus 13

reducing agents 2–3, 9, 15, 122, 189, 193, 202, 225–6, 270, 292

Rhodobacter 252, 256
Rhodocyclus 252
  R. purpureus 258
Rhodomicrobium vannielii 252, 256
Rhodopila globiformis 253
Rhodopseudomonas 253, 256, 257
  R. acidophila 251
  R. palustris 258
Rhodospirillaceae 247, 252–3
Rhodospirillum 253, 256, 258
  R. salexigens 266, 274
16s rRNA restriction patterns 94–8, 247
rocket immunoelectrophoresis 112–13, 116
Rumen fluid 188, 281, 290, 293, 295
Ruminococcus 189

Saccharomyces cerevisiae 151, 161
sapropelic protozoa 288–90
serotyping 21, 101–2
shake cultures 1–2
sodium polyanethol sulphonate (SPS, liquoid) 19, 32
sonication 104
Sphaeromonas 285, 290
Spirogyra 290
spores 39
Staphylococcus aureus 133, 151
Streptococcus 169
  S. faecalis 151
Sulfolobus acidocaldarius 161
sulphate reducing bacteria 201–21

tetanus 13
Tetratrichomonas 285
Thermodesulfobacterium 209
thin layer chromatography 80–5, 87–8
Thiocapsa 249, 257
Thiocystis 249
Thiodictyon 249
  T. elegans 258
Thiopedia rosea 249
Thiorhodaceae 247
Thiospirillum jenense 249
toxins 163–80
  Bacteroides fragilis 179
  bioassays 24, 101, 165–9, 174–7, 179
  C. botulinum 24–6, 163–8
  C. difficile 22, 172–8
  C. perfringens 169–71
  C. tetani 168–9
transformation 154–5
transposons 54, 151–4
Trichomitopsis termopsidis 286–8
Trichomonas vaginalis 279
Trichonympha sphaerica 286, 288
Tritrichomonas 285

vector systems 150–3
Veillonella 32, 43, 194, 196
  V. alcalescens 122, 130, 133, 134, 138
volatile fatty acids 9, 14, 39, 43–9, 66–7, 291

Winogradsky columns 259–60

Zootermopsis 285